Lecture Notes in Applied and Computational Mechanics

Volume 77

Series editors

Friedrich Pfeiffer, Technische Universität München, Garching, Germany
e-mail: pfeiffer@amm.mw.tum.de

Peter Wriggers, Universität Hannover, Hannover, Germany
e-mail: wriggers@ikm.uni-hannover.de

About this Series

This series aims to report new developments in applied and computational mechanics—quickly, informally and at a high level. This includes the fields of fluid, solid and structural mechanics, dynamics and control, and related disciplines. The applied methods can be of analytical, numerical and computational nature.

More information about this series at http://www.springer.com/series/4623

Th. Triantafyllidis
Editor

Holistic Simulation of Geotechnical Installation Processes

Numerical and Physical Modelling

Springer

Editor
Th. Triantafyllidis
Institute for Soil and Rock Mechanics
The Karlsruhe Institute of Technology
Karlsruhe, Baden-Württemberg
Germany

ISSN 1613-7736 ISSN 1860-0816 (electronic)
Lecture Notes in Applied and Computational Mechanics
ISBN 978-3-319-38701-7 ISBN 978-3-319-18170-7 (eBook)
DOI 10.1007/978-3-319-18170-7

Springer Cham Heidelberg New York Dordrecht London

Softcover reprint of the hardcover 1st edition 2015

Printed on acid-free paper

Springer International Publishing AG Switzerland is part of Springer Science+Business Media
(www.springer.com)

Preface

This textbook summarizes the midterm results of the research group GEOTECH dealing with the holistic consideration of geotechnical installation processes. In this sense, the entire process for the realization of a geotechnical construction project starting from a well-defined initial stress and deformation state is taken into account. For every subsequent construction stage, as e.g., the vibro-installation of piles in excavation pits, the changes in stress and deformations caused by the installation process are calculated.

It is well known that the majority of events with unacceptable deformations of nearby structures due to the geotechnical construction activity were caused during the installation of the supporting elements (D-wall panels, pile installation, etc.) and not by the pit excavation or insufficient stiffness of the shoring system. Construction steps like excavation, dewatering, shoring support of the pit, etc. implemented in commercial codes can reliably predict the deformations using appropriate input parameters for soil–structure interaction. However, accompanying measurements in accordance with the observational method in geotechnics on construction sites testify that the deformations of the supporting system in large excavation pits and nearby structures caused by the installation processes are the most significant ones over the entire construction procedure, but unfortunately up to now they are not satisfactory predictable.

In the light of EC 7, it is on the other hand crucial for the serviceability state to estimate the deformations and distortions of the supporting systems taking also into account the effects of construction (installation process).

Within the frame of the envisaged research plan, the research group tackles this problem creating the necessary fundamental knowledge for the improvement of the soil–structure interaction models in order to describe the installation process especially in the case of the vibro-installation of piles.

The research group operates in three levels:

- benchmarking projects with element-like and large-scale model tests for calibration and validation of the developed numerical models

- theoretical fundamental research for the development of high-quality constitutive soil models and contact formulations in combination with efficient numerical implementations and algorithms
- application of the developed theoretical models to boundary value problems with parametric studies of respective geotechnical installation processes and recommendations for further use of the numerical models in practice as well as for the practical optimization of these processes.

In this book, the demonstrator experiments for pile installation as benchmark are presented with very interesting results which can be used either for validation of numerical simulations/calibration of simulation purposes or for further experimental and numerical investigations. Further simulation methods for vibro-injection pile installation with multimaterial flow and large material deformations are also presented.

The mortar contact formulations for soil–structure interaction problems with a hypoplastic material model and the evolution of effective stress around a vibrating pile toe using a combination of a hypoplastic model with an explicit formulation for very high numbers of cycles implementing new numerical strategies are described.

A new hypoplastic formulation and an improved integration for explicit high-cycle loading using hierarchical, enhanced, and assumed strain finite elements is described as well as experimental results on strain response-envelopes of sandy material for monotonic and low-cycle loading processes.

The main target of the research group is the provision of suitable methods for the simulation of geotechnical installation processes based on fundamental research in order to reliably predict the serviceability state of supporting systems and nearby structures.

The editor likes to thank all his colleagues (Prof. Ehlers, Prof. Wriggers, Prof. Savidis, Prof. Rackwitz, Prof. Hettler) and co-workers (Dr. Niemunis, Dr. Osinov, Dr. Huber) for their valuable contributions and their extreme efforts and engagement in order to achieve the high scientific targets of the projects.

Furthermore, I would like to express my thanks to Mrs. Meininger for organization of all the workshops of the research group GEOTECH and her engagement to make those events pleasant.

Finally, all of us like to express our deep gratitude to DFG (German Research Council) for the generous financial support of this very interesting research topic in geotechnical engineering.

Karlsruhe Th. Triantafyllidis
February 2015

Contents

Effects of Soil Deposition on the Initial Stress State in Model Tests: Experimental Results and FE Simulation

J. Vogelsang, H. Zachert, G. Huber and Th. Triantafyllidis

Abstract The knowledge of the initial soil state (stress and density distribution) in geotechnical model tests is indispensable, particularly with regard to FE back calculation of experimental results. Usually, so-called K_0-conditions are assumed, which for many cases do not describe the soil stress state before the experiment begins adequately. Using an exemplary test device we present and discuss different measurement techniques for the interpretation of soil deposition procedures and the evaluation of the initial state. By means of stress and bearing force measurements, the stress state is captured representatively. The soil deformations during the filling of the test device are evaluated with Digital Image Correlation (DIC) methods and the initial density distribution is examined by cone penetration tests (CPT). Afterwards, a simple FE simulation method is presented, which models the soil deposition procedure by a weight increase layer-by-layer. It is shown that the method is suitable to provide a realistic initial soil state. The methods presented can be easily transferred to other geotechnical test devices and can in many cases ensure a better comparability of tests with their simulations.

Keywords Initial soil state · Model test · Back calculation · Cone penetration test · Digital image correlation

1 Introduction

Objectives of geotechnical model tests are various: they reach from simple qualitative investigation to sophisticated, quantitative studies and model validation. A starting point for new test types can be the qualitative investigation of the behaviour of geotechnical structures. Subsequently, model tests are often used to validate

J. Vogelsang (✉) · H. Zachert · G. Huber · Th. Triantafyllidis
Institute of Soil Mechanics and Rock Mechanics, Karlsruhe Institute of Technology, Karlsruhe, Germany
e-mail: jakob.vogelsang@kit.edu

Th. Triantafyllidis (ed.), *Holistic Simulation of Geotechnical Installation Processes*, Lecture Notes in Applied and Computational Mechanics 77,
DOI 10.1007/978-3-319-18170-7_1

calculation methods, soil models or general new numerical developments. This is also the aim of the experiments performed in the research group FOR 1136 funded by the German Research Foundation (DFG). This contribution reports some often neglected but for successful validations essential aspects of the interpretation of initial states in geotechnical test devices. This may help to close the gap between experimental measurements and numerical simulations.

Most geotechnical model tests are performed in box- or tube-shaped test devices. The test material is usually installed in horizontal layers. Such procedure resembles the filling process of a silo. The stress distribution in the sample differs from free field conditions, which are usually assumed to be initial conditions for the simulation of model tests. For the filling case and simple geometries, the stress distribution in the soil body can be estimated with sufficient accuracy using classical silo theories (e.g. Janssen [7]). In 2D-conditions, the vertical stress σ_z is a function of depth z and depends on the soil weight γ, the width of the silo b, the coefficient of lateral earth pressure K and the soil-wall friction angle δ, Eq. (1).

$$\sigma_z = -\frac{\gamma \cdot b}{4 \cdot K \cdot \tan\delta}(1 - e^{-4 \cdot K \cdot \tan\delta \cdot z/b}). \quad (1)$$

In a normalized form, Eq. (1) can be written as:

$$\frac{\sigma_z}{\gamma \cdot b} = -\frac{1}{4 \cdot K \cdot \tan\delta}(1 - e^{-4 \cdot K \cdot \tan\delta \cdot z/b}). \quad (2)$$

where the normalized vertical stress $\sigma_z/(\gamma \cdot b)$ is a function of a normalized (dimensionless) depth z/b.

Figure 1a shows the experimental set-up of an exemplary test device for the investigation of "plane strain" model pile penetration. Similar set-ups are used e.g. by White and Bolton [17] and many other researchers. Figure 1b plots the vertical stress distribution in those test devices assuming plane strain conditions for different soil-wall friction angles δ according to Eq. (1). The coefficient of lateral earth pressure K is set equal to 0.4, which is a reasonable value for many cohesionless soils. Note, that K is lower than the coefficient of earth pressure at rest K_0 due to friction in the interface soil-wall.

Silo effects result in an under-linear increase of stress with depth below the soil surface. As a matter of course, they are more pronounced for higher soil-wall friction angles. In order to minimize wall friction, smooth wall surfaces are often used in experiments ($\delta = 10–20°$). Figure 1b shows that for these low δ values, silo effects have a minor influence for $z/b < 0.5$. For increasing z/b, silo effects become more apparent and should not be neglected for $z/b > 1$. Their impact on the stress distribution is significant for $z/b > 2$. Existing test devices easily reach values of $z/b = 7–15$ in the regions of interest (e.g. [17]). For the Interface test device (ITD), used for the validation of numerical methods in the research group FOR 1136, z/b lies in the range of 1–4. As a result, the stress distribution in the soil body is strongly influenced by silo effects. However, due to different surface roughnesses in contact

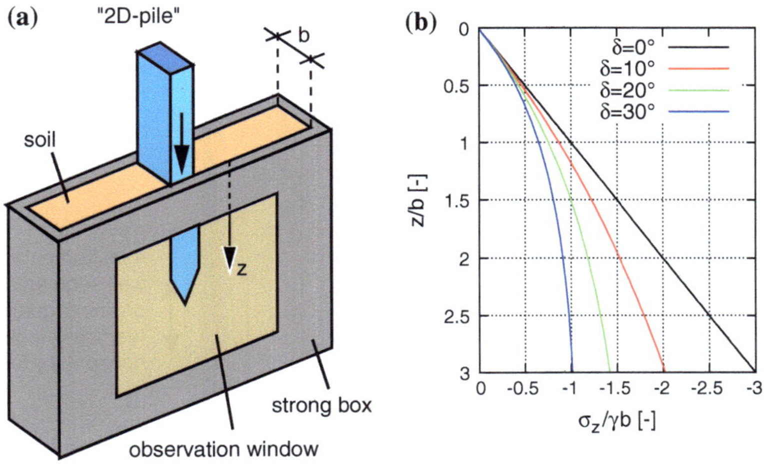

Fig. 1 **a** Schematic test device intending plane strain model pile penetration (after [17]) and **b** normalized vertical stress in a 2D silo problem for different soil-wall friction angles (Eq. (1) according to [7] with $K = 0.4$)

with the sand and more complex geometries, conventional silo equations fail in this special case. The suitability of the test device to provide benchmark experiments is therefore questionable.

Important questions to be answered are:

(a) How can the initial soil state be evaluated experimentally? (stress state and density distribution)
(b) Are simple and reliable numerical methods available to obtain a realistic initial state for the simulation of the experiments?
(c) Taking into account (a) and (b), can the tests in the Interface test device (ITD) still be used as benchmark experiments?

In order to answer these questions, the ITD is briefly presented and available experimental methods for the evaluation of the initial soil state are discussed. Afterwards, a simple method for the simulation of the soil installation is described and validated by comparison with experimental data.

2 Test Device

In the ITD, large scale 1g-model tests can be performed on soil-structure interactions. It was designed to provide high quality experimental data for the validation of constitutive models and new simulation techniques. Details concerning the construction and test results can be found in [14, 15]. The basic concept is schematically

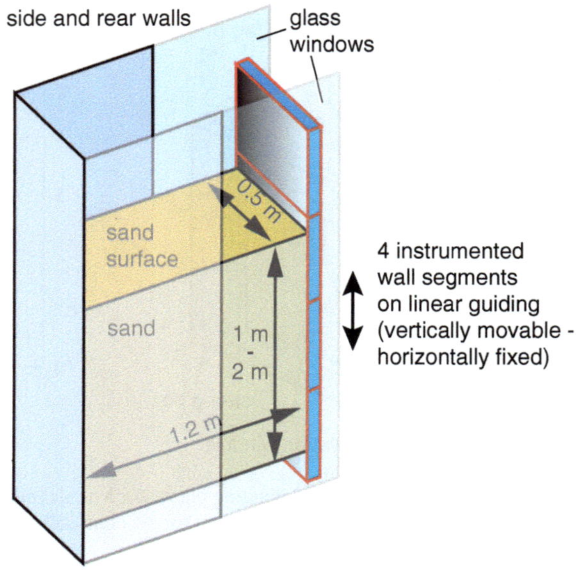

Fig. 2 Schematic construction of the Interface test device

illustrated in Fig. 2. The inner dimensions of the test device are 1.2 × 0.5 × 2.4 m (length × width × height).

In the test, the device is filled with sand up to heights of 1.5–2.0 m. Integral part of the test device are four instrumented wall segments placed on one of the short sides. They can be installed with different surface roughnesses or geometries. In the tests, the wall segments are displaced vertically and the reaction forces in the soil-structure interface are measured. The soil is partly kept visible through observation windows on both sides of the device. The soil displacement and deformations can thus be evaluated with DIC.

The test device was developed intending plane strain conditions. In order to ensure this, the side walls have to be designed with very smooth surfaces and have to be supported by a stiff construction. Within the realms of technical possibility, this was achieved by using glass plates or stainless steel for the inner surfaces. Nevertheless, a "silo effect" occurs in the test device [14]. Below a certain depth of about 0.5–1.0 m under the sand surface a significantly lower stress is observed compared to free field conditions. Silo effects result in a considerable discrepancy from plane strain conditions. Any quantitative simulation of these tests has to take this phenomenon into account.

2.1 Denotation

The chosen coordinate system is shown in Fig. 2. The y-axis corresponds to the vertical direction. The x-direction is perpendicular to the wall segments surfaces.

Table 1 Nomenclature Interface test device

b	Width of test device in x-direction
h_{sand}	Fill height
γ	Averaged unit soil weight (dry)
σ_{y}	Vertical stress
σ_{x}	Horizontal stress in x-direction
$F_{\text{x,ij}}$	Horizontal bearing force j in x-direction on wall segment i
$F_{\text{y,i}}$	Vertical bearing force on wall segment i
I_{D}	Relative density
I_{d}	Pressure-dependent relative density

The origin lies in the intersection of wall segments, bottom plate surface and symmetry plane of the test device. Mechanical sign convention is used (compressive force/stress/strain negative). Table 1 contains some important variables subsequently used for the interpretation and visualization of test results.

2.2 Remarks on Test Material and Installation Method

Good descriptions of requirements and procedures for the preparation of uniform large scale soil samples are provided e.g. in [3, 8]. For dry cohesionless soil, the pluviation method is usually applied, which leads to reproducibly uniform test samples. The soil density can be controlled via variation of free fall height and intensity. The preparation methods for the Interface test device were already extensively described in [15] and are therefore not repeated here. A dry, uniform medium quartz sand was used as test material. Detailed information can be found in [14, 15].

2.3 Initial Density Distribution

As already mentioned, the pluviation method aims to produce a uniform density distribution over the whole soil sample. However, a certain fluctuation of void ratio cannot be avoided. Depending on the quality and thorough execution of the placement method, this fluctuation varies only within a few percent of I_D. In [10], it was shown that a cyclic preloading (before the test) can lead to homogenization soil samples. However, this kind of procedure is not applicable to all test devices and a cyclic pre-loading can influence the soils mechanical behaviour significantly in subsequent tests.

2.4 Uniformity Control Using Cone Penetration Tests (CPT)

There are several methods to examine the density of in situ soils. CPT are widely accepted to be a reliable tool for the determination of the soil density profile and are often used in model tests [3, 6, 16]. However, the interpretation of soil's relative density from the CPT-results is still a challenging task. Especially for small confining pressure, as in 1g-model tests without additional pressure boundaries or surcharge loads, the simple relations between tip pressure and relative density generally used (e.g. DIN 4094-1 [5]) are no longer valid. This is the reason for the use of alternative and more advanced methods for the assessment of the CPT-results (e.g. [1, 4, 12]). In this contribution, the procedure after Cudmani and Osinov [4] is used.

Cone Penetration Tests in model tests are often influenced by boundary conditions of the test device. In order to approach to free field conditions, the cone for CPT in model tests should be designed in relation to the test device's dimensions. According to [9], a relation of sample to cone diameter of $D_{\text{sample}}/d_{\text{cone}} \geq 71$ is sufficient to neglect the effects of the boundaries on the cone resistance. The cone used in this contribution was developed in [18] for a test device shown in [20] with a diameter of 94 cm. The cone dimensions are shown in Fig. 3.

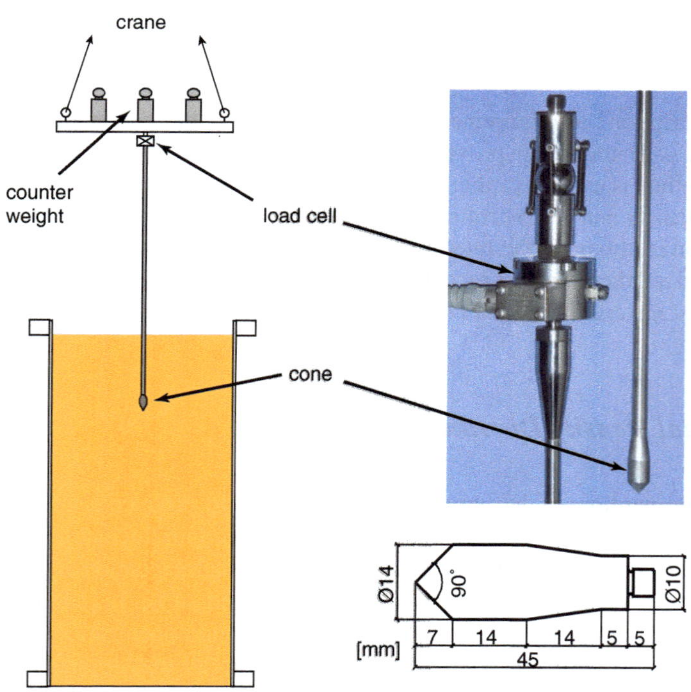

Fig. 3 Scheme of the CPT in the test device and miniature cone for CPT

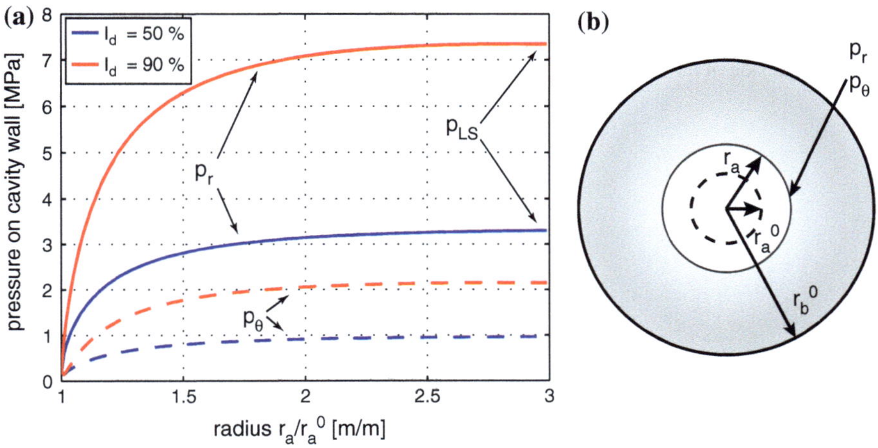

Fig. 4 **a** Development of the expansion pressure. **b** Scheme of the cavity expansion

The penetration force is measured by a load cell, mounted to the head of the sounding rods. It captures tip pressure and skin friction at once without the ability to separate them. The sounding rods were therefore designed with a diameter reduction compared to the cone, which will reduce the skin friction considerably. The rod with the load cell is mounted to a counter weight, which was driven downwards with a constant speed during the CPT. A scheme of the setup is shown in Fig. 3. The resulting tip pressure is then interpreted by the method proposed in [4]. This is a semi-empirical method based on the numerical solution of a spherical cavity expansion problem. An initial cavity (radius r_a^0) in a hypoplastic continuum (radius r_b^0 with $r_\mathrm{b}^0 \gg r_\mathrm{a}^0$, Fig. 4b is expanded until a stationary expansion pressure p_LS is reached (see p_r and p_θ in Fig. 4a). This limit pressure is influenced by the material and its state, in detail the initial density I_d and the initial effective mean pressure p_0, as shown in Fig. 4a for medium dense ($I_\mathrm{d} = 50\,\%$) and dense ($I_\mathrm{d} = 90\,\%$) sand.

The pressure-related relative density I_d is defined by [4] as

$$I_\mathrm{d} = \frac{e_c - e}{e_c - e_d} \tag{3}$$

with the pressure-dependent void ratios e_c and e_d according to the hypoplastic model [2]. The limit pressure p_LS is coupled with the results of calibration chamber tests q_c using a so-called shape factor k_q

$$q_\mathrm{c} = k_\mathrm{q}(I_\mathrm{d}) \cdot p_\mathrm{LS}(p_0, I_\mathrm{d}) \tag{4}$$

with

$$k_\mathrm{q}(I_\mathrm{d}) = 1.5 + \frac{5.8 \cdot I_\mathrm{d}^2}{I_\mathrm{d}^2 + 0.11}. \tag{5}$$

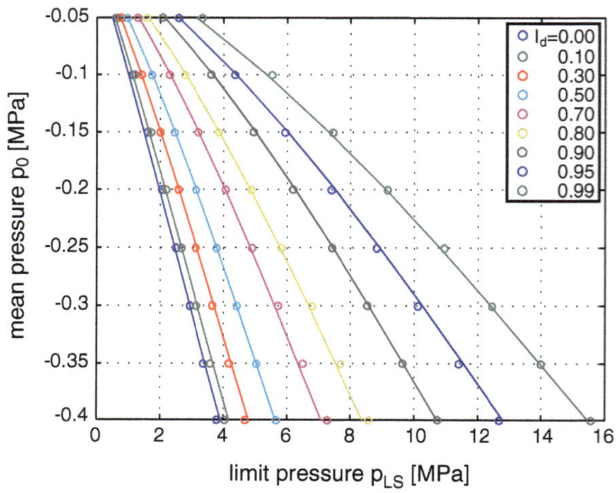

Fig. 5 Limit pressure $p_{\mathrm{LS}}(I_{\mathrm{d}}, p_0)$ in spherical cavity expansion of the test sand

Table 2 Parameters a_i and b_i for Eq. (6) of the test sand

a_1	a_2	a_3	b_1	b_2	b_3
6.433	−3.644	−1.138	0.4961	0.5755	1.172

The connection between p_{LS}, the density I_{d} and the effective mean pressure p_0 is obtained from the numerical calculation of a set of cavity expansions with different initial conditions. This leads to the results shown in Fig. 5.

These data can be described with a surface using the parameters a_i and b_i according to [4]

$$p_{LS} = a\ p_0^b \tag{6}$$

with

$$a = a_1 + \frac{a_2}{a_3 + I_d} \quad \text{and} \quad b = b_1 + \frac{b_2}{b_3 + I_d}. \tag{7}$$

This set of parameters is unique for every soil, in particular for every set of hypoplastic parameters, used for the determination of $p_{\mathrm{LS}}(I_{\mathrm{d}}, p_0)$. With the hypoplastic parameters from Table 3 the parameters a_i, b_i in Table 2 were gained:

With the results of a CPT (q_{c} over depth z) and the corresponding mean pressure $q_{\mathrm{c}}(z)$, Eq. (6) leads to the density I_{d}.

2.5 Deformation During Filling

DIC is usually used to evaluate soil deformations during model tests. Furthermore, the method is also applicable to investigate the soil deformations during the soil deposition process. The filling process is documented photographically with image sequences like the subsequent experiment. These image sequences are then evaluated

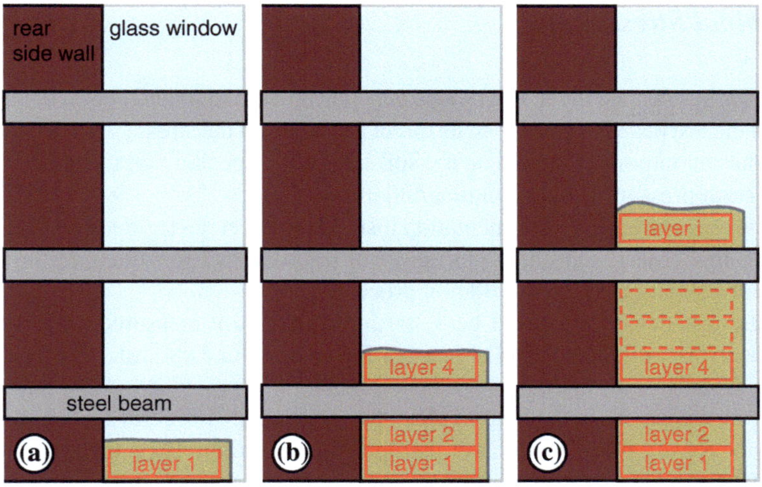

Fig. 6 DIC-evaluation of the sand installation and definition of soil layers: start of DIC-evaluation for **a** layer 1, **b** layer 4 and **c** layer i

with the program JPIV [13]. Therefrom, the compression and settlement of a soil layer due to the deposition of the soil above can be estimated.

Figure 6 illustrates the evaluation procedure in greater detail. Three different filling levels are shown in schematical side views on the test device. Between the horizontal steel beams, the current sand height is visible. Every 3–4 cm a new soil layer is defined for which the evaluation process starts just after its deposition. The deformations of this layer—due to the continuing filling process—are measured indirectly in this way. The soil layers hidden by the steel beams cannot be evaluated when this procedure is used. Their deformation has to be interpolated between the adjacent visible layers above and below.

Note that these displacements or deformations of the distinct layers do not have the same reference time. The deformation of a layer is defined as its deformation with respect to the initial configuration of this layer (configuration of the layer at the time directly after deposition).

For medium dense to dense soil, the observed displacements are very small ($|u_y| < 1$ mm). They result in changes of void ratio of about -0.001. The change in density is more significant for loose soil (up to -0.1 in void ratio). In other words, the lower the relative density is, the more relevant becomes the DIC analysis of the installation process. In every case, the evaluated displacements can serve for the validation of displacements obtained from FE-analysis.

In particular the DIC results can be useful to verify the soil stiffness in simulations, which may differ from the one obtained from laboratory tests due to different stress levels.

2.6 Initial Stress State

The initial stress state in the test device depends on void ratio and geometric boundaries of the test device. Apart from its direct influence on the stress ($\gamma(e) \cdot h$), the void ratio e has an important impact on the soil behavior, especially on the coefficient of lateral earth pressure and on jamming/unjamming effects.

Beside measurement instrumentation installed in a test device for the observation during a model test, additional placement of transducers can sometimes be useful for the investigation of the installation process.

In the test device discussed here, earth pressures are measured with pressure transducers as well as in terms of bearing forces on boundary structures (wall segments). Detailed information on the instrumentation is provided in [15]. Here, only an overview is given. In the bottom plate, two relative pressure transducers are integrated measuring the vertical earth pressure during filling. Their membran area is about 12 cm^2. Thus, the measurement provides only local information. Especially in loose sand, the membrane can be to small to provide representative and reliable results.

The instrumentation of the test device also allows for separate measurement of horizontal and vertical components of earth pressure. On each wall segment i ($i =$ 1, ..., 4), the resulting vertical component, named $F_{y,i}$, is recorded with one load cell and the horizontal one with four load cells $F_{x,1-4,i}$, see Fig. 7. From the equilibrium of forces and momentum we can also estimate the earth pressure distribution. This is done by assuming a sectionwise linear (trapezoidal) distribution in order to obtain a statically determined system, see Fig. 7.

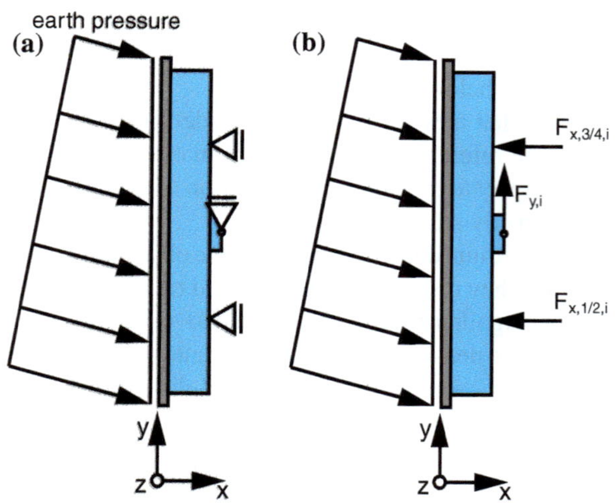

Fig. 7 **a** Static system of wall segment i and **b** denotation of measured bearing forces

3 Simulation of Sand Deposition

Several kinds of simulation techniques can be applied to model the soil deposition process in a test device (FEM, FDM, DEM, etc.). Considering only the deposition process, the most appropriate algorithm should be used. But often, the applied software is already chosen for the simulation of the subsequent experiment. The same applies for the choice of an appropriate constitutive model. In both, simulation of soil installation and subsequent test, only one model should be used in order to ensure a realistic evolution of internal state variables. Advanced constitutive models should be preferred, but basic phenomena can already be modelled by using simple elasto-plastic or even elastic models. However, at least elasto-plastic description of the contact behaviour in the interface soil-test device boundaries is indispensable (e.g. Coulomb friction model).

In this study, the FE software Abaqus is used. The soil behavior is described by a hypoplastic constitutive model (after [19]). The concept of intergranular strain is not necessarily to be applied because the soil is installed in a "virgin" state and no cyclic loading occurs during the installation process. If the model test is simulated subsequently, the intergranular strain can be assumed to be fully mobilized in vertical direction. The used parameters for the hypoplastic constitutive model are taken from [18] and are listed in Table 3.

A simple Coulomb contact description was applied. The friction coefficients were chosen according to the experimental measurements (the instrumentation of the wall segments allows for determination of the interface friction angle). The contact friction angle between sand and the side walls and the wall segments 1 and 3 was set to 12°. The second wall segment had a rough surface (coarse fraction of the test sand bonded on the steel surface). The friction angle between this wall segment and sand was chosen to 38°, which corresponds to the peak friction angle for medium dense sand.

3.1 FE Model of the Test Device

A 3D FE model of the ITD was developed in order to include geometric effects appropriately, Fig. 8. Reduced integration elements (Abaqus keyword C3D8R) were used. Taking advantage of the symmetry of the problem, only the half of the test device was modelled. The side and rear walls, as well as the wall segments, were

Table 3 Used hypoplastic parameters (from [18])

φ_c(°)	e_{d0}	e_{c0}	e_{i0}	h_s (MN/m^2)	n	α	β
32.8	0.568	0.866	0.953	5800	0.28	0.13	1.05

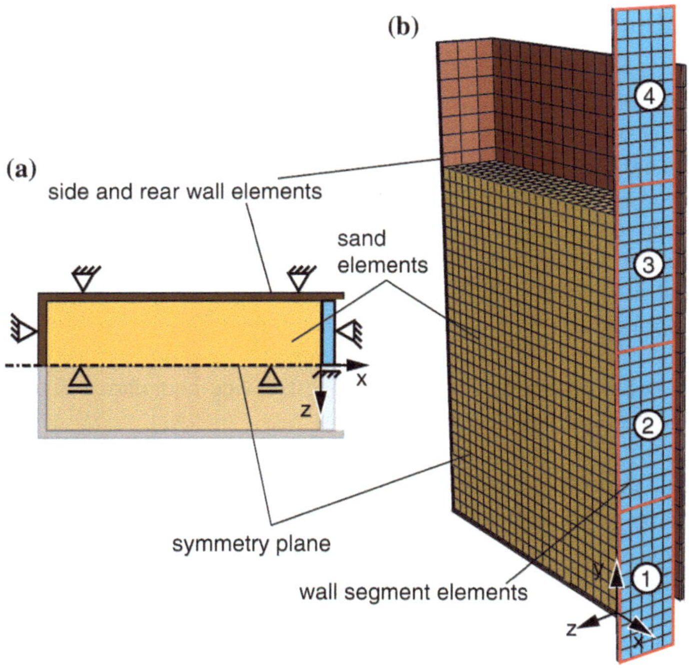

Fig. 8 **a** Top view and **b** 3D view of the FE model

modelled with shell elements (Abaqus keyword S4R). The supporting steel beams were not modelled. The deflection of the side walls was disabled. If this simplification is admissable, has to be verified in the individual case.

3.2 Simulation Steps

The simulation steps are schematically illustrated in Fig. 9. The first simulation step refers to a geostatic equilibrium calculation. The difference to usual simulations is the application of a strongly reduced gravity (e.g. to 1/100 g). This gravity must be in equilibrium with the defined initial stress field so that no (or only very small) displacements occur. Since there is no relative displacement between soil and boundaries, no friction is mobilized.

In the subsequent steps, the gravity is "switched-on" layer by layer from the bottom to the top. Each increase of weight of a layer is associated with one step. The weight-increase leads to a compaction of the lower layers and to a mobilisation of shear stresses on the boundaries. This reproduces the well-known "silo effect". It is also possible to use the Abaqus user subroutine *dload.for* for these kinds of calculation, but the presented method turned out to be more robust.

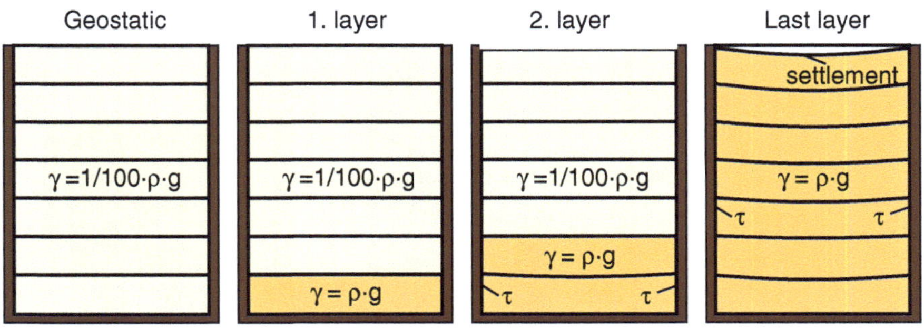

Fig. 9 Schematical illustration of the calculation process

The FE model was created in the *Abaqus scripting interface* and was entirely parameterised. Choosing a geometry and a number of soil layers, all element sets, calculation steps, etc. are then automatically defined. This allows for easy adaption of the FE model to other test device or geometries. The model was also applied for other test devices (e.g. [20]) to study the influence of boundary effects.

4 Comparison of Experimental and Simulation Results

4.1 Vertical Stress Evolution

Figure 10 shows the evolution of vertical stress σ_y on the two pressure transducers PT1 and PT2 as a function of the fill height h_{sand} for an exemplary test and its simulation. Figure 10b illustrates the position of the two pressure transducers with respect to the wall segments. The green line colour in Fig. 10a is attributed to test results and the red one to simulation results. PT1 is plotted in solid line and PT2 in dotted line. For better orientation, the linearly increasing pressure $-\gamma \cdot h_{sand}$ is also shown in black solid line. On the bottom x-axis h_{sand} is given in m and on the upper x-axis h_{sand} is normalized with the width b_z of the test device ($b_z = 0.5$ m).

As predicted by the silo theory, test results and simulation show an increase of vertical stress on both pressure transducers lower than the $\gamma \cdot h_{sand}$-curve. A good accordance between test and simulation can be observed. The evolution of stress on PT1 is very similar. On PT2, the differences between test and simulation are more pronounced with about 1 kN/m^2 for greater sand heights. With regard to a planned FE simulation of the subsequent model test, the agreement of test and simulation can be considered as satisfying.

4.2 Horizontal Stress Distribution

Figure 11 compares the resulting earth pressures on the wall segments for the same test. In the centre of the figure, the three wall segments interacting with sand are shown

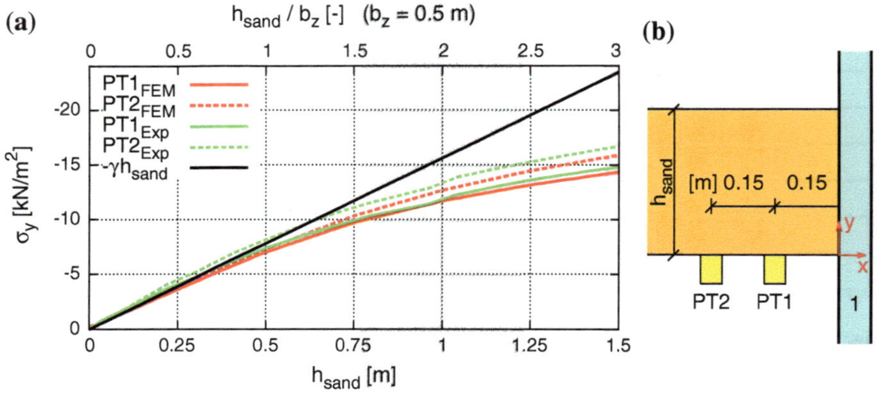

Fig. 10 **a** Vertical stress evolution $\sigma_y = f(h_{sand})$ on PT1 and PT2 measured an experiment (internal name V13-4) and obtained from FE simulation and **b** location of pressure transducers PT1 and PT2 in schematic side view (zoom)

schematically. The region filled with sand in shown with yellow background on their left. The second segment has a rough surface (coarse fraction of test sand bonded on the front sheet). In the upper right part of the figure, the static system of a wall segment is illustrated. The horizontal stress distribution is plotted in the left half of the figure (x-range bottom left) and horizontal bearing forces on the right (x-range top right). The horizontal stress distribution obtained from force and momentum equilibrium (assuming a section-wise trapezoidal shape) is plotted in green. The simulation results are plotted in red as the average over the width of the wall segments. For better orientation, the K_0-line is plotted in black (using $K_0 = 0.37$). Test and simulation results can be quantitatively compared considering the horizontal bearing forces. They are displayed in the right part of Fig. 11 as bar graphs for each horizontal bearing.

Figure 11 shows a very good agreement of test and simulation results. Globally, the earth pressure distribution is similar. Differences between experiment and simulation can be seen around the transition from one segment to another. The smooth curve obtained from the simulation shows a qualitatively expected and plausible shape, e.g. with reduced earth pressure under the rough second segment and directly above the bottom plate. The interpreted experimental results can per definitionem not provide such detailed information, but principally the stress distributions are well comparable.

The bar graphs in the right part of Fig. 11 are rather suited for quantitative comparison since they do not imply any assumptions. They prove the good accordance between test and simulation in an impressive way. All bearing forces are reproduced very well by FE simulation, except the upper one on the first segment. Even here, the difference is lower the 20 %, which is still an acceptable error tolerance for geotechnical simulations. Note that for these results no soil parameter fitting had to be performed. The good accordance is probably contributed to very reliable measurements. The wall segments are exposed to earth pressures over quite large areas (0.5 m $\times$ 0.6 m) so local inhomogeneities have a lower impact on the results.

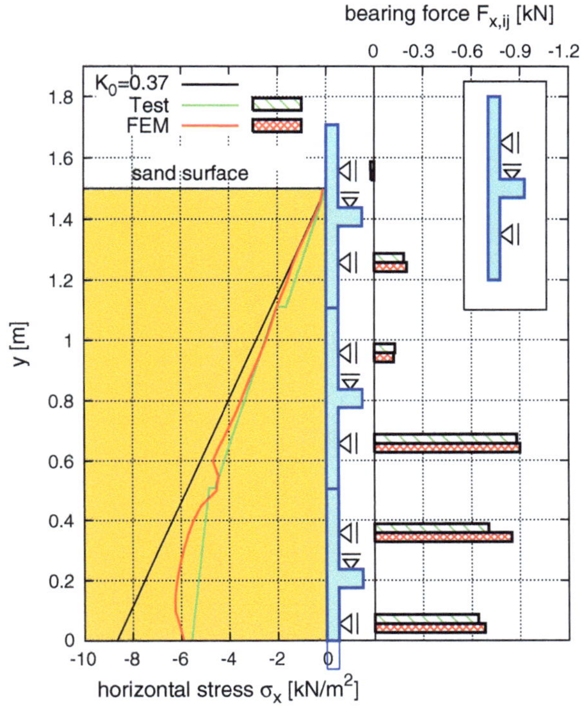

Fig. 11 Horizontal earth pressures and bearing forces measured in an experiment (internal name V13-4) and obtained from FE simulation

Furthermore, the well-known good performance of hypoplastic constitutive models for monotonic loading cases is once more proven.

4.3 Density Distribution

In order to examine the density distribution in the model test, CPT are conducted in recent experiments. The results of three CPT are depicted in Fig. 12 for an experiment with the internal name V18-2. The three CPT are named S18-21, S18-22 and S18-23. For the test presented before, no CPT were performed.

The whole force, measured by the load cell is interpreted as cone tip resistance q_c. An estimation of skin friction with a friction angle $\delta = 1/3\varphi' = 13°$ yields only to a low impact of about 1 % and was therefore neglected. Furthermore, Fig. 12 contains the vertical effective stress σ_y calculated as linear increasing with depth ($\sigma_y = \gamma \cdot h$) as well as by taking silo effects into account (see Fig. 10).

The tip resistances q_c in Fig. 12 show a certain scatter but CPT S18-22, which is close to the instrumented wall segments, gives the best and quite smooth results. This confirms the experimental impression, that the pluviation is of higher quality near to the wall segments (S18-22) than in the rear part of the test device (near S18-23), where the sand dropper sometimes stood still when it was refilled by the

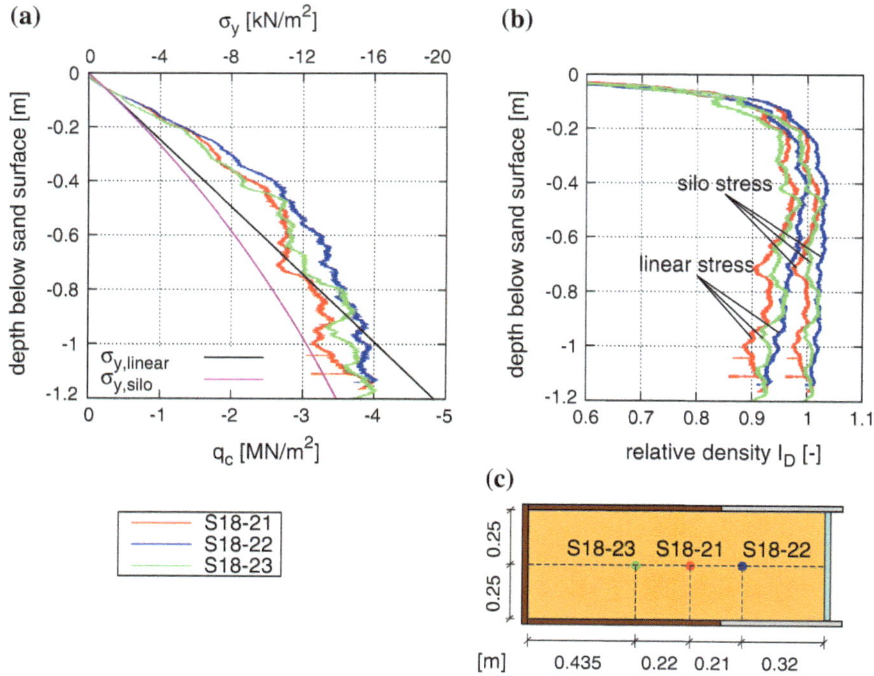

Fig. 12 **a** q_c of three CPT in the model test and assumed vertical effective stress σ_y (linear distribution with depth and with silo effect), **b** relative density I_D calculated with linear stress distribution with depth and with consideration of the silo effect and **c** location of the CPT in schematic top view

sand reservoir. The relative density I_D obtained from these CPT is evaluated with the method described in Sect. 2.4 with the linear stress distribution versus depth and also with the problem-adapted stress distribution incorporating the silo effect (see Fig. 12b). Only with the realistic stress distribution, considering the silo effect, the calculated density is almost homogeneous over depth. A linear stress distribution with depth (which is mostly adopted for interpretation of results!) gives an apparently non plausible decreasing relative density with depth.

The absolute value of $I_D \approx 1$ seems to be too high. The averaged relative density, calculated with soil weight and sample volume, is $I_D = 0.81$. This difference results probably from boundary effects. The cone was designed for a diameter of the sample $D_{sample} = 94$ cm. The space between the walls in this test device was only 50 cm, which results in a ratio of $D_{sample}/d_{cone} = 35.7$. As the walls are quite stiff, the measured cone tip resistance q_c is higher than it would be in free field conditions. This leads to the evaluated higher relative density, compared to mean density that was measured. Another reason for the discrepancy of I_D may also be found in non-optimized soil parameters for this special application.

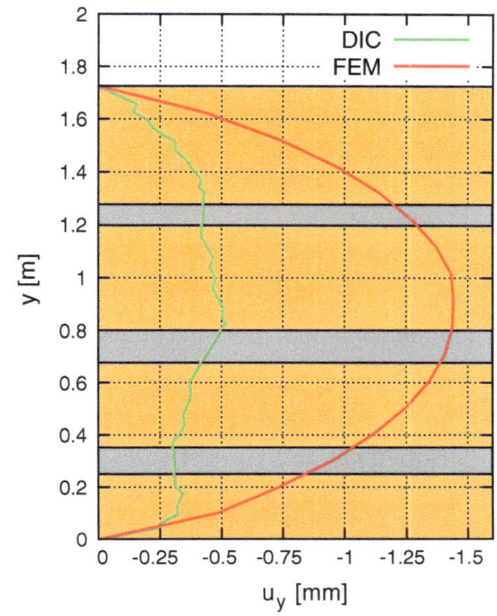

Fig. 13 Comparison of vertical displacements during filling process evaluated by DIC and FEM simulation

4.4 Displacements During Filling Process

Figure 13 shows the displacements occuring during the filling process in test V18-2 (internal denotation) compared with the associated simulation using 17 soil layers á 0.1 m height. The configuration of this test was a 2D-pile (for details see [14]). This configuration is more challenging for numerical simulation due to the non-planar wall geometry.

Figure 13 shows a qualitative shape as explained above for the test results as well as for the simulation. The displacements at the bottom plate and at the sand surface are zero per definition. The qualitative curve shape is similar in test and simulation. The maximum displacement is reached at a height of about 1 m. However, the maximum settlement in the simulation is three times as large as in the experiment. This is probably contributed to a non representative soil stiffness obtained by the hypoplastic equation for this low stress level. The stress level in the experiment ranges from 0 to $-15\,\text{kN/m}^2$. Constitutive equations are usually calibrated for stress levels ranging from -50 to $-500\,\text{kN/m}^2$. Specific parameter calibration considering low stresses may provide better results. The change of void ratio related to this settlement is very small and can be considered to be negligible for the simulation of the subsequent model test. Taking into account the soils small strain stiffness in the constitutive model might enhance the prediction of displacement. However, this point is not discussed in this contribution. The satisfying results of the stress distribution in the soil were considered to be sufficient for the simulation of the subsequent test.

5 Summary and Outlook

In this paper, some aspects related to the evaluation of the initial soil state in geotechnical model tests are discussed. Methods for the quantitative description of initial soil state are shown. Numerical modelling of the deposition process with the Finite Element Method is shown to be a solution for treating silo effects in geotechnical model test devices.

For the simulation of installation procedures a 3D FE model was used. The soil behavior was described by a hypoplastic constitutive model without the concept of intergranular strain. The sand deposition is modelled by a layered increase of weight from near to zero to the unit soil weight starting from the bottom of the soil body. The principal results can be summarized as follows:

(a) Horizontal and vertical stress measurements during the soil installation in a model test device are indispensable for the evaluation of the initial soil state. The sample homogeneity can be proven using CPT, if advanced interpretation methods are applied. Even more effort has to be made, when boundary effects influence the interpretation of CPT results. DIC can be an useful tool for the evaluation of settlements and soil deformation during the deposition process.
(b) A realistic back-calculation of the initial soil state in model tests is possible with reasonable effort using a 3D FE-model, a well-established constitutive model (here hypoplastic) and simple Coulomb contact description. Good results can only be achieved if contact friction angles experimentally determined and fully mobilized. The presented measurement and simulation methods are adaptable to other test devices and should be considered when quantitative validation of numerical simulation is intended.
(c) Experiments influenced by silo effects can provide experimental data for benchmark boundary value problems as long as the soil state can be modelled realistically with reasonable effort and can be validated with experimental data. The presented Interface test device was shown to meet these requirements and is therefore suited to provide experimental data for benchmark tests.

The contribution focuses on the preparation of large scale samples of cohesionless soil. The basic problem exists also for tests in clayey or silty soil. Whether the presented methods can be applied to those soils has to be investigated in the particular case. Usually, moist tamping compaction methods are applied in those soils, whose incorporation in the simulation is challenging.

Another important point only briefly mentioned before is the beneficial interaction of experiment and simulation, even for the experimentalist. Interpretation of experimental measurements can be strongly supported by numerical modelling of the process. The improved interpretation of CPT-data is only one example. Some effects occurring in the experiments can easily be studied in FE models, such as the influence of boundaries on the test results or the evaluation of soil deformations via DIC. The presented methods can also be applied to the design process of a test device. Pre-dimensioning with FEM can be a cost-efficient tool replacing expensive pre-tests. Their advantage is the "overall instrumentation" of an FE model. Stress and deformation can be plotted for every point in the model. This allows e.g. appropriate placement of transducers and avoids faulty design. This approach is only meaningful if high quality material models are available describing the soil behaviour realistically.

Important issues of future work are an improved consideration of boundary effects in the evaluation of soil density via CPT and the simulation of soil deposition for other test configurations resp. for sloped soil deposition (used for loose samples). Furthermore, the elevated soil stiffness at small strains has to be considered in order to investigate its influence in this case.

Acknowledgments The work presented in this paper was supported by the German Research Foundation (DFG) as part of the research project "Central project of the researchers group FOR 1136 with demonstrator experiments". The authors acknowledge the financial support.

References

1. Ahmadi, M.M., Karambakhsh, P.: K_0 determination of sand using CPT calibration chamber. In: 2nd International Symposium on Cone Penetration Testing. California (2010)
2. Bauer, E.: Calibration of a comprehensive constitutive equation for granular materials. Soils Found. **36**, 13–26 (1996)
3. Choi, S.-K., Lee, M.-J., Choo, H., Tumay, M.T., Lee, W.: Preparation of a large size granular specimen using a rainer system with a porous plate. Geotech. Test. J. **33**(1), 45–54 (2009)
4. Cudmani, R., Osinov, V.: The cavity expansion problem for the interpretation of cone penetration and pressuremeter tests. Canadian Geotech. J. **38**(2), 622–638 (2001)
5. DIN 4094–1:2002: Baugrund, Felduntersuchungen Teil 1: Drucksondierung. Juni (2002)
6. Fretti, C., Lo Presti, D.C.F., Pedroni, S.: A pluvial deposition method to reconstitute well-graded sand specimens. Geotech. Test. J. **18**(2), 292–298 (1995)
7. Janssen, H.A.: Versuche über Getreidedruck in Silozellen. Zeitschrift Verein deutscher Ingenieure, Band XXXIX. 1045–1049 (1895)
8. Maier, C.: Herstellung trockener Großproben mit verschiedenen Lagerungsdichten. Diplomarbeit am Institut für Bodenmechanik und Felsmechanik des Karlsruher Instituts für Technologie (2011)
9. Mayne, P.W., Kulhawy, F.H.: Calibration chamber database and boundary effects correction for CPT data. In: Proceedings of the 1st International Symposium on Calibration Chamber Testing, pp. 257–264. Potsdam and New York (1991)
10. Niemunis, A., Wichtmann, T., Triantafyllidis, Th.: Spatial stress fluctuations: acoustic evidence and numerical simulations. In: Pande and Pietruszczak (eds.) Numerical Models in Geomechanics (Numog X), pp. 159–166. Taylor & Francis, London (2007), ISBN 978-0-415-44027-1
11. Rebstock, D.: Verspannung und Entspannung von Sand entlang von Baukörpern. Dissertation (2011), http://digbib.ubka.uni-karlsruhe.de/volltexte/1000023891
12. Schmertmann, J.H.: An updated correlation between relative density Dr and Fugro-type electric cone bearing qc. DACW 39–76 M6646, Waterways Experiment Station, USA (1976)
13. Vennemann, P.: JPIV-software package for particle image velocimetry (2007), http://www.jpiv.vennemann-online.de
14. Vogelsang, J., Huber, G., Triantafyllidis, Th, Schindler, U.: Pfahlpenetration in nichtbindigem Boden: Großmaßstäbliche Modellversuche und Nachrechnungen. Pfahlsymposium, Braunschweig (2013)

15. Vogelsang, J., Huber, G., Triantafyllidis, Th: A large scale soil-structure interface testing device. Geotech. Test. J. **36**(5), 613–625 (2013). doi:10.1520/GTJ20120213, ISSN 0149-6115
16. Walker, B.P., Whitaker, T.: An apparatus for forming uniform beds of sand for model foundation tests. Géotechnique **17**(2), 161–167 (1967)
17. White, D.J., Bolton, M.D.: Displacement and strain paths during plane-strain model pile installation in sand. Géotechnique **54**(6), 375–397 (2004)
18. Wienbroer, H.: Umlagerung von Sand durch Wechselbeanspruchung. Diss., Veröffentlichungen des Instituts für Bodenmechanik und Felsmechanik des Karlsruher Instituts für Technologie, Heft 174 (2011)
19. von Wolffersdorff, P.-A.: A hypoplastic relation for granular materials with a predefined limit state surface. Mech. Cohesive-Frictional Mater. **1**, 251–271 (1996)
20. Zachert, H., Wienbroer, H., Triantafyllidis, Th.: Experimentelle Untersuchung der Verformungsakkumulation eines Flachgründungsmodells von Offshore-Windenergieanlagen. Bautechnik 88, Heft 11 (2011)

Demonstrator Experiments on Significant Effects During Pile Installation

J. Vogelsang, G. Huber and Th. Triantafyllidis

Abstract Benchmark tests for the numerical simulation of pile installation require clearly defined boundary value problems with corresponding experimental data. These experiments have to provide quantitative information on the soil deformations and stresses. Large scale model tests in dry, granular soil were carried out for this purpose. The interface testing device, that is used for the tests, allows the investigation of selected aspects of pile penetration. The normal and shear forces on the pile structure are measured. The displacements in the surrounding soil zone can be evaluated via Digital Image Correlation (DIC). The test results concentrate on the interface behavior between the soil and the pile and the evolution of stresses and deformations around the pile tip. For rough pile surfaces the occurrence of dilatancy effects in the pile-soil interface is shown. The localization of deformations in the post-peak phase is analyzed for monotonic and cyclic test paths. The influence of the pile driving mode on the evolution of stresses around the pile tip is demonstrated.

Keywords Pile installation · Large scale model test · Benchmark test · Soil-structure interaction

1 Introduction

The research unit FOR 1136 has the objective to provide numerical tools for the simulation of pile driving processes. The main focus is on the installation of vibro-injection piles, in German RI-piles (Rüttelinjektionspfähle). Such piles are schematically illustrated in Fig. 1. RI-piles are H-section steel profiles that are vibrated into the ground. The collar, a welded flat-bar placed above the pile tip, displaces the soil and creates a cavity above. Simultaneously to the pile installation, cement grout is

J. Vogelsang (✉) · G. Huber · Th. Triantafyllidis
Institute of Soil Mechanics and Rock Mechanics, Karlsruhe Institute of Technology, Karlsruhe, Germany
e-mail: triantafyllidis@kit.edu

Th. Triantafyllidis (ed.), *Holistic Simulation of Geotechnical Installation Processes*, Lecture Notes in Applied and Computational Mechanics 77,
DOI 10.1007/978-3-319-18170-7_2

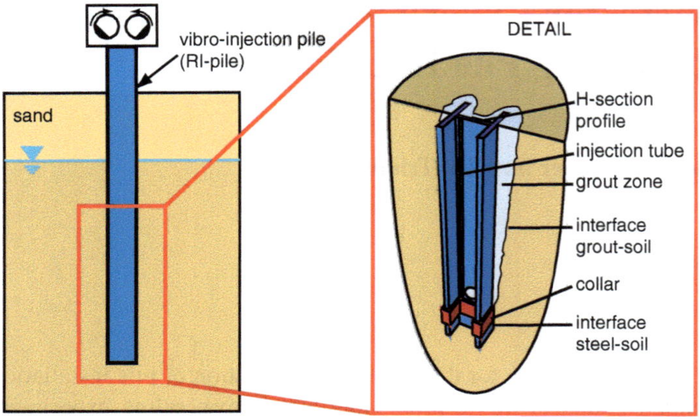

Fig. 1 Schematic illustration of RI-pile installation

injected into this cavity and encloses the steel profile. When cured, the grout offers a corrosion protection in combination with a rough contact to the surrounding soil.

These processes are influenced by several effects that have to be taken into account in simulations. Pile driving is a high-cyclic, dynamic process that often takes place in water saturated soil. This causes important liquefaction effects around the pile. The penetration of the pile structure induces large soil deformations that have to be handled using adequate simulation techniques. The interface between the soil and the pile structure has to be considered as well as the interaction between soil and cement grout. The influences of these aspects are difficult to separate, for real pile installations as well as in experiments. Therefore, demonstrator experiments on selected aspects have to be performed. The experimental investigations discussed here concentrate on the pile penetration in dry soil without grout injection. In this contribution, the requirements for these tests are defined, the test devices are illustrated and representative results are shown.

Many geotechnical model tests dealing with penetration of pile-like structures have been performed, mostly in dry granular soil in order to exclude pore water effects. E.g. [2] or [6] present experiments with small scale piles penetrated into dry sand. Usually the overall axial/vertical pile force is measured in relation to the penetration depth. In [2] the tip resistance and skin friction forces were extracted from the pile head force. Soil displacements are rarely evaluated in these kind of tests. Other experiments using glass windows installed in symmetry planes allow the observation of soil displacements, e.g. the model test performed in [1]. Another problem is the measuring of the stress distribution. In [9] a possible solution is presented, however, without evaluation of the soil displacements. The combination of extensive stress measurements and evaluation of the soil displacements in one experiment is a complex challenge. Such experiments were therefore only performed very rarely.

Often, the purpose of geotechnical model tests is a qualitative prediction on the behavior of real pile structures. As shown e.g. in [14], the quantitative application on real structures of results obtained from small scale tests is doubtful. Scaling effects can significantly limit the transferability of the test results. Some field tests that enable the application of test results on real pile structures have been performed. Examples are presented in [8] or [4]. Those tests eliminate scaling effects and are performed in the in-situ soil layers. The instrumentation is usually limited on the axial pile head force. In [8], in addition, a sophistically instrumental model pile allowed the evaluation of tip resistance and skin friction.

To provide experimental data for boundary value problems that can be back-calculated with numerical methods is another purpose of geotechnical model tests. Such experiments have to reproduce similar effects as the real process that is to be investigated. The test conditions have to be kept as simple as possible and an extensive instrumentation has to provide quantitatively reliable results on the evolution of stresses and deformations. It is useful to focus the experimental set-up on selected aspects (e.g. contact zone between the soil and the structure or liquefaction effects) and to exclude effects of secondary interest. These effects can rarely be separated in small scale tests. Large scale tests that simulate only parts of a real structure in a realistic scale are one possibility to solve this problem. Such model tests are presented in this paper.

A large scale interface testing device is used to investigate the soil-structure interaction for monotonic and cyclic test paths. A sophisticated instrumentation enables the measuring of shear and normal stresses in the interface for smooth and rough structural surfaces. The soil displacements are observable and can be evaluated by DIC.

The interface testing device is also used for tests on the penetration of pile-like structures under plane strain conditions. The test configuration and the instrumentation allow the observation of skin friction and tip resistance. Idealized tests simulating a 2D pile penetration are performed for the validation of FE-techniques. In the tests, the difference between the so-called cavitational and non-cavitational pile driving [2] and its influence on the soil deformation is shown.

2 Interface Test Device

The interface test device used for the tests has already been presented in detail in [10, 13]. Nevertheless, for a better comprehension of the following recent results, we will recapitulate the essential aspects.

2.1 Basic Idea

The basic idea of the interface test device is the transition of an idealized pile structure into plane structures that are investigated separately. In Fig. 2a an idealized pile

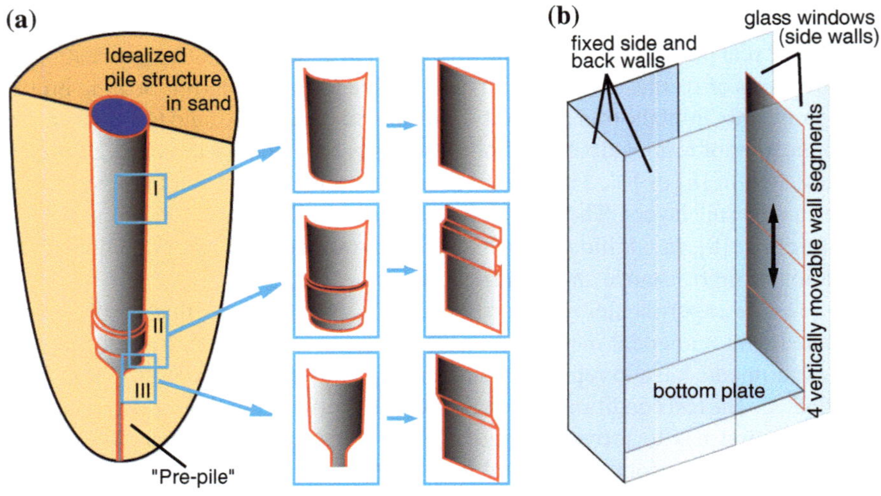

Fig. 2 Interface testing device—basic idea. **a** Transition of sections of an idealized pile structure to 2D wall elements. **b** Schematic design of the interface test device

structure is shown. The idealized pile is simplified to a radial symmetric structure that is penetrated into sand. This simplified pile has skin sections (detail I), structural elements imitating the collar of an RI-pile (detail II) and a pile tip (detail III).

As in many numerical simulations the pile has a "pre-pile" with a small diameter compared to the pile. The pile itself represents only an expansion of this pre-pile. In the interface test device parts of this idealized radial symmetric structure are cut out and "unfolded" to plane elements. This transition is illustrated in Fig. 2a. The plane pile elements are integrated in the interface test device as wall segments, Fig. 2b. The wall segments are guided and can be moved vertically between two glass windows placed on the side. The back part of the test device is designed with stiff construction plates. A bottom plate forms the lower boundary of the box and the sand filling. In the tests, the wall segments interact with the sand and are displaced vertically performing various test paths. The sand displacements can be observed through the glass windows on the sides. The forces acting between the sand and the wall segments are measured by load cells integrated in the supports of the wall segments. In that way, both, displacements and stresses in the zones of interest, are evaluated.

2.2 Design

Figure 3a shows a side view on the interface test device. Some design details are shown schematically in Fig. 3b. More technical details are discussed in [13].

A stiff outer steel construction minimizes the deformations of the test device and their influence on the experiments. From the outside, the back part of the sand body

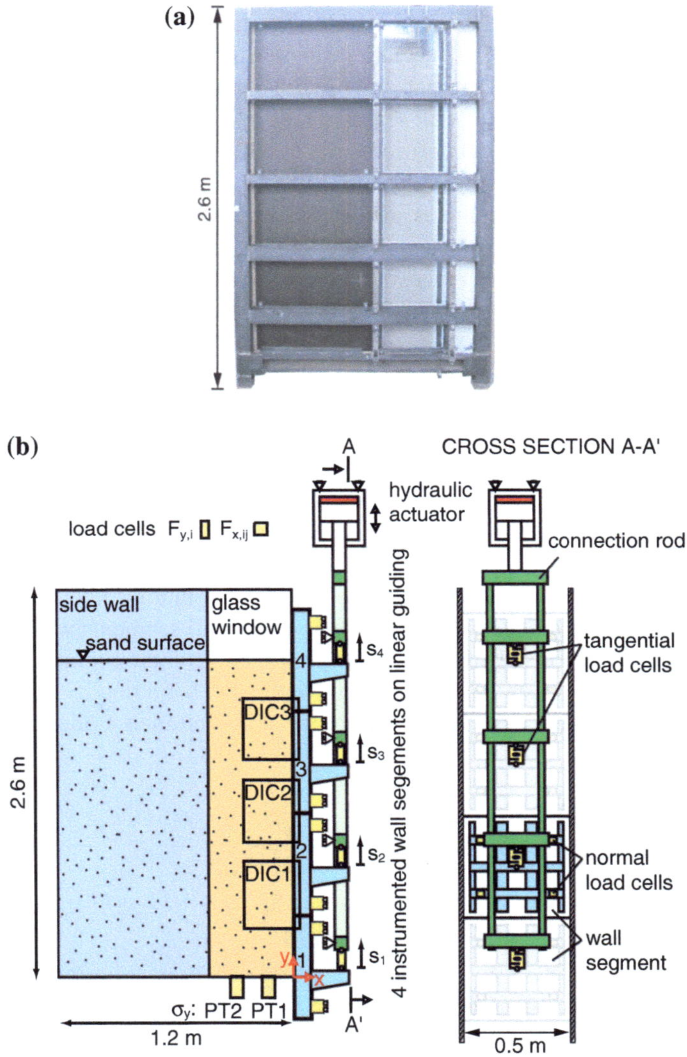

Fig. 3 Overview—interface test device. **a** Side view and **b** schematic side view and view on the back side of the wall segments

is hidden by side walls, only the first part in vicinity to the wall segments is visible through glass windows (on the front and on the back side). Between the horizontal steel beams five fields with free view on the sand body can be used to take image sequences during the tests. Usually, the three middle windows are used (named DIC1, DIC2 and DIC3 in Fig. 3b).

The base area of the sand filling is 1.2 m × 0.5 m. Sand heights up to 2.0 m are possible. Usual sand heights range between 1.5 and 1.7 m. The movable wall is divided into four instrumented segments (numbered 1–4 from bottom to top). The wall segments are based on a stiff steel frame in order to minimize their deformations. The front plates of the wall segments are modifiable and can be adapted depending on the test configuration. Each wall segment has load cells for the measuring of the normal and the tangential (vertical) force acting on the front plate, see below. The wall segments are coupled separately to a ladder-like connection rod. The view on the back side of the wall segments in Fig. 3b illustrates their position relative the connection rod. The second segment is highlighted. A hydraulic actuator is placed on top of the test device and moves the connection rod in vertical direction. The maximum possible wall displacements are about 200 mm.

2.3 Test Configurations

Various test configurations are possible in the interface test device. Three basic configurations for the simulation of pile driving can be distinguished and were investigated, Fig. 4.

Configuration I focuses on the cyclic and monotonic interface behavior between the soil and the pile. In this configuration four plane wall segments with modifiable surface roughness are installed. In the tests presented here, the segments 1, 3 and 4 have smooth surfaces of stainless steel, while segment 2 has a rough surface (sanded steel). The wall segments represent a part of the skin section of a pile.

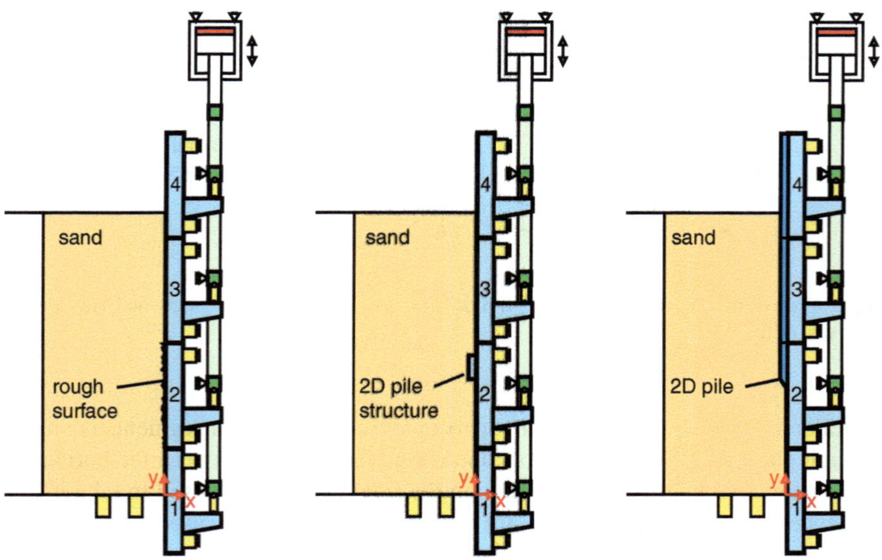

Fig. 4 Basic test configurations in the interface test device

Configuration II investigates the influence of structural geometries more similar to real pile structures. The segments are installed with smooth steel surfaces. A pile structure imitating the collar of RI-piles is mounted on segment 2. The tip resistance on this structure can be evaluated. In addition to this, its influence on the skin friction on the upper segments can be observed as well.

Configuration III represents a 2D pile and is particularly suited for the validation of numerical simulations of the pile penetration problems. Segment 2 contains an inclined pile tip. Segment 3 and 4 are shifted into the test device and form a complete pile in 2D in combination with segment 2.

2.4 Instrumentation

The test device provides a sophisticated instrumentation. The basic instrumentation installed in the test device is presented in [10]. The more advanced instrumentation of the test device which has been added in a later stage of development is discussed in [13]. Here, an overview of the instrumentation components is provided, Fig. 3.

Tangential Forces $F_{y,i}$. Each wall segment is attached individually to the connection rod by one tangential load cell (HBM RSCA). In the case of plane wall segments, the tangential load cell generally measures the resulting shear force on the front plate and furthermore a weak systemic friction force of the Teflon sealings on the glass windows. In the case of segment 2 in the test configurations II and III, a tip force under the pile tip is also included in the tangential force.

Normal Forces $F_{x,ij}$. In the basic version of the test device [10], the measuring of the normal forces on the wall segments was not possible. However, for the usage as demonstrator experiments for the validation of contact formulations, the knowledge of normal stresses in the contact zone is crucial, but its measurement in granular soil is complex. The measuring elements have to be very stiff in order to avoid a significant influence on the earth pressure. Due to high quality and stiffness requirements in combination with geometric constraints, the usage of commercial products was excluded. Thus, self-designed and fabricated load cells are used. Their construction is discussed in detail in [13]. Four load cells per segment are integrated between the steel frame and the guidings. Usually, the sum of all four load cells $F_{x,ij}$ per segment is shown in the results ($F_{x,i}$: resulting normal force on segment i).

Displacements s_i. The non-uniform force distribution on the wall segments leads to different deformations of the tangential load cells, ergo varying displacements of the wall segments. Therefore, the displacements are measured separately with one absolute magnetostrictive position transducer (Novotechnik TLM 300) per segment.

Vertical Stress σ_y. Two pressure transducers measure the evolution of the vertical stresses on the bottom plate. They are integrated centrically into the bottom plate with a horizontal distance to the wall segments of 0.15 m (PT1) and 0.30 m (PT2). Pressure transducers with very stiff front membrane are used (VEGABAR 52).

2.5 Evaluation of Soil Displacements

Image Sequences. The first part of the side walls is constructed with glass windows. Thus, between the horizontal steel beams the contact zone between the wall and the soil is kept visible throughout the tests. During the tests three digital cameras take image sequences in the fields DIC1, DIC2 and DIC3, Fig. 3. The arrangement of the cameras on the side of the device is illustrated in detail in [13].

The obtained image sequences are preprocessed digitally before the DIC evaluation. The images are rotated in order to precisely place the wall in a vertical orientation. Only the zone of interest (sand without steel beams) is evaluated subsequently. The edges of the wall segments (where the displacements s_i are measured) are kept in this zone in order to allow a control of the calculated displacements.

Digital Image Correlation (DIC). The freeware code JPIV [12] is used to evaluate the incremental displacement fields by comparing every pair of consecutive images in the recorded image sequences. Many recommendations can be found in the literature for the settings in DIC evaluations [15, 16]. The settings chosen here are in accordance with these. The vector spacing (width and height of the search patch) is set to 12–16 pixel ($4-5 \times d_{50}$). Thus, the search patch contains enough pixel and sand grains for a good precision and it enables the visualization of localized deformations (shear band width about $10 \times d_{50}$, see below).

After the JPIV evaluation invalid vectors are marked by a "normalized median test" and replaced by the median of the neighboring vectors. Invalid vectors are marked if their displacement differs too much compared to the neighboring vectors (details in [12]). Invalid vectors can occur in areas of the image that provide only low texture. In this particular application the problematic zones are on the edges of the wall segments and especially the teflon sealings. In the evaluations presented, the number of invalid vectors was very low (maximum 3 per image evaluation). If appropriate settings are used invalid vectors in the sand zone do not occur.

From the JPIV analysis ASCI-files containing the incremental displacements (image n and n+1) are obtained. Subsequently, these displacements are summated to the cumulated displacements and the deformations are calculated.

3 Test Results

3.1 Configuration I: Soil-Pile Interface Behavior

The soil-pile interface behavior was investigated with a series of tests using the first configuration (plane wall segments). In this test configuration, the segments 1, 3 and 4 were installed with smooth steel surfaces. Segment 2 has a rough (sanded) surface. Monotonic tests were performed as well as cyclic tests with different test paths. Here, a cyclic test is compared to a test performing a monotonic wall displacement. The initial conditions were similar in both tests with about 1.5 m sand height and an initial relative density of $I_D \approx 65\,\%$. The initial conditions and test paths are schematically illustrated in Fig. 5a, b.

A wall displacement of 57 mm downwards is imposed in the monotonic test. In total, the cyclic test performing an alternating wall displacement reaches also 57 mm wall displacement. The alternating displacement was carried out using the following progression sequence: first 11 cycles of 3 mm down and 1 mm up and afterwards 7 cycles of 5 mm down and 1 mm up.

The Fig. 5c–e compare the results of both tests. The evolution of the tangential force on the second wall segment is shown in Fig. 5c. In the cyclic test (blue curve) the tangential force increases over 3 cycles and reaches its maximum at 9 mm displacement. In the pull-out phases the tangential force vanishes in all cycles, except the first. In all other cycles the wall displacement of 1 mm is not enough for a reversion of the shear stresses. In the monotonic test (red curve) the tangential force increases slightly stronger on the first 6 mm. After 6 mm displacement the curve kinks, cuts the peak of the cyclic curve and behaves like an envelope in the residual test path.

Both tests show a peak in the curve of the normal force, Fig. 5d. The peak occurs after 9 mm in the monotonic and after 11 mm in the cyclic case. The normal force in the monotonic test envelopes the curve of the cyclic test. The normal force does not vanish in the pull-out phases because in granular soil a minimum earth pressure has to be maintained.

Since normal force $F_{x,2}$ and tangential force $F_{y,2}$ act on the same area, the mobilized friction angle on the segments surface δ_{mob} can be calculated as follows:

$$\delta_{mob} = \arctan(F_{y,2}/F_{x,2}). \tag{1}$$

The cyclic test shows a global maximum of $\delta_{mob} = 39°$ after 7 mm displacement. Furthermore, in all subsequent cycles a small peak is mobilized. The residual value of δ_{mob} is about 35°. The monotonic test shows an earlier peak after 4 mm displacement. Its maximum is about 38°. The residual value ranges between 33 and 36° and corresponds approximately to the residual values of the cyclic test. In neither case a critical state is reached ($\varphi_{c,sand} = 33°$ from drained triaxial tests).

The results shown in Fig. 5 become more clear when the soil displacements obtained from DIC are considered. The displacement vectors of the monotonic test are plotted in Fig. 6a for a small horizontal cross section in the interface zone between the rough wall segment 2 and the sand. The accumulated displacements are shown for the monotonic test at the peak of the mobilized friction angle (s = −3.5 mm), after the peak (s = −6.0 mm about at the kink of $F_{y,2}$), at the global minimum (s = −8.5 mm about at the peak in $F_{x,2}$) and after a large displacement of s = −20 mm. For the cyclic test the accumulated displacements are shown at similar wall positions in Fig. 6b. The position s = −3.5 mm is now before the peak, s = −5.8 mm about at the peak and s = −8.5 mm just after it. A displacement of s = −20.4 mm still corresponds to a position on the residual path.

Figure 6a reveals that the displacement field is smooth until the peak of δ_{mob} is reached. The displacements constantly decrease at larger distances from the wall

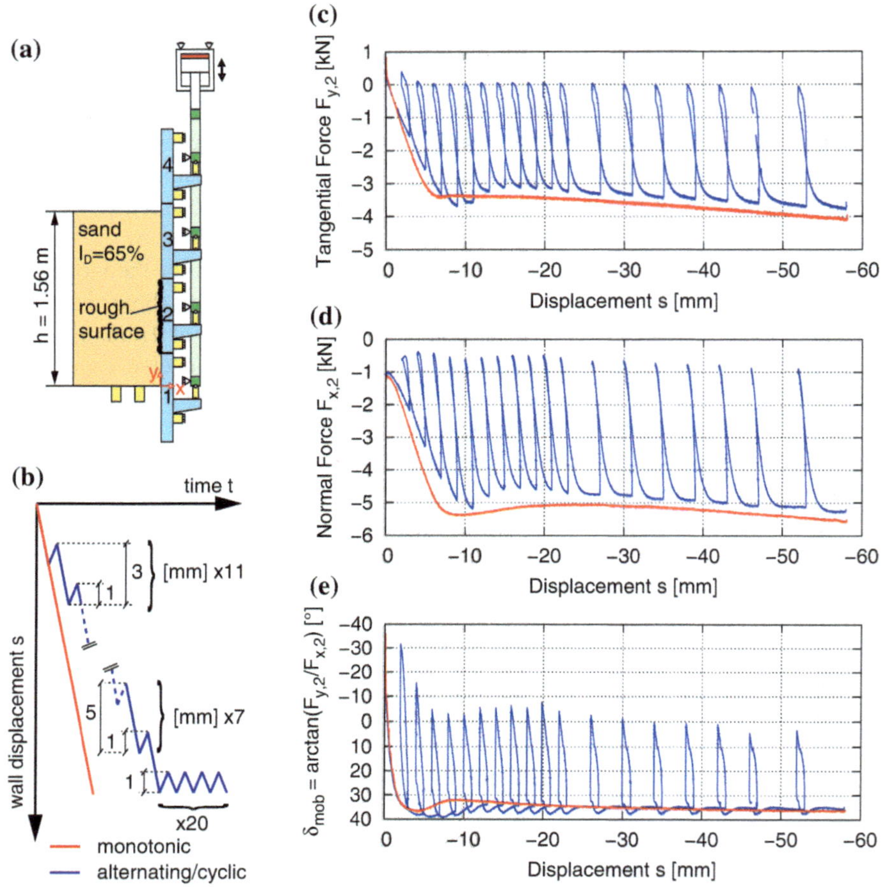

Fig. 5 **a** Initial conditions, **b** test paths and **c–e** tangential force $F_{y,2}$, normal force $F_{x,2}$ and mobilized friction angle $\delta_{mob} = \arctan(F_{y,2}/F_{x,2})$ on 2nd segment over wall displacement

segment. The deformations are more or less homogeneous. After the peak, an increase of the displacements near the segment surface can be observed. The deformations start to localize in a shear band in front of the wall segment. After 8.5 mm wall displacement, the zone with large gradients is already more pronounced. It becomes evident after a large displacement of 20 mm. The zone of large localized deformations consists in a band of about 5–6 mm in width. This corresponds to approximately 9–11 times the mean grain diameter d_{50} (d_{50} = 0.55 mm). The slight drift of the displacement vectors to the left, beginning from the fourth vector (from the right), indicates a dilative soil behavior in the shear band. The post peak behavior with the beginning of strain localization was already reported, e.g. for triaxial tests in [5, 11] or for various model tests in [7].

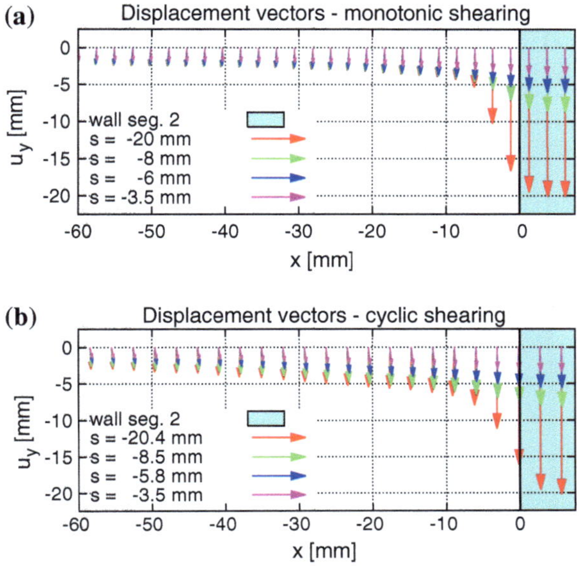

Fig. 6 Displacement vectors for **a** the monotonic and **b** the cyclic test in a horizontal cross section at y = 955 mm

The cyclic test shows similar displacement vectors but the localization occurs later, Fig. 6b. After 8.5 mm wall displacement, the deformations are still homogeneous. Also for the cyclic test, the strain localization corresponds to the peak of δ_{mob}. For 20 mm displacement the shear band in front of the wall segment can clearly be distinguished. The width of the shear band is similar to the monotonic case. With about 7 mm in width, it is a bit larger than in the monotonic case. This width corresponds to approximately 12–13 times the mean grain diameter d_{50}.

In the case of the cyclic test, after the wall displacement sequence downwards a cyclic wall displacement with constant amplitude of 0.5 mm was performed. 20 cycles were performed. Figure 7 plots the averaged tangential (shear) stress over the normal stress for selected cycles. The averaged stresses are the tangential and normal forces $F_{\mathrm{y},2}$ resp. $F_{\mathrm{x},2}$ divided by the surface area of the interface zone (0.3 m^2). This presentation is equivalent to the averaged $\tau-\sigma$-plane in the interface between the soil and the wall. The two critical state lines (CSL) are plotted with $\varphi_{\mathrm{c}} = 33°$. In the bottom left corner of each subfigure, the test path and the current position are indicated schematically. The current cycle is highlighted in red. The previous test path is illustrated in grey with continuous line and the subsequent test path with a grey dotted line. The force path is presented analogously. Small arrows illustrate the curve orientation.

Figure 7a shows the last phase large displacements downwards of the previous test path. Both, tangential and normal stress, increase significantly and exceed the CSL. In the first cycle, the tangential stress drops to zero, Fig. 7b. The end of the first cycle corresponds to the preceding phase of large displacement downwards. In Fig. 7c a strong relaxation of both stresses can be observed for the second cycle. The cycle ends on the CSL. After five cycles a relaxation can still be observed, however, at a

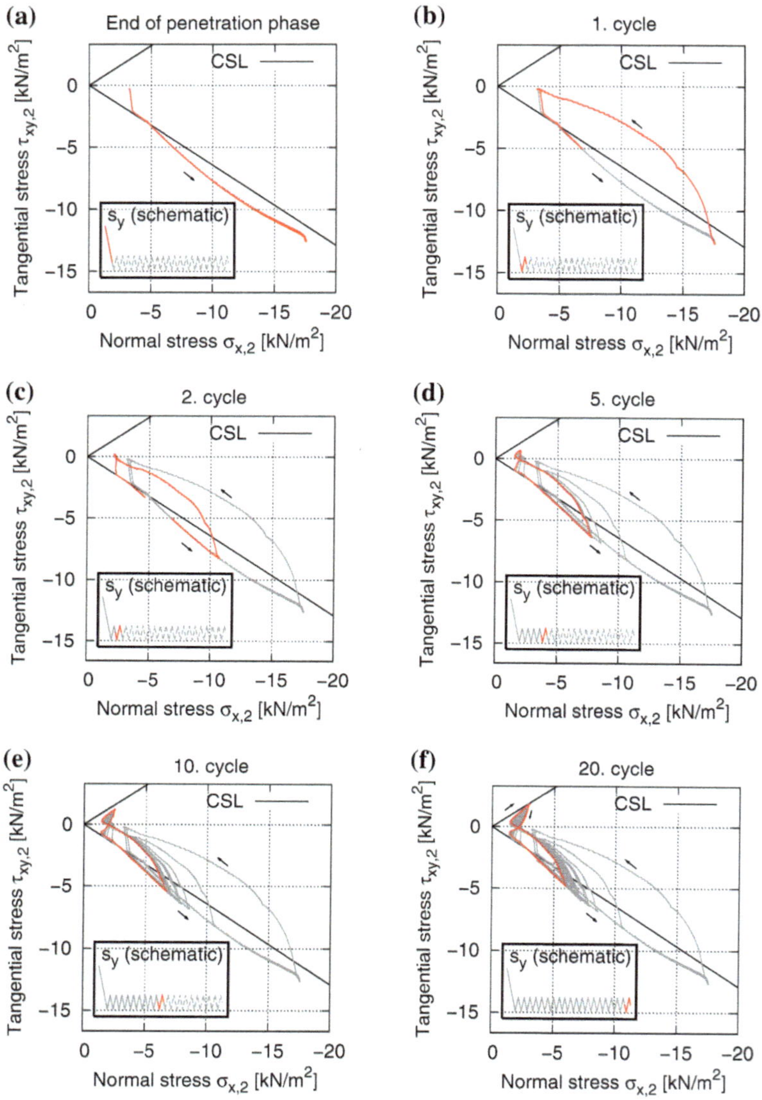

Fig. 7 Tangential stress $\tau_{xy,2}$ over normal stress $\sigma_{x,2}$ for cyclic wall displacement with $s_{ampl} = \pm 0.5$ mm

lower rate. In the pull-out phase, the tangential stress starts to reverse and a butterfly curve shape forms. In the following cycles, the penetration phases are similar to the preceding cycles with only slight relaxation. However, in the pull-out phases, a stronger stress reversion takes place (Fig. 7e, f).

Figure 7 clearly reveals the soil-like behavior of rough interfaces. Contractancy and dilatancy effects have to be modeled by contact formulations if the interaction is of primordial importance for the simulation. For pile penetration problems this is assumed to be the case in the region close to the pile tip and along the pile shaft (in the case of non-injection piles).

3.2 Configuration II: Estimation of the Tip Resistance

The results of a test using configuration II are shown in order to illustrate the influence of the pile driving mode for dense sand. The initial conditions and the test path are given in Fig. 8a. The test path corresponds to an alternating penetration of 57 mm in total. After 20 mm, two small cycles with 0.5 mm movement upwards are performed. In comparison to these small cycles, three larger cycles (5 mm upwards) are performed after 40 mm wall penetration. The presented results concentrate on the effects around the 2D pile tip installed on the second segment.

Figure 8b shows the tangential force $F_{y,2}$ on the second segment (with pile tip) over the wall displacement. The arrows indicate the curve progression. After about 10 mm wall displacement, the tangential force increases almost asymptotically. The two small cycles lead to a slight reversion of the tangential force up to about 0.25 kN. The response after the pull-out phases is stiff. The asymptotic envelope curve is already reached 4 mm after the last reversal point. The larger cycles are significantly different. In the pull-out phases a value of 0.25 kN is rapidly reached and remains constant throughout the whole phase. After the pull-out phase, the tangential force almost instantaneously changes direction to a negative value of −0.4 kN. Subsequently, the force remains approximately constant for 2 mm of penetration before it restarts to increase. The increase is much slower compared to the small cycles. The soil behavior is significantly softer. 4 mm additional displacement in these cycles are not enough to reach the asymptotic curve of the previous penetration.

Evidently, the evolution of the tangential force is a result of the soil behavior under the 2D pile tip (collar). The tip force cannot be measured directly in the current configuration but it can be evaluated from the measured forces.

The force distribution on wall segment 2 is illustrated in Fig. 9 for displacements downwards and upwards.

The tangential force $F_{y,2}$, measured by the tangential load cell, is usually directed downwards in the penetration phases and upwards in the pull-out phases. The normal force $F_{x,2}$ is always a compressive force and enables the mobilization of friction on the wall segments surface. $F_{x,2}$ is measured via four normal load cells $F_{x,2} = \Sigma F_{x,2,i}$ ($i = 1{-}4$). The resulting friction force is named F_τ and can be estimated as follows:

$$F_\tau = \int^{A_{seg}} \tau \, dA = F_{x,2} \cdot \tan(\delta_{mob}), \qquad (2)$$

wherein A_{seg} is the surface area of the segment and δ_{mob} is the mobilized friction angle steel-soil. For large unidirectional displacements (>1 mm), this friction angle can be assumed to be fully mobilized ($\delta_{mob} = \delta_{max}$). The segments surface partly

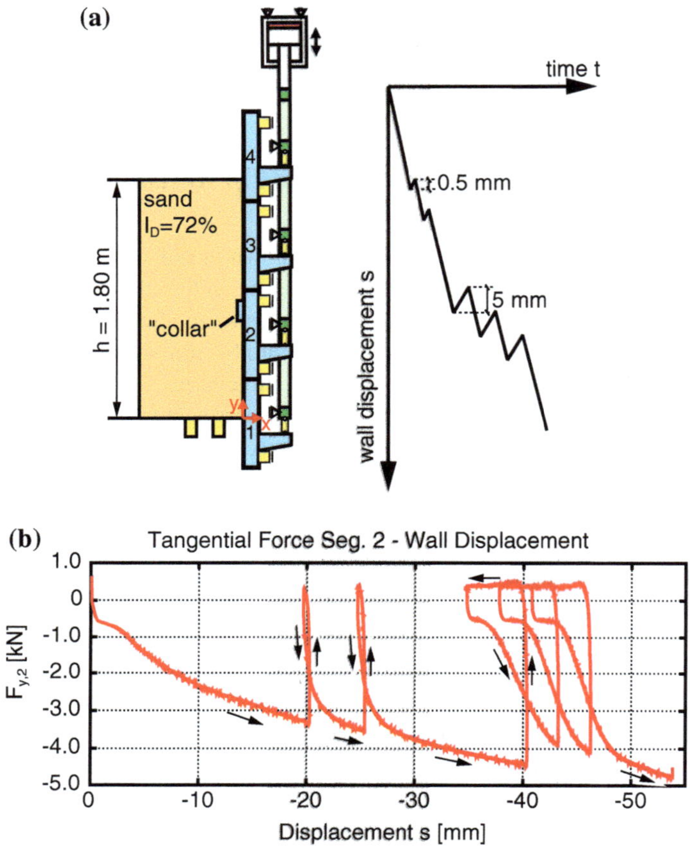

Fig. 8 V14-4—Alternating penetration in dense sand. **a** Initial conditions and test path and **b** tangential force $F_{y,2}$ on 2nd segment over displacement

Fig. 9 Force distribution on segment 2 for displacement **a** downwards and **b** upwards

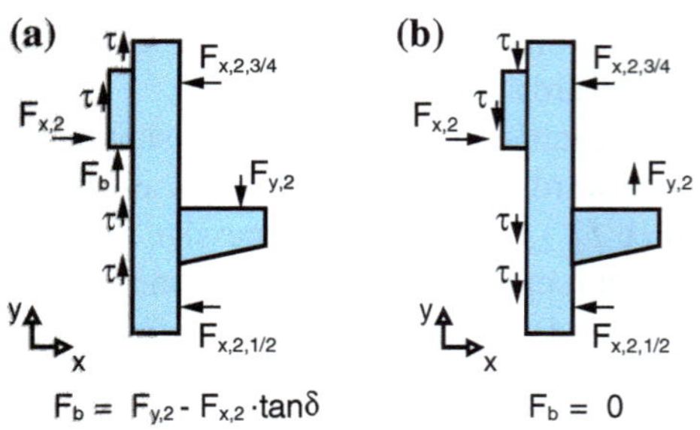

consists of stainless (front plate) and partly of common steel (collar). For the stainless steel surfaces, δ_{max} can be measured on the smooth segments 1 or 3. Usually, values of about 13° are obtained. Taking the higher roughness of common steel into account, δ_{max} was estimated at 16° here. The friction force is directed against the wall displacement, thus, against $F_{y,2}$. The last force to be considered is the tip force F_b. This force is always directed upwards and vanishes for large displacement in the pull-out phases.

Assuming that the friction angle is fully mobilized, the tip force can be estimated accurately. Taking the distinction of cases (displacements down- or upwards) into account, the absolute value of the tip force F_b can be calculated as follows:

$$|F_b| = |F_{y,2}| - |F_{x,2}| \cdot \tan(|\delta_{max}|) \quad (\Delta s \downarrow) \tag{3}$$

$$|F_b| = |F_{x,2}| \cdot \tan(|\delta_{max}|) - |F_{y,2}| \quad (\Delta s \uparrow). \tag{4}$$

Problems occur at reversal points in the test path. Just after changes of direction of the wall displacements, the mobilized friction angle is unknown. A characteristic line for its evolution after reversal points cannot be specified for all cases. Amongst others, it strongly depends on the stress state and the soil density. This problem is (yet) not solved at the moment. Thus, the tip force can only be roughly interpolated in the phases about 0.3–0.4 mm wall displacement after reversal points.

Figure 10 presents the calculated tip force obtained from Eqs. 3 and 4 for test V14-4. The sections calculated with Eq. 4 are plotted in red, those calculated with Eq. 4 in blue and the interpolated section in green.

The calculated tip force in Fig. 10 is comparable to the tangential force $F_{y,2}$ in Fig. 8b. However, there are important differences. In the small cycles, the tip force does not vanish. A minimum value of -0.4 kN remains in the top position of both cycles. 2D pile tip and sand are always in contact. In the new penetration phases, the tip force first increases rapidly. Exceeding the lowest previous position, the response gets softer and the curve approaches the asymptotic penetration envelope curve. In the larger cycles in contrast, the tip force evolutes in a different way. A complete stress relief under the 2D pile tip is reached after a displacement upwards of about 1.5 mm. A loss of contact between pile tip and sand was observed in these phases during the test. In the new penetration phases, there is still no tip force for about 2 mm displacement. This penetration is necessary to close the gap between pile tip and sand surface. Afterwards, the tip force slowly restarts to increase with a much softer response than in the small cycles.

From Fig. 10 can be concluded that the evaluation of the tip force following Eqs. 3 and 4 works correctly in the given case. A strong argument in favour of this conclusion is that in the phases without contact between tip and sand zero tip force is calculated with Eq. 3 as well as with Eq. 4. For large unidirectional displacements the calculation can be considered to work adequately in any case.

The calculated tip force can now be compared with other penetration tests using pile-like structures. [8] for example presented in-situ experiments using an instrumented pile. For smaller scales, similar effects were observed and discussed in [2, 3]. The occurrence of two basic pile driving modes was shown, named cavitational and non-cavitational pile driving. The schematic illustration of these two modes is given in Fig. 11. Figure 11a illustrates the case of non-cavitational and Fig. 11b the case of cavitational pile driving.

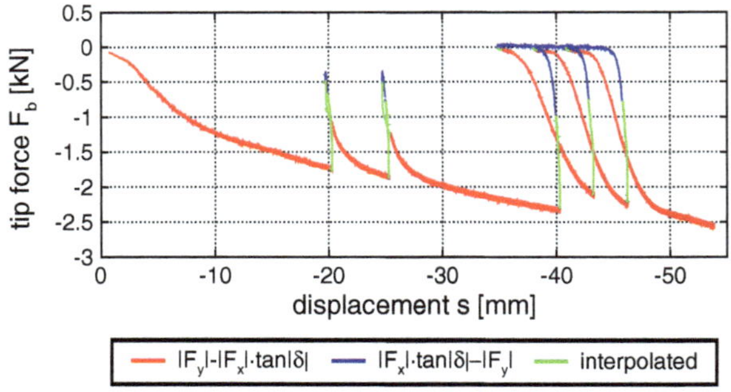

Fig. 10 Estimation of the tip force F_b in test V14-4

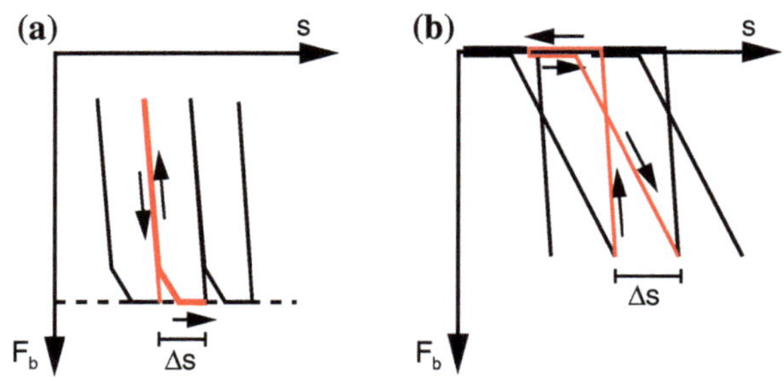

Fig. 11 **a** Non-cavitational and **b** cavitational pile driving [2]

Comparing Figs. 10 and 11, it can easily be seen that in the small cycles non-cavitational and in the large cycles cavitational pile driving occurs. For non-cavitational pile driving even in the highest point pull-up phase of the pile a small tip force remains. The stress ratio of the soil elements under the pile tip does not reverse. The deformations in the pull-up phase are relatively small. Therefore, the soil behavior in the penetration phases is very stiff and the tip force rapidly reaches a maximum value. In the test results, this maximum value is not observed but the approaching to a straight increasing line. This difference can be explained considering the boundary conditions in the test device where the pile tip approaches to the rigid bottom plate during the penetration. In the larger cycles in Fig. 10 as well as in Fig. 11b the tip force, ergo the vertical stress in the soil elements under the pile tip, vanishes in the phase of the displacement upwards. The stress ratio reverses and large soil deformations occur. This leads to a much softer response in the subsequent penetration phase. In the experiments and field tests in [2, 3], the soil zone under the pile tip was not visible and the soil deformations could not be evaluated. The test results presented here show that the attribute "cavitational" for this driving mode was justified because a cavity actually occurs under the pile tip in certain conditions.

3.3 Configuration III: 2D Cavity Expansion

The test configuration II is not explicitly suitable for simulation using FEM because the 2D pile tip provides edges of 90° and for simulations smoothed geometries are

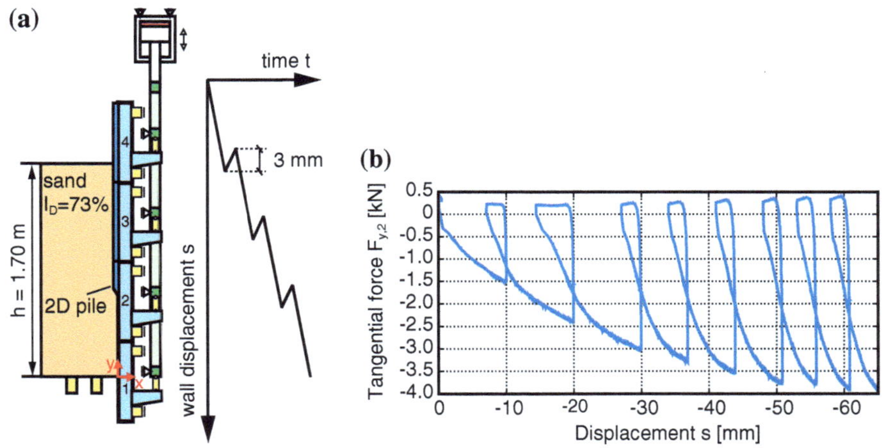

Fig. 12 **a** Initial conditions and test path V15-4 and **b** tangential force $F_{y,2}$ over wall displacement

desirable. Configuration III fits better in these requirements. The pile tip is inclined, sharp edges are avoided and the surfaces of the wall segments are smooth. Thus, the tests performed with this configuration are particularly suitable for simulations. This paper only gives an exemplary overview of the results obtained from DIC analysis. The test path and the initial conditions are shown in Fig. 12a. The test path (Fig. 12a) is an alternating penetration of 65 mm. In the pull-out phases, the displacement upwards is 3 mm, except in the second cycle (about 4 mm). Figure 12b shows the tangential force $F_{y,2}$ on segment 2 over the wall displacement.

Compared to the test presented using configuration II, the evolution of the tangential force $F_{y,2}$ in Fig. 12b indicates a rather cavitational pile penetration. The tangential force increases in the penetration phases and reaches about 3.8 kN after 60 mm wall displacement. The envelope curve in these phases is not linear, but increases gradually. After a strong amount in the beginning of the test, the tangential force increases slowly at the end of the test. In the pull-out phases, a small positive value is reached rapidly. With restarting penetration, the tangential force increases slowly over about 8 mm displacement before it reaches the envelope curve. This behavior is similar to the cavitational pile driving mode shown above. Note that with this test configuration, the inclined pile tip never loses contact to the sand at any time. However, the contact pressure vanishes and the behavior is similar to the cavitational pile driving.

Figure 13 shows the soil displacements obtained from DIC analysis after (a) 20 and (b) 50 mm wall penetration. The contour plots including isolines show the magnitude of the displacements in three areas between the horizontal steel beams. For better orientation, the steel beams that hide parts of the glass windows are illustrated in grey and the initial positions of the four wall segments in blue.

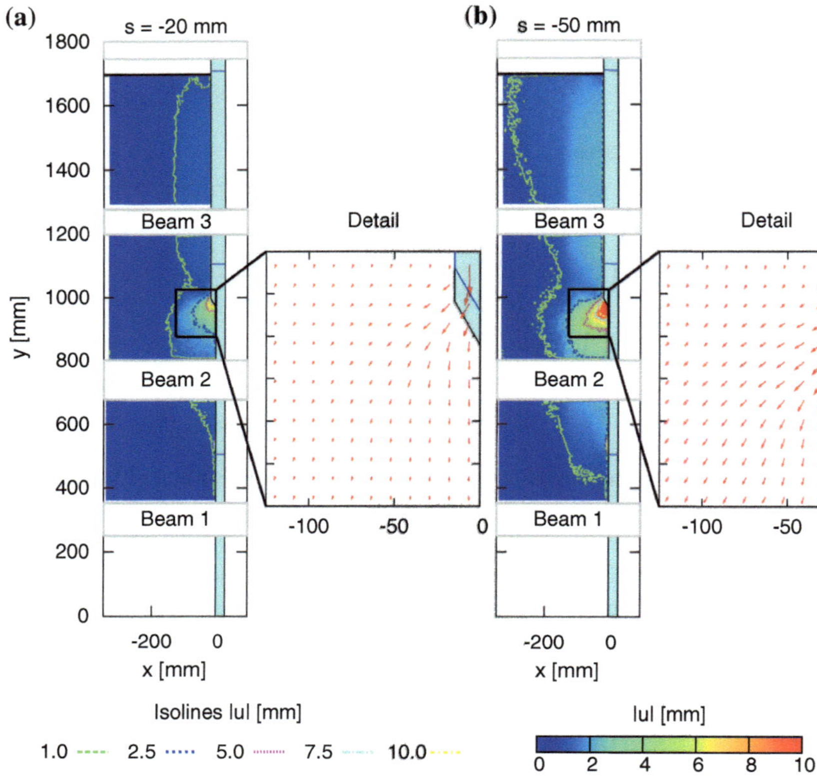

Fig. 13 Displacement magnitudes and vectors for **a** $s = -20\,\mathrm{mm}$ and **b** $s = -50\,\mathrm{mm}$

Beside these information on the displacement magnitudes, vector plots indicate the direction of the displacements in the zone of interest around the pile tip. In order to give a clear overview, not all displacement vectors are shown.

Figure 13a indicates a bulb-like zone of large soil displacements close to the pile tip. This zone extends over about two times the pile diameter from the pile axis. The largest soil displacements are about 8 mm. The complete zone with displacements induced by the pile penetration is about four times as wide as the pile diameter. From the vector plot detail it can be seen that the pile tip displaces the soil particles laterally downwards in the zone adjacent to the inclined pile tip. Under the pile, the displacements are more directed downwards. After 50 mm of penetration, the affected soil zone spreads over the whole width of the glass window. The zone with large displacements has widened too and is more than four times as wide as the pile diameter. The largest displacement is about 25 mm. The 1 mm isoline shows a distinct necking slightly above the pile tip. This can be explained by the penetration of pile volume and the dilative soil behavior in vicinity to the pile tip. Both cause slight displacements upwards above.

4 Summary and Conclusions

Representative results of large scale model tests on important aspects of pile penetration are shown. The tests are designed as boundary value problems for the validation of numerical simulations. An innovative large scale test device is used to perform cyclic and monotonic penetration tests of planar and non-planar wall structures. These structures imitate pile penetrations in dry sand under laboratory conditions (near to plane strain). The distribution of tangential and normal forces on the pile structure is measured. The soil displacements are obtained from DIC analysis. Thus, a holistic impression of the test is provided, which allows an extensive comparison of experiments and simulations.

Tests using three different configuration are discussed. The first test configuration investigates the soil-pile interface using a planar wall structure displaced along an adjacent sand body. These tests clearly show the soil behavior of rough soil-pile interfaces. The influence of the loading on the development of shear zones is pointed out. In monotonic tests, the localization of shear deformations occurs earlier than in cyclic penetration tests. The results represent an important database for the development and the validation of constitutive soil models and contact formulations.

The test configurations II and III concentrate on the soil behavior around pile tips. The occurrence of the pile driving modes, cavitational and non-cavitational pile driving, is demonstrated in the model tests. It is shown that, for flat-ended pile tips, in the case of the cavitational pile driving mode, a cavity forms under the pile tip and in the case of the non-cavitational pile driving not. However, for the acute-angled pile tip in test configuration III no cavity is observable although the pile driving modes are investigated. This leads to the assumption that the cavity is not required for the cavitational pile driving but only the vanishing of the contact stress under the pile tip.

From the test results, the general suitability of the tests to provide similar processes to real pile driving is demonstrated. The wide range of application of the test device and the abilities of the instrumentation for the evaluation of stresses and soil displacements have been pointed out. The experimental database is thus available for Benchmark tests of numerical simulations. In the research unit FOR 1136 the other subprojects can now use these test data for the validation of their work.

Acknowledgments The work presented in this paper was supported by the German Research Foundation (DFG) as part of the research project "Central project of the researchers unit FOR 1136 with demonstrator experiments". The authors acknowledge the financial support from the German Research Foundation (DFG).

References

1. Aubram, D.: *An Arbitrary Lagrangian-Eulerian Method for Penetration into Sand at Finite Deformation*. In: Veröffentlichungen des Grundbauinstitutes der Technischen Universität Berlin, vol. 62. (Shaker, Aachen, 2013)
2. Cudmani, R.O., Huber, G., Gudehus, G.: Zyklische und dynamische Penetration nichtbindiger Böden. Contribution to the Workshop "Boden unter fast zyklischer Belastung", Bochum (2000)
3. Cudmani, R.O.: Statische, alternierende und dynamische Penetration in nichtbindigen Böden. Diss., Veröffentlichungen des Instituts für Bodenmechanik und Felsmechanik der Universität Fridericiana in Karlsruhe, vol. 152 (2001)
4. Dührkop, J., Grabe, J.: Verbesserter Vertikallastabtrag durch konische Pfähle—Feldversuche und Vorschlag für ein Bemessungsverfahren. Bautechnik **85**(11), 748–751 (2008)
5. Finno, R.J., Rechenmacher, A.L.: Effects of consolidation history on critical state of sand. J. Geotech. Geoenviron. **129**(4), 350–360 (2003)
6. Grabe, J., Dührkop, J., Henke, S., Kinzler, S., König, F.: Pfähle mit veränderlichem Querschnitt—konische Pfähle, Fertigteilpfähle mit Fußaufweitung und Flügelpfähle. In: Tagungsband zum Pfahl-Symposium 2007, Mitteilungen des Instituts für Grundbau und Bodenmechanik der TU Braunschweig, vol. 84, pp. 131–155 (2007)
7. Gudehus, G., Nübel, K.: Evolution of shear bands in sand. Géotechnique **54**(3), 187–201 (2004)
8. Huber, G.: Vibrationsrammen: GroßmaSSstäbliche Versuche. Contribution to the Workshop "Vibrationsrammen", Karlsruhe (1997)
9. Kempfert, H.-G., Thomas, S.: Pfahltragverhalten infolge zyklisch axialer Belastung- Versuchsergebnisse und Modellbildung. Tagungsband zur 31. Baugrundtagung der DGGT in München, pp. 255–261 (2010)
10. Rebstock, D.: Verspannung und Entspannung von Sand entlang von Baukörpern. Diss., http://digbib.ubka.uni-karlsruhe.de/volltexte/1000023891
11. Rechenmacher, A.L., Saab, N.A.: Digital Image Correlation (DIC) to evaluate progression and uniformity of shear bands in dilative sand. In: 15th ASCE Engineering Mechanics Conference. Columbia University, New York (2002)
12. Vennemann, P.: JPIV-software package for particle image velocimetry. http://www.jpiv.vennemann-online.de (2007)
13. Vogelsang, J., Huber, G., Triantafyllidis, Th: A large scale soil-structure interface testing device. Geotech. Test. J. **36**(5), 613–625 (2013). doi:10.1520/GTJ20120213, ISSN 0149-6115
14. Walz, B.: Der 1g-Modellversuch in der Bodenmechanik—Verfahren und Anwendung. Vortrag zum 2. Hans Lorenz Symposium, Veröffentlichung des Grundbauinstituts der Technischen Universität Berlin, Heft 40, pp. 13–26 (2006)
15. Westerweel, J.: Digital particle image velocimetry—theory and application. Dissertation, Delft University (1993)
16. White, D.J., Take, W.A.: Particle image velocimetry (PIV) software for use in geotechnical testing. CUED/D-SOILS/TR322 report (2002)

On Soil Deformation and Stress Redistribution Around Pressed-in and Vibrated Displacement Pile Tips

J. Vogelsang, G. Huber and Th. Triantafyllidis

Abstract The experimental study compares soil displacement trajectories around model pile tips obtained from Digital Image Correlation (DIC) for different penetration modes. Monotonic and cyclic quasi-static penetration under plane strain-resembling conditions in dry sand and vibratory model pile penetration in saturated sand are investigated. The comparison results agree well although the penetration mode and the degree of saturation differ considerably. In the experiments, the soil below the pile tip is first pushed downwards as the pile approaches and is then moved more and more sidewards. A slight uplift of the grains is observed when the pile tip has passed. Subsequently, a clear trend of the soil adjacent to the pile shaft to move towards the pile is measured in the case of quasistatic cyclic and vibratory penetration. This trend is considered to be an indicator for "*friction fatigue*", the degradation of shaft friction at a certain depth as the pile penetrates further. A discussion on the comparability with numerical results and on the influence of disturbing boundary effects concludes this contribution.

Keywords Pile penetration · Vibratory pile driving · Friction fatigue

1 Introduction

The application of Digital Image Correlation (DIC) to geotechnical model tests on pile penetration has been an important improvement for understanding related effects. Substantial efforts have been made on the evaluation of soil displacements

J. Vogelsang (✉) · G. Huber · Th. Triantafyllidis
Institute of Soil Mechanics and Rock Mechanics, Karlsruhe Institute of Technology, Karlsruhe, Germany

Th. Triantafyllidis
e-mail: triantafyllidis@kit.edu

Th. Triantafyllidis (ed.), *Holistic Simulation of Geotechnical Installation Processes*, Lecture Notes in Applied and Computational Mechanics 77,
DOI 10.1007/978-3-319-18170-7_3

and deformation around pile tips with small diameter, e.g. [15]. However, most work concentrates on small, non-instrumented model piles penetrating into dry material in monotonic penetration mode. The measurement of stresses acting on the pile is usually performed in closed test devices (strong boxes or calibration chambers) without evaluation of soil deformation [3, 4, 16]. Important phenomena, like e.g. *friction fatigue*, were measured in such tests. These effects were also investigated via DIC in tests with non-instrumented piles by consideration of characteristics of displacement and strain paths [15]. The attribution of these optical observations to stress measurements in the same test is still a challenging task.

The aim of this paper is to contribute to this research work and to provide reliable experimental data for the validation of Finite Element (FE)-simulation techniques. Therefore, 1g-model tests were performed to investigate the soil mechanical effects around pile tips during pressed-in and vibrated-in penetration modes in coarse grained material. A cohesionless medium quartz sand was used in all tests. Two test devices were used: a large scale test device with 2D-resembling conditions for pressed-in sheet-like piles and a half cylinder-shaped device for rectangular vibro-driven piles. Both devices allow the evaluation of soil displacements and deformations around the pile tip via DIC. In this contribution displacement trajectories and principal strains (resp. strain directions) will be analysed. Major subjects of this study were penetration mode and pile tip geometry. Quasi-static monotonic and cyclic penetration were investigated using the large scale test device. Vibratory penetration was performed in the second test device. Both will be quantitatively compared in terms of displacement and strain paths. Remarks on the transferability of the test results on real displacement piles and the limitations of model testing will be made. A comparison of test results and effects of pile driving observed in FE simulations will conclude this contribution.

2 Experimental Methods and Setup

The interface test device with possible test configurations is presented in detail e.g. in [8, 12, 16]. The setup used in this study is illustrated schematically in Fig. 1. The so-called 2D-pile is a wall structure with 0.5 m width placed between two glass observation windows. This 2D-pile is divided into four instrumented segments in order to evaluate the force distribution on the structure. In the second segment, the "pile tip" is located. Thus, the lowest segment and the lower part of the second segment represent a kind of pre-pile. Segments three and four represent the pile shaft. This assembly resembles the zipper-technique, that is often used in FE simulations of pile penetration. The glass windows have relatively smooth surfaces and are reinforced horizontally. This experimental setup tries to achieve plane-strain conditions (which is actually not completely successfull, see [13] and Sect. 6.2). The 2D-pile is penetrated into the sand using an hydraulic actuator mounted on top of the test device. This actuator moves a connection rod that is coupled individually to each wall segment. The 2D-pile is 15 mm wide ($b_{\text{pile}} = 30$ mm) and has a tip inclination of 30°.

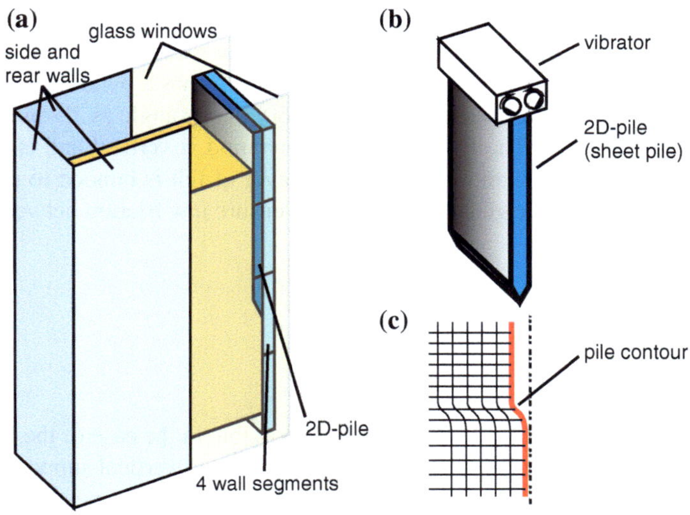

Fig. 1 **a** Schematic interface test device, **b** sketch of a 2D-pile (*sheet pile*) and **c** corresponding FE-mesh with pre-pile

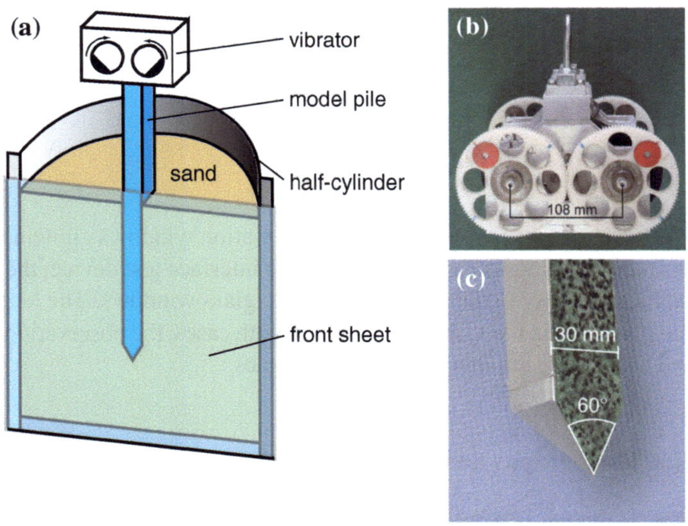

Fig. 2 **a** Test device for vibro penetration, **b** vibrator and **c** detail pile tip

In the half-cylinder test device, model tests with water saturated sand are possible. Model piles with small equivalent diameter can be penetrated into the sand by vibro-driving. They slide along the acrylic glas front sheet, so that the pile tip and one face are visible throughout the whole test. A schematic illustration of this test device is shown in Fig. 2a. The model piles are driven by a small vibrator, Fig. 2b.

The model pile shaft is an aluminium tube with rectangular cross sections of 30×30 mm and 2 mm wall thickness. The tip geometry was matched to the 2D-pile tip in order to provide comparable geometric effects. It has a pyramidal shape with the same width as the shaft and a cone angle of 60° (same angle as 2D-pile tip) resp. 45° (back), see Fig. 2c. The front face of the pyramid is vertical and is in contact with the front sheet (observation window). A layer of felt is bonded to the surface in contact with the acrylic glas front in order to ensure low friction between the pile and the window.

2.1 Instrumentation

Substantial force and stress measurements are possible in the case of the large scale test device. They include measurement of horizontal and vertical support forces on the wall segments. A detailed discussion on the evaluation of tip resistance and earth pressure distribution is provided in [12]. The displacement is measured separately for each wall segment.

In the half-cylinder, only pile head force and pile displacement are measured. The load cell is placed between vibrator and pile. The displacement is measured with a rope pulley transducer.

2.2 Evaluation of Soil Deformation

Both test device have glas resp. acrylic glas observation windows allowing the evaluation of soil deformations during the tests. In the interface test device, the front and rear side walls in vicinity to the wall segments are glass windows. The half-cylinder has a acrylic glas sheet in the symmetry plane. In both cases, the observation windows are assumed to behave similarly to symmetry planes.

2.2.1 Digital Photography and Post-processing

Three digital cameras are used in interval mode for the observation of quasi-static tests. The dynamic model pile penetration is documented using a high frequency video camera. Information on cameras and picture recording are given in Table 1.

Representative sections of the test are subsequently extracted from the image sequences. If necessary, the image sequences are post-processed. Post-processing includes fine rotation with respect to known vertical or horizontal axes. Cropping the regions of interest is done inside the DIC software JPIV [10]. Figure 3 shows image material taken from two test using both test devices. It can be seen that a random pattern is painted on the felt in order to provide a sufficient contrast for DIC.

Table 1 Camera information

	Pressed-in piles	Vibro-driven piles
Camera	Nikon coolpix P7700	Basler CMOSIS ace acA2000-340km
File type	xxx.jpg	xxx.seq
Interval (s)	30	0.007
Images per min	2	8400
Size (Pixel)	2816 × 2112	2040 × 1088
Scale (Pixel/mm)	≈15	≈10
Scale (Pixel/d_{50} (mm))	≈8	≈5
Color depth (bit)	24	8

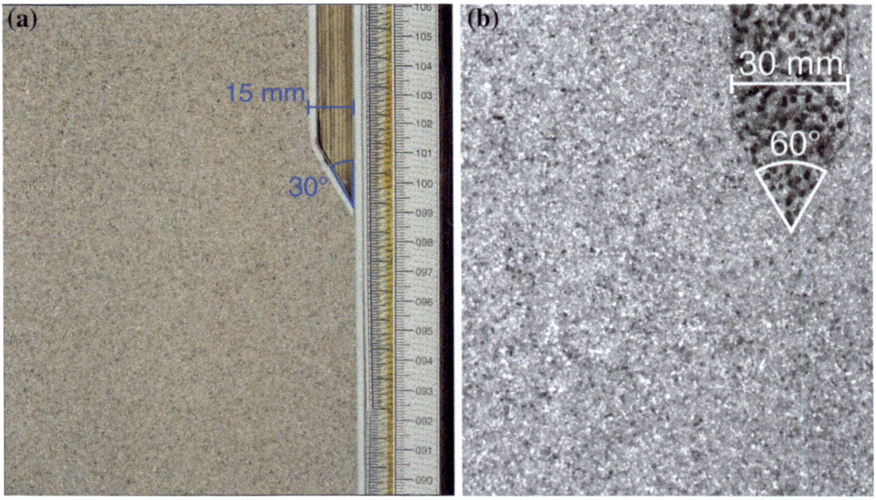

Fig. 3 Image material from **a** 2D-pile penetration and **b** 3D-vibrodriven pile penetration

A correct evaluation of the pile displacement is important, because it is the only part of the image where the quality of the evaluation can be proved.

2.2.2 Digital Image Correlation

JPIV [10], a freeware code for Particle Image Velocimetry, is used for the evaluation of the displacement field between two subsequent images. Summation of the displacements and strain calculation is done afterwards. The evaluation procedure is described in [8, 16].

2.3 Test Material, Sample Preparation, and Uniformity Control

A medium silica (quartz) sand is used in the experiments. Some important granular properties are given in [11, 14]. The sand is installed in the test device by using a sand rainer system (described in detail in [8, 16]). Before the test, the uniformity of the soil sample is controlled by CPTs. The set-up, evaluation procedure and first results are discussed in [13].

2.3.1 Dry Conditions

Dry samples for the large scale interface test device are obtained by air pluviation of dry material. Details on the pluviation system and methods can be found in [8, 16]. Note that the initial stress state before the test is influenced by silo effects. An investigation of these effects is provided in [13].

2.3.2 Saturated Conditions

Saturated samples in the half-cylindrical test device are obtained by pluviation of dry material into water. This procedure results in relative densities of about 40 %. Higher densities can be achieved by excitation of the test device with slight hammer blows against the side walls. The densification was stopped when a relative density of about 70 % has been reached. This method is considered to provide high degrees of saturation [17]. However, subsequent densification of the soil body is not applicable in all test devices. In addition to this, the uniformity of the sample has to be ensured.

3 Nomenclature, Coordinates, and Sign Conventions

A schematic illustration of the model pile embedded in sand is shown in Fig. 4 with the relevant geometries and the chosen coordinate system. The pile diameter in the view plane is denoted b_{pile}. The width of the 2D-pile corresponds to $b_{\text{pile}}/2$. A local coordinate system is defined relative to the initial pile position. The origin is located in the symmetry axis/plane of the pile on the vertical level of the pile shoulder. Coordinates downwards and to the left are defined negative. In the deformed configuration, the current vertical distance between an arbitrary sand element and the pile shoulder is named Δy (also negative towards bottom and left), Fig. 4c. A sand element below the pile tip has thus a negative Δy. According to [15], this coordinate will be presented in a normalized form: $2\Delta y/b_{\text{pile}}$. For selected points on the displacement trajectory, the deformed element shape will be plotted in the directions of principal strain. This configuration is schematically illustrated in Fig. 4d.

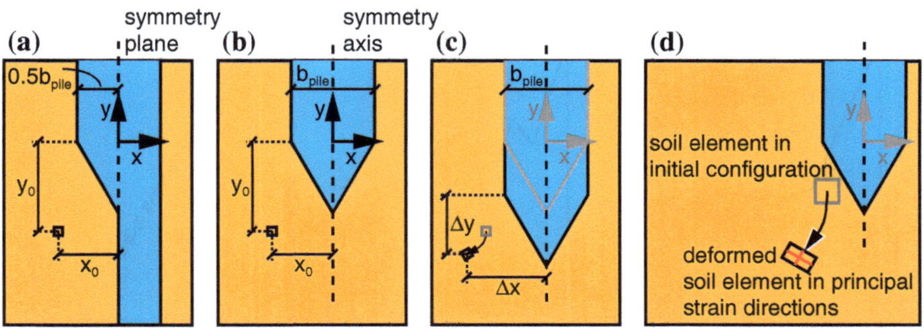

Fig. 4 Coordinate system relative to pile tip: **a** 2D-pile, **b** 3D-pile, **c** coordinates relative to current pile position and illustration of displacement trajectory and **d** element shape in deformed configuration in directions of principal strain

Table 2 Information concerning the compared tests

Test	Dim.	Tip shape	Penetration mode	I_D (%)	S_R (%)	h_{sand} (m)	$\sigma_{y,0}$ (kN/m^2)
V18-1	2D	2D-30°	Monotonic	69	0	1.7	≈10
V15-4	2D	2D-30°	Cyclic (pseudo-dynamic)	69	0	1.7	≈10
Vib-01	3D	3D-30°	Vibro-driven	69	≈100	0.9	≈3–4

4 Displacement and Deformation Around Pile Tips

Three model tests will be compared: V18-1, V15-4 and Vib-01 (internal names). Some relevant information concerning these tests is listed in Table 2. The test paths are illustrated schematically in Fig. 5. Note that the "time" axis is strongly stretched for the vibro-test compared to the two quasi-static tests.

A good comparability of all three tests was achieved due to the similar pile tip shape and almost equal initial soil density of $I_D = 69\,\%$. Important differences were the stress level, the water content, the penetration mode and the dimensionality of the process. The soil displacements and deformations will be compared for all three tests. A possible evolution of pore pressures in test Vib-01 is not considered in this study.

4.1 Displacement Trajectories and Deformation Paths

Figure 6a shows displacement trajectories of selected points in vicinity to the 2D-pile tip during monotonic penetration into dry sand. The axes on the left resp. bottom are the coordinates with respect to the pile symmetry plane in horizontal and the

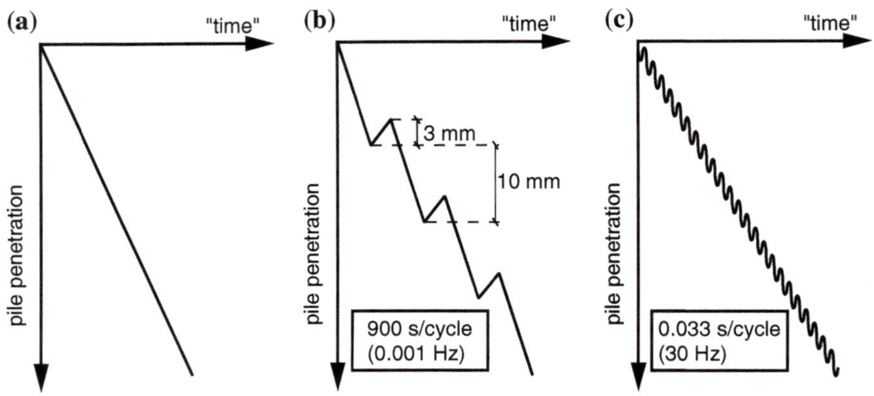

Fig. 5 Schematic test paths in **a** V18-1, **b** V15-4 and **c** Vib-01

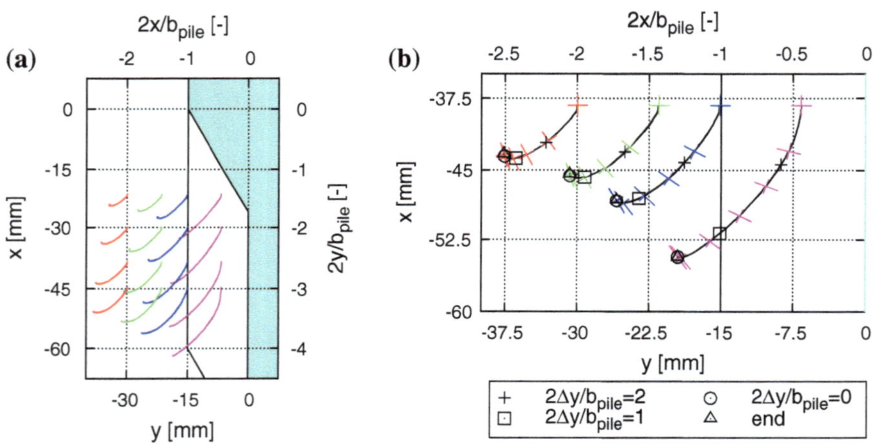

Fig. 6 Selected displacement trajectories during monotonic plane strain model pile penetration evaluated via DIC: **a** overview, **b** detailed figure with deformed elements in principal strain directions; $I_{D0} = 69\,\%$ and $S_R = 0\,\%$

pile shoulder in vertical direction, see Fig. 4. These coordinates are normalized with $b_{\text{pile}}/2$ on the top and right axis. The initial pile position is shown in blue color. The end position is indicated by solid lines. The region below the pile tip (vertical projection) corresponds to coordinates $x \in (-b_{\text{pile}}/2, b_{\text{pile}}/2)$. Displacement trajectories for sand elements in vertical columns (same initial horizontal position) are plotted in the same line color. Figure 6b is a detailed view on four soil elements in a horizontal cross section. The current normalized vertical position of the soil element with respect to the current position of the pile shoulder is indicated for four selected moments: $2\Delta y/b = 2$, $2\Delta y/b = 1$, $2\Delta y/b = 0$ and for the end of the test. $2\Delta y/b = 0$ is the moment where the soil element passes the vertical level of the pile shoulder. In addition to the trajectories, the current shape of each element is plotted as a cross in the principal strain directions, see Fig. 4d. The cross represents the current minimal and maximal width of the element in the view plane.

Figure 6 indicates that the sand is pushed downwards and to the side as the pile tip passes the sand elements. The vertical displacement downwards is larger in the regions under the vertical projection of the pile tip. The outer sand elements have a predominant horizontal displacement component. Some sand elements show a slight uplift towards the end of the penetration process. This detail will be taken into consideration later on.

The two lower rows show very similar displacement trajectories. They can be treated as representative displacement patterns due to pile penetration. The sand regions above do not undergo the whole history of the pile penetration process. The third row (counted from the top) is therefore shown separately in a detailed view in Fig. 6b. Figure 6b also illustrates the chronology of the process and the occurring deformations. As the pile approaches vertically towards the sand elements the horizontal displacement component becomes more and more dominant. The further pile penetration causes only very small displacements. These displacements are oriented upwards for the sand elements outside of the pile projection. The displacement upwards start before the pile shoulder has passed the sand elements. For the sand elements in direct vicinity to the pile tip, the displacements are purely downwards.

The deformation mode resulting from pile penetration was often reported: Below the pile tip, the deformation regime is roughly vertical compression/horizontal extension and beside the pile tip the opposite, vertical extension/horizontal compression. A deformation pattern like this was observed in experiments (e.g. in [15]) and is also obtained qualitatively from FE analysis [1]. The direction of minimal width (maximal compression) is oriented approximately tangentially to the displacement trajectory as long as an element is under the pile projection. When the pile tip passes an element, a counter-clockwise rotation of directions of principal strains with respect to the trajectory can be seen.

Figure 7 shows the same curves for similar points during an alternating or pseudo-dynamic pile penetration in dry sand. The pile penetration consists in movements downwards interupted by phases of 3 mm pile displacement upwards, see Fig. 5. The effective penetration per "cycle" is 10 mm.

The displacement trajectories during cyclic penetration in Fig. 7a are very similar to the displacement patterns during monotonic penetration. In Fig. 7b, the cycles can be analysed more clearly. The cycles are like loops attached to the monotonic path. For small displacements upwards, a positive vertical displacement occurs due to the stress relief. Subsequently, the soil elements move to the side towards the symmetry axis. In the new penetration phase, the monotonic path is rapidly reached and followed.

In Fig. 7a, a problem related to DIC evaluation of pile penetration processes can be seen. The element in the second row and right column (highest pink trajectory) shows a strong displacement downwards as the pile tip has already passed this element. This displacement pattern is anormal and is in distinct contrast to other elements starting from the same horizontal position. It can be seen, that the considered soil element ends up close to the pile and actually overlapes the pile shaft, which is physically

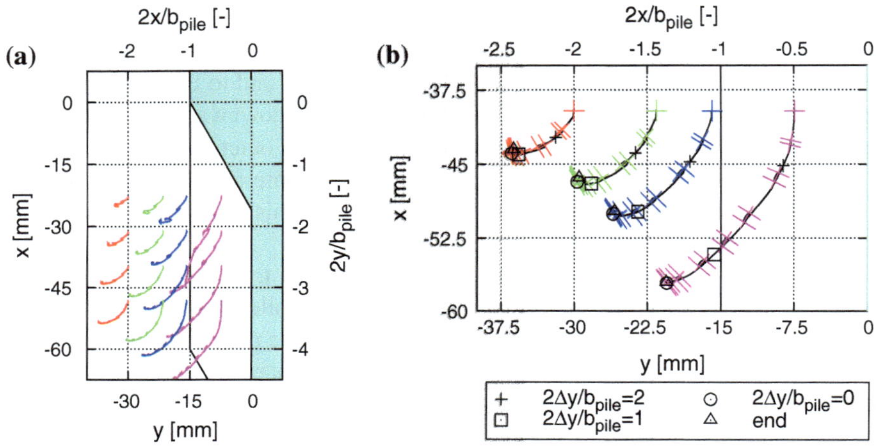

Fig. 7 Selected displacement trajectories during pseudo-dynamic plane strain model pile penetration evaluated via DIC: **a** overview, **b** detailed figure with deformed elements in principal strain directions; $I_{D0} = 69\,\%$ and $S_R = 0\,\%$

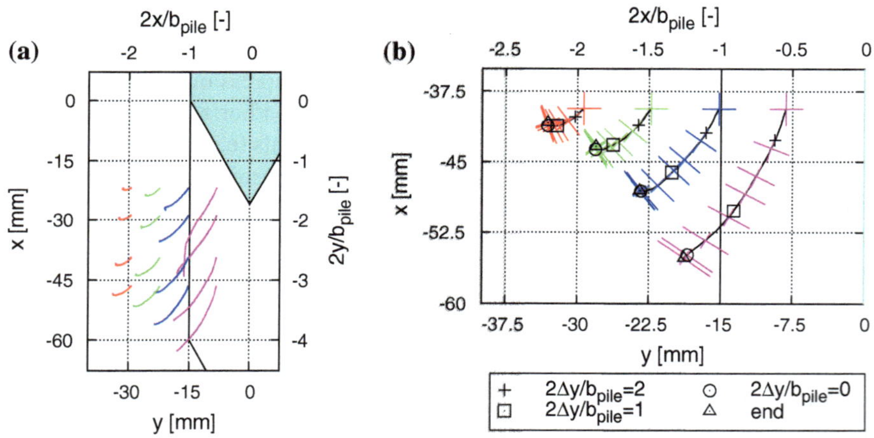

Fig. 8 Selected displacement trajectories during vibrodriven model pile penetration evaluated via DIC: **a** overview, **b** detailed figure with deformed elements in principal strain directions; $I_{D0} = 69\,\%$ and $S_R \approx 100\,\%$

not possible. This behaviour results from an erratic DIC evaluation. When a sand region is located too close to the pile tip, it may happen that parts of the test patch contain sections of the pile. These sections disturb the evaluation of displacements and the soil displacement contains erroneously portions of the pile displacement (rigid body). Thus, the evaluation of regions in vicinity to sand-structure interfaces have to be treated very carefully.

Figure 8 shows similar displacement trajectories for the vibro-driven 3D-pile penetration. Globally, the displacements are similar the quasistatic 2D-pile penetration. Quantitative comparison will be made in Fig. 9.

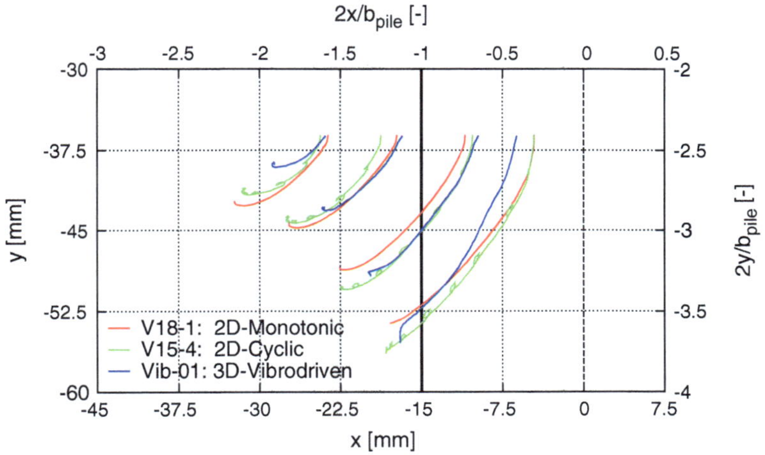

Fig. 9 Comparison of displacement trajectories for monotonic, cyclic and vibrodriven model pile penetration

One detail stands in contrast to Figs. 6 and 7: the sand elements near the pile shaft tend to move downwards along the shaft when the pile tip has already passed these elements. This calculated displacement was verified and does not result from an erratic DIC evaluation.

Figure 9 compares the displacement trajectories in the three penetration modes for selected points. Note that the positions of the selected points are slightly different because the size of the DIC patches differ. The vertical offset was eliminated in order to start from one given vertical level. The horizontal position was not altered.

Figure 9 shows a qualitatively similar displacement mechanism for all three penetration modes. The sand elements in the vertical projection of the pile are pushed down- and sidewards with a predominant vertical component. They end up next to the pile shaft as the pile has passed. The outer sand elements are moved more to the side and show a slight uplift as the penetration continues. The two plane strain tests are in very good agreement. The vertical displacements downwards of the sand elements below the pile tip is slightly stronger in the cyclic test. Comparing the 2D tests and 3D vibrodriven test, important differences can be seen. The horizontal displacements for the 3D vibrodriven case are significantly lower than in the 2D cases. This difference is more pronounced for the sand element outside of the pile projection and results at least in parts from the dimensionality of the problem. In plane strain conditions, a horizontal displacement of a sand element is not necessarily connected to a volume change. The 3D case is comparable to a rotationally symmetric process, where a horizontal displacement results in a strain in the offplane direction (circumferential). Thus, a smaller horizontal displacement of the outer sand elements is sufficient to compensate the additional penetrated pile volume and dilation effects. Figure 10 illustrates the dimensionality effects observed in the displacement trajectories.

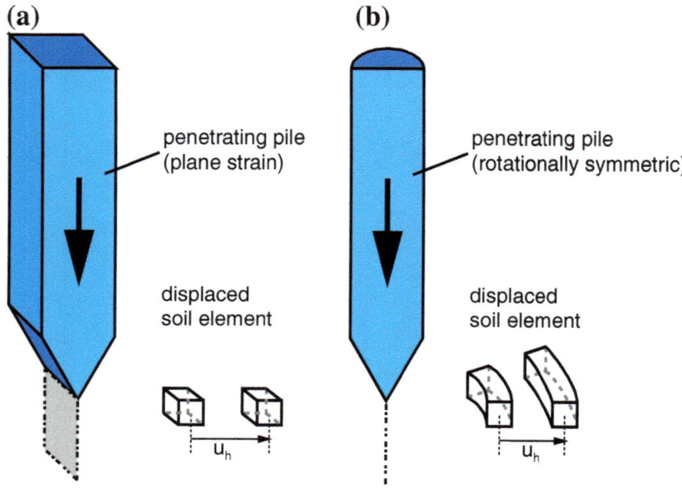

Fig. 10 Illustration of the dimensionality effect during pile penetration

Another important difference between the quasi-static and the dynamic test is the displacement of sand elements in vicinity to the pile shaft when the tip has passed. As already discussed, these sand elements show the clear tendency to follow the pile displacement downwards.

Towards the end of the outer displacement trajectories, a small but significant detail between monotonic and cyclic/vibro-driven penetration can be seen: the outer sand regions tend to move towards the pile shaft, when the pile tip is located below these elements. This important detail is discussed in the next chapter.

5 Stress Redistribution Around Pile Tips

The phenomenon of "friction fatigue" (also called the "h/R effect") is known to be a governing factor in the penetration process of a pile as well as in the evaluation of initial capacity of driven piles [7, 16]. "Friction fatigue" is defined as the decrease of shaft friction at a given depth as the pile penetrates further. It was shown to occur during cyclic and dynamic penetration. For monotonic penetration, "friction fatigue" is not to be observed [9]. "Friction fatigue" corresponds to a significant relaxation of horizontal (radial) stress in the sand regions adjacent to the pile shaft and is at least partly explained by the soil behaviour under cyclic loading. Thus, the name "friction fatigue" is not completely correct as the cause is a relief or relaxation of radial stress and the effect is a decrease of ultimate shaft friction (assuming a constant contact friction angle). Therefore we prefer to use the term "stress relaxation" in the following.

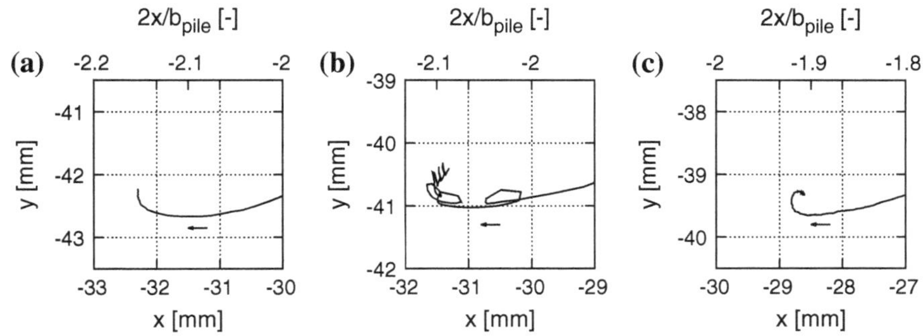

Fig. 11 End (*tail*) of displacement trajectory for **a** V18-1, **b** V15-4 and **c** Vib-01

5.1 Experimental Evidence of Stress Relaxation Along the Shaft

The investigation of relaxation effects requires the measurement of the evolution of displacements and forces resp. stresses and strains. During this study only displacements have been measured. Therefore the results can only used in an indicative manner and further investigation is needed.

The trend at the end (or tail) of displacement trajectories of elements near to the pile shaft is considered to be an indicator for the occurence of stress relaxation [16]. Figure 11a–c shows a detailed view on this section of the trajectory for the three tests. The sand elements in the third row of the first column are compared (red curves in Figs. 6, 7 and 8). The small arrows indicates the direction of displacement.

The global shape of the trajectories in Fig. 11a–c is similar: the sand is pushed sidewards and then undergoes a small uplift. However, the displacement trend towards the end contrasts sharply. During monotonic penetration (Fig. 11a) the sand element remains at a constant horizontal position. The distance between sand element and pile shaft does not change. During cyclic and vibro-driven penetration, a clear trend of displacement towards the pile shaft is observed.

6 Discussion of the Results

6.1 Comparison with Other Model Tests and FE Simulation

The presented results are in good agreement with other model tests on pile penetration. The global penetration mechanism is found similar, although different pile diameter, tip inclination and penetration modes were investigated. The observations in DIC analysis agree with the hypothesis of displacement indicators for friction fatigue and offer a plausible explanation for effects related to cyclic pile penetration.

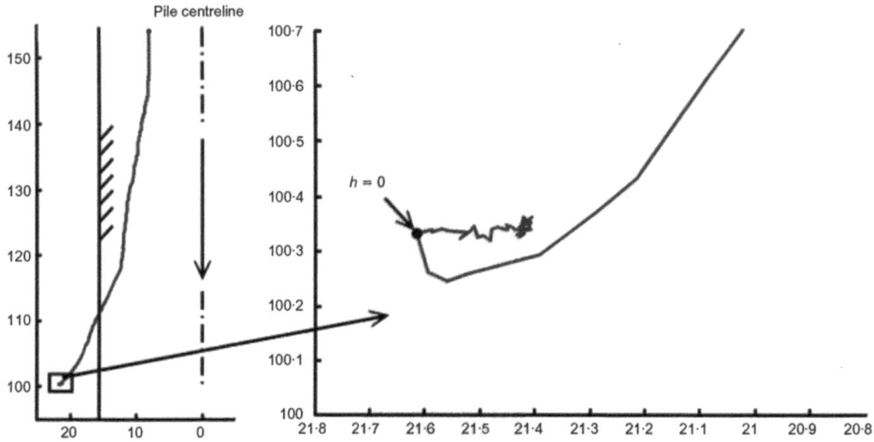

Fig. 12 Measured displacement trajectory for a sand element in direct vicinity to a penetrating model pile tip (from [15], h = Δy)

White and Bolton [15] reported similar trajectories during monotonic model pile penetration and related those to the mechanism of stress relaxation, see Fig. 12.

An explanation offers the behaviour of soils under cyclic loading conditions. Following [15], the mechanism can be described as follows: The soil direct adjacent to the pile shaft is sheared cyclically due to the cyclic motion of the pile shaft. As a consequence, it tends to contract so that the outer sand regions can move towards the pile shaft. The reaction of the sand located at the pile shaft should be a combined contraction/relaxation mode. Contraction leads to the outer displacements towards the pile shaft and relaxation leads to a degradation of shaft friction ("friction fatigue"). This interpretation is in accordance to the experimental observations in [12] on the soil behaviour adjacent to a cyclically displaced wall segment with rough surface. It was found that a jamming (opposite case of relaxation) is accompanied by a displacement trend away from the soil-wall interface. In the 3D-case, circumferential arching occurs. The circumferential horizontal stress can exceed the radial horizontal stress. It is not easy to perform measurements of this stress redistribution.

In Fig. 12, it can be seen that a displacement towards the pile tip can occur even during monotonic pile penetration. In the tests presented here, such displacement was only observed during cyclic and vibro-driven penetration. One reason could be that the performed penetration depth was not sufficient.

Although a vast number of numerical simulations of pile driving are available, comparisons with the presented test results are not easy to be drawn. Either different geometries are modelled or displacement patterns are not analysed explicitly. However, some basic analogies can be seen. Large compressive strains essentially in vertical direction and extension in horizontal direction are usually observed below the pile tip, e.g. [1]. The effect of "friction fatigue" was also observed in numerical simulations, e.g. [2]. More verification of numerical work using existing model test data is necessary in this context. It is the aim in the research unit FOR1136 to use the presented model tests for the validation of FE simulations.

6.2 Disturbing Boundary Effects

The interaction between side walls resp. observation windows and soil was neglected until now. The displacement trajectories were interpreted as representative for the soil flow around the pile tip. However, two principal effects of interaction that actually occur have to be distinguished: the mobilization of friction in the interface side wall-soil and the deflection of the side walls due to increasing horizontal earth pressure. The parasitic impact of the observation windows is not easy to evaluate but has to be taken into account in the interpretation in a qualitative manner. The qualitative influence of both effects can be summarized as follows:

- *Friction on the side walls*
  Friction between observation window and soil inhibits soil displacements along the window and leads to off-plane shear deformation of sand elements near to the window. The soil displacements and deformations can be larger in the inner part of the soil body.
- *Deflection of the side walls*
  Deflection of the observation windows leads to strains in the off plane direction. For the 2D-case, volumetric strain can no longer be evaluated assuming plane strain conditions. This deformation cannot be evaluated with planar DIC.

The influence of frictional effects between glass and soil is considered small due to the low contact friction angle ($\approx 10°$). By observation of the free sand surface at the top, no 3D effects or arching were optically identifiable. Nevertheless, in [12] it was shown that arching effects do have an impact on the experimental results. The deflection of the side walls was measured for a monotonic test and lies in the range of 0.3 mm. Therefrom, an off-plane strain of 0.1–0.5 % can be calculated. Occuring strains during the test results presented here, reach a few % (up to more than 20 %) and are therefore predominant compared to the off-plane strain.

FE back calculation of the experiments can take account for these boundary effects if the side walls are incorporated in the model. This leads to an important improvement of comparability between test and simulation and can even enhance the interpretation of the experiment.

7 Summary and Outlook

In this contribution, the soil displacements and strains occurring during model pile penetration were investigated. Effects of penetration mode and dimensionality (plane strain-like or 3D) were studied using two test facilities: Quasi-static monotonic and cyclic plane strain penetration was performed in a large scale interface test device and vibrodriven pile penetration was performed in a half-cylinder test device. Soil deformations were evaluated using DIC. Representative displacement trajectories of sand elements with different horizontal distances to the pile were compared. The overall penetration mechanism is similar although pile penetration mode, tip shape and degree of saturation differ. A marked dimensionality effect was demonstrated. The evaluated horizontal soil displacements were significantly larger in plane strain pile penetration than in 3D-vibro-driven penetration. Indicators for stress relaxation were shown to appear during cyclic plane strain pile penetration and during vibrodriven pile penetration. Future investigation will focus on the link between DIC evaluation of soil deformation and stress measurements on the model pile.

The tests presented include the effect of pile penetration. In subproject 6 of the research group FOR 1136, the soil liquefaction around vibrating pile tips is investigated numerically [5, 6]. However, in these simulations, the pile vibrates around a constant position, the effect of penetration is neglected. In order to compare simulations and results, future tests will be adapted to that boundary value problem. A modified pile with flat ended tip will be used for that purpose.

Acknowledgments The work presented in this paper was supported by the German Research Foundation (DFG) as part of the research project "Central project of the researchers group FOR 1136 with demonstrator experiments". The authors acknowledge the financial support.

References

1. Dijkstra, J., Broere, W., van Tol, A.: Experimental investigation into the stress and strain development around a displacement pile. In: Proceedings of the sixth European Conference on Numerical Methods in Geotechnical Engineering, pp. 595–600 (2006)
2. Grabe, J., Mahutka, K.-P.: Finite-Elemente-Analyse zur Vibrationsrammung von Pfählen. Bautechnik 82, Heft 9 (2005)
3. Klotz, U., Coop, M.R., Taylor, R.N.: Zur Verteilung von Spitzendruck, Mantelreibung und Radialspannung bei Installation von Rammpfählen in Sanden. Pfahlsymposium, pp. 201–221, Braunschweig (2001)
4. Lehane, B.M., White, D.J.: Lateral stress changes and shaft friction for model displacement piles in sand. Can. Geotech. J. **42**, 1039–1052 (2005)
5. Osinov, V.A., Chrisopoulos, St., Triantafyllidis, Th.: Numerical study of the deformation of saturated soil in the vicinity of a vibrating pile. Acta Geotechnica (2012). doi:10.1007/s11440-012-0190-7
6. Osinov, V.A., Triantafyllidis, Th.: Modelling soil liquefaction around a vibrating pile. The influence of soil permeability. 3rd Workshop "Holistic simulation of geotechnical installation processes", Karlsruhe (2014)
7. Randolph, M.F., Dolwin, J., Beck, R.: Design of driven piles in sand. Géotechnique **44**(3), 427–448 (1994)
8. Rebstock, D.: Verspannung und Entspannung von Sand entlang von Baukörpern. Diss. (2011). http://digbib.ubka.uni-karlsruhe.de/volltexte/1000023891
9. Rimoy, S.P.: Ageing and axial cyclic loading studies of displacement piles in sands. Diss, Imperial College London (2013)
10. Vennemann, P.: JPIV-software package for particle image velocimetry (2007). http://www.jpiv.vennemann-online.de

11. Vogelsang, J., Huber, G., Triantafyllidis, Th., Schindler, U.: Pfahlpenetration in nichtbindigem Boden: Großmaßstäbliche Modellversuche und Nachrechnungen. Pfahlsymposium, 409 Braunschweig (2013)
12. Vogelsang, J., Huber, G., Triantafyllidis, Th.: Demonstrator experiments on significant effects during pile installation. 2nd Workshop "Holistic simulation of geotechnical installation processes", Karlsruhe (2013)
13. Vogelsang, J., Zachert, H., Huber, G., Triantafyllidis, Th.: Effects of soil deposition on the initial stress state in model tests: experimental results and FE simulation. 3rd Workshop "Holistic simulation of geotechnical installation processes", Karlsruhe (2014)
14. Vogelsang, J., Huber, G., Triantafyllidis, Th: A large scale soil-structure interface testing device. Geotech. Test. J. **36**(5), 613–625 (2013). doi:10.1520/GTJ20120213, ISSN 0149-6115
15. White, D.J., Bolton, M.D.: Displacement and strain paths during plane-strain model pile installation in sand. Géotechnique **54**(6), 375–397 (2004)
16. White, D.J., Lehane, B.M.: Friction fatigue on displacement piles in sand. Géotechnique **54**(10), 645–658 (2004)
17. Wienbroer, H.: Umlagerung von Sand infolge Wechselbeanspruchung. Diss. Veröffentlichungen des Instituts fr Bodenmechanik und Felsmechanik (IBF) am Karlsruher Institut fr Technologie (KIT), Heft 174, Karlsruhe (2011)

Modelling of Soil Structure Interaction by Applying a Hypoplastic Material Behaviour Within Mortar Contact Formulation

P. Dziewiecki, C. Weißenfels and P. Wriggers

Abstract The main goal of the project is the realistic simulation of pile installation processes. By considering this processes prediction of soil behaviour using numerical simulation is used. The boundary conditions, here the external loads, are the contact forces between soil and structure. For correct prediction of the external loads a suitable contact and friction model is required. During the relative movement of a pile or generally a body with a rough surface within sand, a shear zone actually develops within the sand, directly to the contacting surfaces. Thus the interaction behaviour between sand and pile results in varying coefficient of friction, which is assumed as a quantity dependent on the stress state within sand body near to the contact surface. This assumption leads to an extension of the classical formulation of friction laws used within the contact mechanics framework related to the inelastic material behaviour. As constitutive law a hypoplastic material model is used, which represents volume changing effects of sand due to loading, which are specific for granular media. The discretisation of the contact constraints based on mortar method will be described. A robust hypoplastic model will be depicted. A proposed projection procedure for calculating the coefficient of friction exploiting the mentioned localisation of the contact surfaces and thus the analogy of simple shear and triaxial test behaviour of sand will be described. For the validation of the finite element model the results are compared with experimental data obtained within a specific large scale shear test.

Keywords Soil structure interaction · Contact mechanics · Mortar method · Hypoplastic material

P. Dziewiecki (✉) · C. Weißenfels · P. Wriggers
Institute of Continuum Mechanics, Leibniz Universität Hannover, Hanover, Germany
e-mail: dziewiecki@ikm.uni-hannover.de

Th. Triantafyllidis (ed.), *Holistic Simulation of Geotechnical Installation Processes*,
Lecture Notes in Applied and Computational Mechanics 77,
DOI 10.1007/978-3-319-18170-7_4

1 Introduction

Installation of construction elements like vibro-driven piles in building grounds may lead to strong influencing of the soil state. Because of this strong interaction the behaviour of the soil have to be predicted by using numerical simulation. Though the boundary conditions and the loads are determined by the contact forces between soil and structure. Therefore there is a need for precisely understanding of the mechanical system consisting of the installed pile and the surrounding medium soil. Especially the surrounding soil exhibits a highly non-linear behaviour. Considering this mechanical system, see Fig. 1, at first it is important to take the contact between sand and the moving pile into account. Implementing contact interaction between soil and structure the friction behaviour has to modelled realistically. Considering a rough surface between sand and pile, contact is actually taking place between soil and soil. This assumption can be made, because the actually unbonding takes place not exactly at the soil-structure interface but approximately at a distance of a few grain sizes of it, see Fig. 2. According to this fact correct prediction of the loads occurring requires a suitable model for contact and friction behaviour. Thus a physically motivated projection scheme has to be used, exploiting the equivalency of material behaviour in triaxial loading and direct shear test. On the other hand it is important to apply a suitable constitutive stress-strain relation for the surrounding medium soil. It is well known, that the mechanical behaviour of the soil is strongly influenced by its non-homogeneous fabric. The soil consists usually of sand grains, movement and interaction of which during loading cause volumetric phenomena like dilatancy and contractancy. In order to describe these phenomena a robust hypoplastic model is

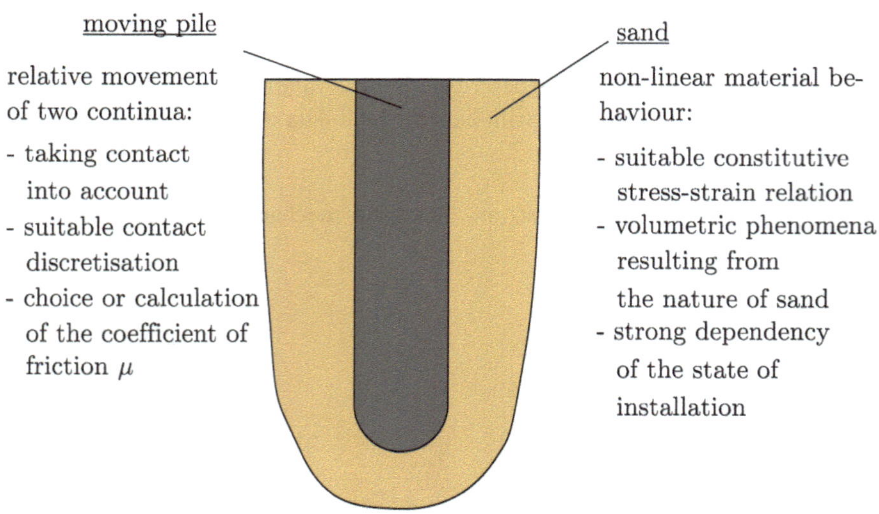

Fig. 1 Classification of the problems

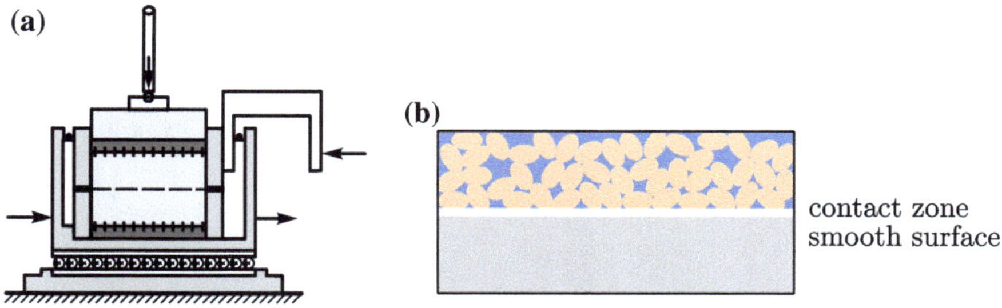

Fig. 2 **a** Direct shear test. **b** Idealisation of contact between sand and a surface

applied, suitable for capturing the specific inelastic behaviour. Since the assumption of contact between sand and sand is made, a hypoplastic model is also used within the projection procedure for obtaining the coefficient of friction.

2 Material Behaviour

Since the soils is a granular material, several mechanical phenomena characteristic for this kind of material are occurring. Due to this they exhibit three characteristic mechanical phenomena. The first one is the dependency of the behaviour on the magnitude of the hydrostatic pressure p. The second one is the possible volume change caused by the reordering of the grain skeleton. Thus the considered sample can increase its volume (*dilatancy*) or decrease it (*contractancy*) due to shearing. Third important property of granular geotechnical materials is the change of the stiffness caused by a change of loading direction. However applying classical plasticity models the mentioned volume change can occur only for changing plastic strains. A hypoplastic material model of type proposed by Niemunis [2], which captures the important phenomena occurring in the large scale shearing test, is used, Eq. (3). The stress of the soil is given in a rate form

$$\dot{\boldsymbol{\sigma}} = \underline{L} : (\dot{\boldsymbol{\varepsilon}} - Y^{w} \boldsymbol{n} \, \|\dot{\boldsymbol{\varepsilon}}\|), \tag{1}$$

where $\underline{L} = \lambda_E p \underline{J}$ is elasticity tensor, which depends on p, with Poisson ratio $\nu = 0$, $\underline{J}$ a 4th order unit tensor and λ_E the barotropy parameter of the model. The quantity Y is an interpolation of the degree of non-linearity between elasticity and plasticity, Fig. 3. The degree of non-linearity Y is defined as follows

$$\begin{aligned} Y &= \frac{(B + M p_R)(p_R - p) + \sqrt{B^2(p - p_R)^2 + (2B/M + p_R) p_R \|\boldsymbol{s}\|^2}}{p_R(2B + M p_R)} \text{ for } p < p_R, \\ Y &= \frac{\|\boldsymbol{s}\|}{\sqrt{(B + Mp)^2 - B^2}} \text{ for } p > p_R, \end{aligned} \tag{2}$$

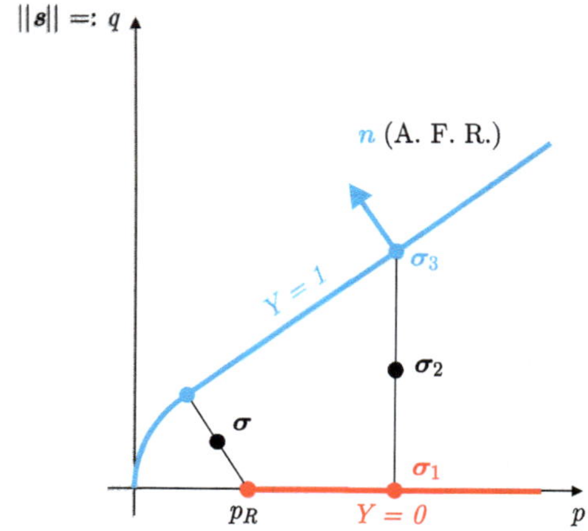

Fig. 3 Used yield surface in $p-q-$space

with constants $M = \sqrt{8/3}\sin\varphi/(3+\sin\varphi)$, $B = \sqrt{24}\,c\cos\varphi/(3+\sin\varphi)$, where φ is the friction angle, c the cohesion of the soil and s the deviatoric part of the stress tensor $\boldsymbol{\sigma}$. The value p_R is a threshold, from which on the used yield surface is rounded for avoiding tension stresses within the soil, see Fig. 3. The degree of non-linearity Y is in the limit case of elasticity 0 and in the case of plasticity 1. The exponent w in Eq. (1) is used for biasing of Y. It determinates the influence of the inelasticity interpolated by Y. For taking the dependency of the yield surface on the hydrostatic pressure into account a Drucker-Prager model with rounded cone end is applied:

$$f^{(\mathrm{DP})}(\boldsymbol{\sigma}) \equiv -\mathrm{M}(\varphi)p - B(\varphi, c) + \sqrt{\|s\|^2 + (B(\varphi, c))^2} = 0. \tag{3}$$

3 Contact Between Sand and Pile

Proper treatment of the contact problems requires a suitable discretisation. Mortar method turned out to be more robust, than node to segment method. According to this the mortar method is applied in this work. Next very important point is the exact as possible constitutive modelling of the mechanical interaction between the contact partners, especially the friction behaviour. Ahead of a detailed explanation of used approaches, some basic definitions used will be explained. Considering current configuration $\varphi(B)$ one can distinguish three types of boundaries. Dirichlet boundary $\varphi(\partial B^{\boldsymbol{u}})$ where displacements can be applied. Neumann boundary $\varphi(\partial B^{\boldsymbol{\sigma}})$. In case of contact a third type has to be considered, boundary $\varphi(\partial B^{c})$ at which in case of contact pressure and tangential forces can be transferred, see Fig. 4. Because of the interaction between the contacting bodies the weak form of the balance have to be extended by additional contact terms

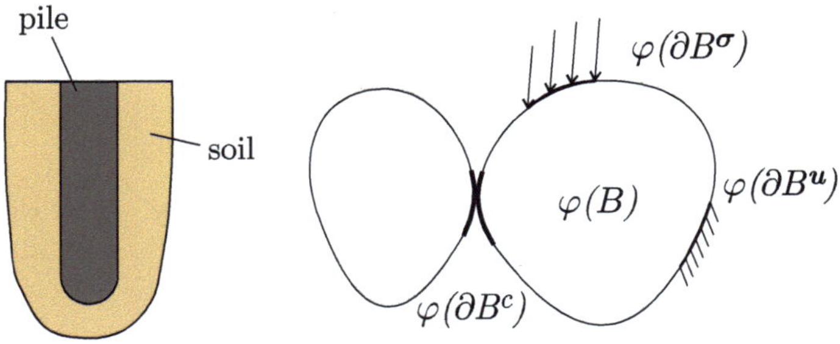

Fig. 4 Capturing the interaction in meanings of contact mechanics framework

$$
\begin{aligned}
G(\boldsymbol{u},\boldsymbol{\eta}) = & \int_{\varphi(B)} (\delta\boldsymbol{\varepsilon}\cdot\boldsymbol{\sigma} - \boldsymbol{\eta}\cdot\rho_0\boldsymbol{b})\mathrm{d}v \\
& - \int_{\varphi(\partial B^{\sigma})} \boldsymbol{\eta}\cdot\boldsymbol{t}\mathrm{d}a \\
& - \int_{\varphi(\partial B^{c})} (t_{\mathrm{N}}\delta g_{\mathrm{N}} + \boldsymbol{t}_{\mathrm{T}}\cdot\delta\boldsymbol{g}_{\mathrm{T}})\mathrm{d}a.
\end{aligned} \tag{4}
$$

where the terms $t_{\mathrm{N}}\delta g_{\mathrm{N}}$ and $\boldsymbol{t}_{\mathrm{T}}\cdot\delta\boldsymbol{g}_{\mathrm{T}}$ represent the virtual work of contact forces. For solving this extended problem additional constraints have to be defined. First one regards to the first contact term in Eq. (4) and is called the **non-penetration condition**

$$
\begin{cases}
g_{\mathrm{N}}\lambda_{\mathrm{N}} = 0 \quad (\text{with } \lambda_{\mathrm{N}} \hat{=} t_{\mathrm{N}}) \\
g_{\mathrm{N}} \geq 0 \\
\lambda_{\mathrm{N}} \leq 0
\end{cases} \tag{5}
$$

which indicates that either the normal gap g_{N} on $\varphi(\partial B^{c})$ is greater than 0, and hence the contact pressure λ_{N} equal to zero. In the other case the regions are in contact, $g_{\mathrm{N}} = 0$ which causes contact pressure λ_{N}. Furthermore a second constraint, the **stick-slip condition**, has to be fulfilled

$$
\begin{cases}
\dot{\gamma}\, f = 0 \\
\dot{\gamma} \geq 0 \quad (\text{where } \dot{\gamma}\,\Delta t = \Delta_t\gamma = \Delta_t\boldsymbol{g}_{\mathrm{T}}) \\
f \leq 0 \quad (\text{with Coulombs friction law } f^{c} = \|\boldsymbol{\lambda}_{\mathrm{T}}\| - \mu\lambda_{\mathrm{N}} \leq 0)
\end{cases} \tag{6}
$$

which means, that either $f < 0$ and the bodies are in stick, where $\dot{\gamma} = 0$ or, in other case, tangential slip occurs, $\dot{\gamma} > 0$, and the slip criterion $f \overset{!}{=} 0$ has to be fulfilled. Considered together Eqs. (5) and (6) imply Kuhn-Tucker conditions, which are treated algorithmically in the meaning of an active set strategy.

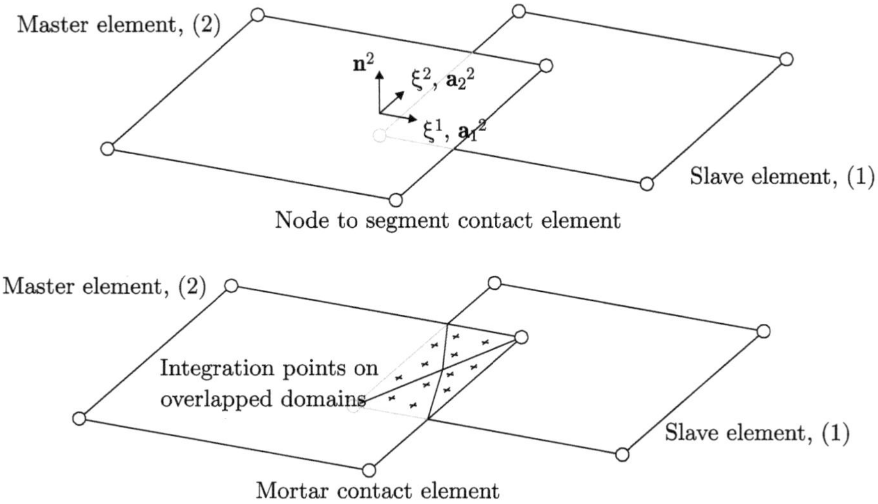

Fig. 5 Comparison of node-to-segment and mortar contact elements

3.1 Mortar Method

Node to segment is a widely used technique for contact discretisation. However the analyzed mechanical problems with all its components may become highly non-linear. In this kind of cases node to segment techniques seems to induce unstable solution behaviour or even fails at all. This behaviour exhibits stronger for coarse discretisation and large time steps. In opposite to this, the mortar method seems to be a more robust approach for more complex problems, see [3, 4]. While within the node to segment method only one slave node and the associated master element yield contributions to the stiffness matrix, the mortar procedure takes into account complete patch of pallets consisting of overlapping areas of the master and associated slave elements, see Fig. 5. Due to this, the quantities are averaged over a mortar patch. As an example showing the improved performance an uncomplicated, but significant simulation setup is chosen, Fig. 6, [7]. Within this test a solid block is placed on another one while assuming friction between contacting surfaces. Both blocks are of Neo Hookian material type

$$\boldsymbol{\sigma} = \frac{\lambda}{2J}(J^2 - 1)\boldsymbol{g}^{-1} + \frac{\mu}{J}(\boldsymbol{b} - \boldsymbol{g}^{-1}), \tag{7}$$

where $J = \det \boldsymbol{F}$ is the determinant of the deformation gradient $\boldsymbol{F}$, $\boldsymbol{b} = \boldsymbol{F}\ \boldsymbol{F}^{\mathrm{T}}$ the left Cauchy-Green strain tensor, $\boldsymbol{g}^{-1}$ the contravariant metric tensor in the current configuration and Lamé constants λ and μ [1] (while engineering constants used are $E = 1.0$, $\nu = 0.3$). The lower block is supported at the bottom. The lower and the upper blocks are discretized by $5 \times 5 \times 5$ and $4 \times 4 \times 4$ elements respectively. The

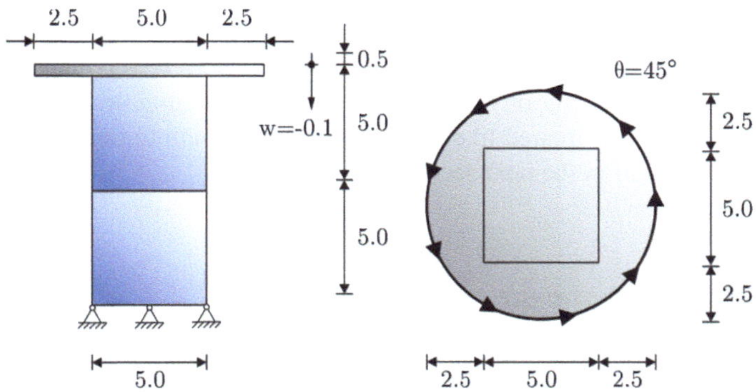

Fig. 6 Geometry of the rotational test

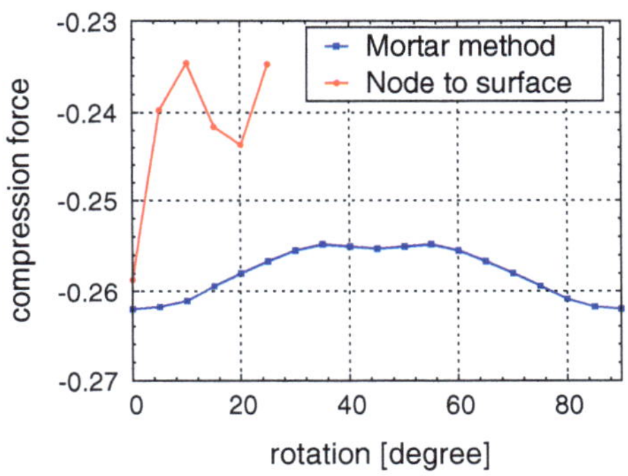

Fig. 7 Comparison of results of node to segment and mortar discretisation

upper block is then rotated by 90° in 5° steps. Comparing the results, Fig. 7, one can see that the node to segment method leads to an unstable solution behaviour, which doesn't complete the full rotational movement. In contrast to this, computation using mortar method runs stable and results in a smooth axial force curve.

3.2 Calculation of the Stress Dependent Friction Coefficient μ

Considering soil-structure interaction the tangential stress behaviour is one of the most influenting values. It is given by the coefficient of friction μ. Taking into account the changing behaviour of soil during loading, new friction laws have to be applied. In reality the coefficient of friction μ depends on different mechanical quantities. In several applications friction is dependent on the tangential slip rate $\dot{g}_{\mathrm{T}}$, the normal pressure λ_{N}, the temperature Θ or on the surface roughness $z(x, y)$. A detailed overview can be found in [8]. A detailed consideration of Finite-Element implementation in [9]. Here different modelling approach is assumed where the coefficient of friction depends on the actual stress: $\mu = \mu(\boldsymbol{\sigma})$ [7]. The proposed approach exploits the analogy of triaxial and simple shear test given in the Mohr-Coulomb and Coulomb friction, see Fig. 2. By applying a suitable numerical procedure the contact layer is treated like a thin limit case of a continuum layer, which is derived in more detail in the next section.

3.3 Numerical Treatment

A contact area can be treated as a limit case of a thin continuum layer, see [7], which is used for applying of the described projection procedure. For this purpose it is more convenient to redefine the contact contribution terms to the weak form (4)

$$\int_{\varphi(\partial B^c)} (\underbrace{t_N \delta g_N}_{} + \underbrace{\boldsymbol{t}_T \cdot \delta \boldsymbol{g}_T}_{}) \mathrm{d}a \tag{8}$$

in terms of kinematical quantities (Fig. 8). This redefinement is based on descriptions of shell kinematics. As a first step the position vector and the test function based on the intrinsic coordinates are introduced

$$\begin{aligned} \boldsymbol{x}(\xi^\alpha, \xi) &= x^1(\xi^{1\alpha}) + \frac{\xi}{h}\left[x^2(\xi^{2\alpha}) - x^1(\xi^{1\alpha})\right] \\ \eta(\xi^\alpha, \xi) &= \eta^1(\xi^{1\alpha}) + \frac{\xi}{h}\left[\eta^2(\xi^{2\alpha}) - \eta^1(\xi^{1\alpha})\right] \end{aligned} \tag{9}$$

where indices $(\bullet)^1$ and $(\bullet)^2$ are according to slave or master surface, respectively, and h is the intrinsic height of the contact layer, see Fig. 9. With terms in Eq. (9) following metric tensors and their virtual counterparts can be derived

$$\begin{aligned} \boldsymbol{g}_\alpha &= \frac{\partial \boldsymbol{x}}{\partial \xi^\alpha} = \left(1 - \frac{\xi}{h}\right)\boldsymbol{a}^1_\alpha + \frac{\xi}{h}\boldsymbol{a}^2_\alpha, \qquad \boldsymbol{g}_3 = \frac{\partial \boldsymbol{x}}{\partial \xi} = \frac{1}{h}\left[\boldsymbol{x}^2 - \boldsymbol{x}^1\right] \\ \boldsymbol{G}_\alpha &= \frac{\partial \boldsymbol{X}}{\partial \xi^\alpha} = \left(1 - \frac{\xi}{h}\right)\boldsymbol{A}^1_\alpha + \frac{\xi}{h}\boldsymbol{A}^2_\alpha, \qquad \boldsymbol{G}_3 = \frac{\partial \boldsymbol{X}}{\partial \xi} = \frac{1}{h}\left[\boldsymbol{X}^2 - \boldsymbol{x}^1\right] \\ \delta\boldsymbol{g}_\alpha &= \frac{\partial \boldsymbol{\eta}}{\partial \xi^\alpha} = \left(1 - \frac{\xi}{h}\right)\boldsymbol{\eta}^1_{,\alpha} + \frac{\xi}{h}\boldsymbol{\eta}^2_{,\alpha} = \left(1 - \frac{\xi}{h}\right)\delta\boldsymbol{a}^1_\alpha + \frac{\xi}{h}\delta\boldsymbol{a}^2_\alpha \\ \delta\boldsymbol{g}_3 &= \frac{\partial \boldsymbol{\xi}}{\partial \xi} = \frac{1}{h}\left[\boldsymbol{\xi}^2 - \boldsymbol{\xi}^1\right] \end{aligned} \tag{10}$$

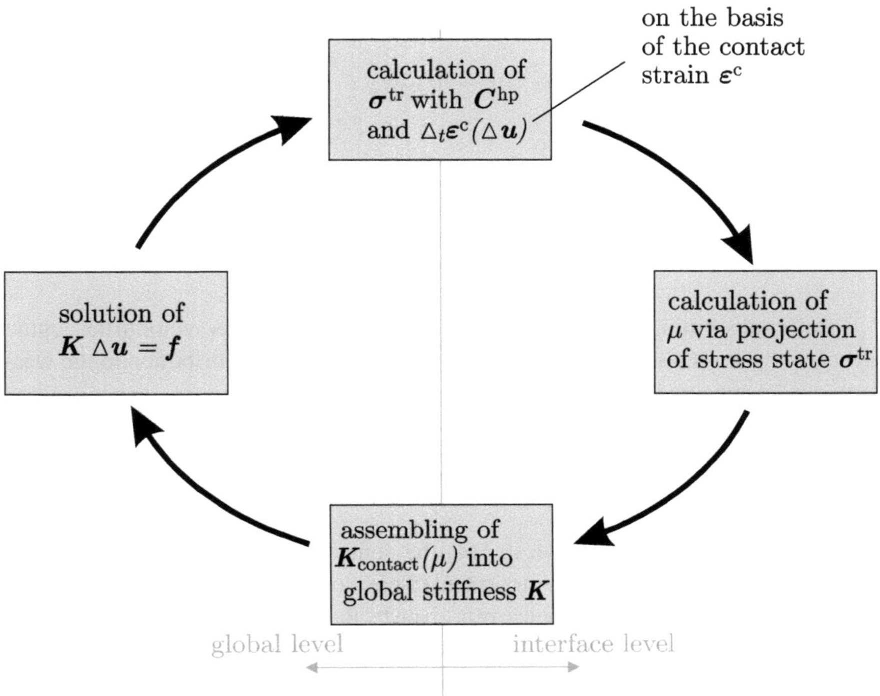

Fig. 8 Computation of the coefficient of friction μ

which can be used for obtaining components of Almansi strain tensor

$$
\begin{aligned}
2\bar{e}_{\alpha 3} &= \lim_{h\to 0}\left(\bar{g}_{\alpha 3} - \bar{G}_{\alpha 3}\right) = \left[x^2 - x^1\right]\cdot a_\alpha - \left[X^2 - X^1\right]\cdot A_\alpha \\
2\delta\bar{e}_{\alpha 3} &= \lim_{h\to 0}\delta\bar{g}_{\alpha 3} = \left[\eta^2 - \eta^1\right]\cdot a_\alpha + \left[x^2 - x^1\right]\cdot \eta_{,\alpha} \\
\bar{e}_{33} &= \lim_{h\to 0}\frac{1}{2}\left(\bar{g}_{33} - \bar{G}_{33}\right) = \frac{1}{2}\left[x^2 - x^1\right]\cdot n \\
\delta\bar{e}_{33} &= \lim_{h\to 0}\frac{1}{2}\delta\bar{g}_{33} = \left[\eta^2 - \eta^1\right]\cdot n
\end{aligned}
\tag{11}
$$

with $\bar{(\bullet)} := \int_0^h (\bullet)\ \det z\ d\xi$ the quantity $(\bullet)$ integrated over the height h. The kinematical terms δg_N and $\delta \boldsymbol{g}_T$ can be redefined using components of virtual Almansi strain tensor $\delta \boldsymbol{e} = [\delta e_{11},\ \delta e_{22},\ \delta e_{33},\ 2\delta e_{12},\ 2\delta e_{23},\ 2\delta e_{13}]^T$

$$\delta g_{\mathrm{N}} = (\boldsymbol{\eta}^2 - \boldsymbol{\eta}^1) \cdot \boldsymbol{n} = \delta e_{33}$$
$$\delta g_{\mathrm{T}\alpha} = (\boldsymbol{\eta}^2 - \boldsymbol{\eta}^1) \cdot \boldsymbol{a}_\alpha = 2\delta e_{\alpha 3} \tag{12}$$

while the in-plane terms are zero

$$e_{\alpha\beta} = 0$$
$$\delta e_{\alpha\beta} = 0, \quad (\alpha, \beta = 1, 2). \tag{13}$$

Finally substituting components t_{N} and $\boldsymbol{t}_{\mathrm{T}}$ of the contact stress vector in the contact layer by $\boldsymbol{t} = \boldsymbol{\sigma} \cdot \boldsymbol{n}$ a solid shell type formulation of contact contribution to the weak form can be specified as follows

$$\int_{\varphi(B^c)} \delta \boldsymbol{e} \cdot \boldsymbol{\sigma} \mathrm{d}v = \lim_{h \to 0} \int_{\gamma^c} \int_0^h \delta \boldsymbol{e} \mathrm{d}\xi \cdot \sigma \mathrm{d}a = \int_{\varphi(\partial B^c)} (t_{\mathrm{N}} \delta g_{\mathrm{N}} + \boldsymbol{t}_{\mathrm{T}} \cdot \delta \boldsymbol{g}_{\mathrm{T}}) \mathrm{d}a. \tag{14}$$

According to the analogy shown in (14) the contact area can be treated like very thin continuum layer, where the strain tensor has now the form $\boldsymbol{\varepsilon}^{\mathrm{c}} = [0,\ 0,\ g_{\mathrm{N}},\ 0,\ \triangle_t g_{\mathrm{T}1},\ \triangle_t g_{\mathrm{T}2}]^{\mathrm{T}}$. The stress response $\boldsymbol{\sigma}$ is given by the hypoplastic material behaviour (see Sect. 2) with the increment of strain ε^{c}. Since the properties of the contact behaviour strongly depend on the state of contact bodies, here in particular of the non-linear material behaviour of sand, the crucial point is the derivation of the stress dependent coefficient of friction $\mu = \mu\,(\boldsymbol{\sigma})$. Using the observation, that performing a simple shear test the actually unbonding takes place in a few grain diameter distance from the moving surfaces, analogy related to the friction angle φ obtained from the triaxial test can be assumed. Following this assumption the common used Coulombs friction law

$$f^c = \|\bar{\boldsymbol{t}}_{\mathrm{T}}\| - |\bar{\lambda}_{\mathrm{N}}| \, \tan \varphi = 0, \tag{15}$$

which in three dimensional case leads to the definition of the Mohr-Coulomb yield criterion, here formulated using invariants of stress tensor and its deviatoric part [5]

$$f^m = \sqrt{\mathrm{II}_s} \cos(\Theta) + \left[\frac{1}{3} \mathrm{I}_\sigma - \sqrt{\frac{\mathrm{II}_s}{3}} \sin(\Theta) \right] \sin \varphi = 0, \tag{16}$$

with I_σ, II_s invariants of stress tensor $\boldsymbol{\sigma}$ and its deviator $\boldsymbol{s}$ respectively, and Θ the Lode angle. The result is conform with the assumed equivalency of direct shear and triaxial test, Fig. 10, in case of sand contacting a surface. The Eq. (16) can be solved for the friction angle φ, and thus for the coefficient of friction μ, as follows:

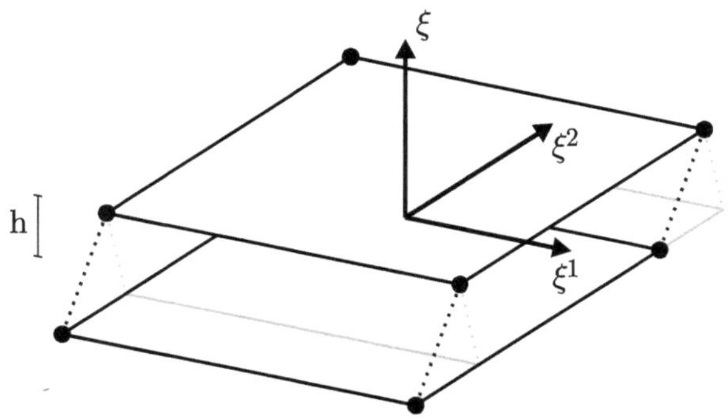

Fig. 9 Solid shell element with intrinsic coordinate system used for defining the kinematics of the contact area

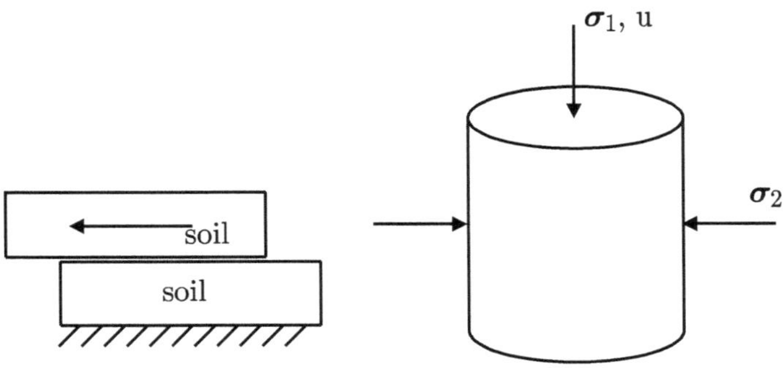

Fig. 10 Principle of direct shear and triaxial test

$$\mu = |\tan \varphi| = \left| \tan \left(\arcsin \left(\frac{\sqrt{\mathrm{II}_s} \cos(\Theta)}{\frac{1}{3}\mathrm{I}_\sigma - \sqrt{\frac{\mathrm{II}_s}{3}} \sin(\Theta)} \right) \right) \right|. \tag{17}$$

By applying the current strain increment $\Delta_t \varepsilon$ a trial stress state $\boldsymbol{\sigma}^{\mathrm{tr}}$ is obtained. Evaluating the Mohr-Coulomb criterion $f^{\mathrm{mc}} = f^{\mathrm{mc}}(\boldsymbol{\sigma}^{\mathrm{tr}})$ it is possible to calculate μ and contact stiffness contributions $\boldsymbol{K}_{\mathrm{contact}}$ to the global stiffness $\boldsymbol{K}$, see Fig. 8.

4 Application to Pull Out of a Wall Experiment

For the verification of the capability of the proposed simulation model a pull out test is performed. The experimental test was conducted by the Central Project within the DFG research unit 1136, which provides the experimental data, see also [6]. The geometry of the model is shown in Fig. 11. The setup consists of a sand body which is bounded at the bottom by a plate. Three sides are supported perpendicularly to

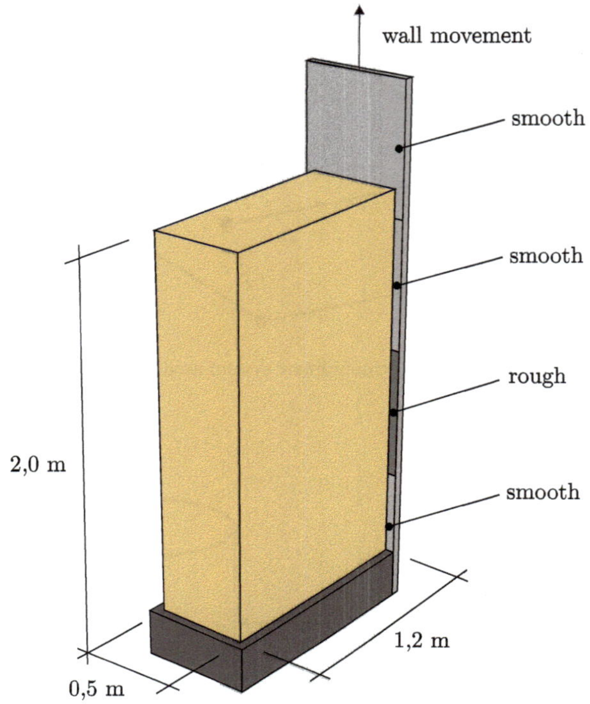

Fig. 11 Pull out test geometry

their surfaces. There is no friction between these three walls and sand. The fourth side has contact with moveable wall section. The wall is divided into four segments. Each segment has different roughness. The surfaces of the lowest and of the two upper segments are smooth ($\mu_1 = 0.2$). The surface of the middle segment ist rough, with the coefficient of friction $\mu = \mu(\boldsymbol{\sigma})$ calculated like described in Sect. 3. For the sand and the contact, for the constitutive behaviour of the rough contact surface, hypoplastic material model described in Sect. 2 is applied, with material parameter set

Setting the exponent $w = 2$ the inelasticity is interpolated by Y quadratically, see Eq. (1) and Fig. 3. The plate at the bottom and the wall itself are made of the same material, which is assumed to be linear elastic with modulus $E = 210 \cdot 10^9 \frac{N}{m^2}$ and Poisson ratio $\nu = 0.3$. After first step of loading with 1g, in the second the wall is moved 100 mm upwards. In Fig. 12 the tangential force between wall and sand body for the rough surface is shown. Its according to expectations, that the resistance within the sand against the movement at the beginning rises. Later with increasing plastic deformations the resistance within the sand is releasing, which results in a kind of softening in the tangential force curve. The quantitative difference between the experimental and the calculated curve is caused by using not fitted material parameters. The values used here, Table 1, are an example set of parameters.

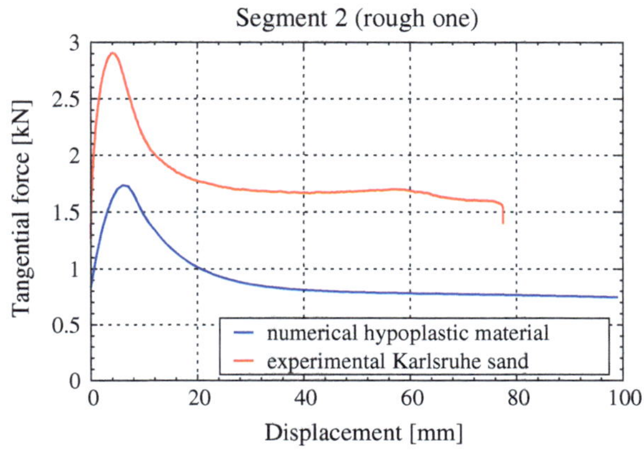

Fig. 12 Tangential force in pull out of a wall simulation

Table 1 Material data used for the simulation of the pull out test

Cohesion c	Friction angle φ	Exponent of Y w	Barotropy coefficient λ_E
5	30	2	100

5 Conclusion and Outlook

Within this contribution a method for considering stress dependent tangential contact behaviour between sand and a rough structure was shown. For modelling the granular medium behaviour as well for the interface modelling a robust hypoplastic material law was used. Basic procedures within computational contact mechanics as well as incorporating the inelastic contact body and thus non-linear, stress dependent, evolution of coefficient of friction have been explained. The proposed approach exploiting the analogy of friction angle in sand to friction within simple shear test shows good qualitative ability to model results obtained experimentally. In conjunction with the mortar method for contact discretisation, the robustness of which has been shown on a significant example, the proposed model seems to be a well promising simulation set up. For a better simulation of installation in sand, two important aspects have to be taken into account. On one side a more precisely identification of material parameters is required. On the other hand an even more sophisticated constitutive model have to be used, which incorporates the changing structure of sand due to volume change phenomena.

Acknowledgments The authors would like to thank to the German Research Foundation DFG for the financial support as well as to other partners of the research group FOR 1136 Modellierung von geotechnischen Herstellungsvorgängen mit ganzheitlicher Erfassung des Spannungs-Verformungs-Verhaltens im Boden (GeoTech, Holistic Simulation of Geotechnical Installation Processes) for pleasant interdisciplinary cooperation.

References

1. Holzapfel, G.A.: Nonlinear Solid Mechanics. A Continuum Approach for Engineering. Wiley, Chinchester (2000)
2. Niemunis, A.: Extended Hypoplastic Models for Soils. Institut für Grundbau und Bodenmechanik der Ruhr-Universität Bochum, Bochum (2003)
3. Puso, M.A., Laursen, T.A.: A mortar segment-to-segment contact method for large deformation solid mechanics. Comput. Methods Appl. Mech. Eng. **193**, 601–629 (2004)
4. Puso, M.A., Laursen, T.A.: A mortar segment-to-segment frictional contact method for large deformations. Comput. Methods Appl. Mech. Eng. **193**, 4891–4913 (2004)
5. Schad, H.: Nichtlineare Stoffgleichungen für Böden und ihre Verwendung bei der numerischen Analyse von Baugrundaufgaben. Institut für Grundbau und Bodenmechanik, Universität Stuttgart, Stuttgart (1979)
6. Vogelsang, J., Huber, G., Triantafyllidis, Th.: A large scale soil-structure interface testing device. Geotech. Test. J. **36**(5), 1–13 (2013)
7. Weißenfels, C.: Contact methods integrating plasticity models with application to soil mechanics. Institute of Continuum Mechanics, Leibniz Universität Hannover, Hannover (2013)
8. Wriggers, P.: Computational Contact Mechanics. Springer, Berlin (2006)
9. Wriggers, P., Reinelt, J.: Multi-scale approach for frictional contact of elastomers on rough rigid surfaces. Comput. Methods Appl. Mech. Eng. **198**, 1996–2008 (2009)

Vibro-Injection Pile Installation in Sand: Part I—Interpretation as Multi-material Flow

D. Aubram, F. Rackwitz and S.A. Savidis

Abstract The installation of vibro-injection piles into saturated sand has a significant impact on the surrounding soil and neighboring buildings. It is generally characterized by a multi-material flow with large material deformations, non-stationary and new material interfaces, and by the interaction of the grain skeleton and the pore water. Part 1 in this series of papers is concerned with the mathematical and physical modeling of the multi-material flow associated with vibro-injection pile installation. This model is the backbone of a new multi-material arbitrary Lagrangian-Eulerian (MMALE) numerical method presented in Part 2.

Keywords Multi-material flow · Large deformations · Mixture · Soil mechanics · Sand · Averaging · Homogenization · Closure law · Arbitrary Lagrangian-Eulerian

Symbols and Operators

$\cdot$	Single contraction of tensors
$:$	Double contraction of tensors
$\otimes$	Tensor product
$\langle\cdot\rangle$	Spatial average
$[\![\cdot]\!]$	Jump across an interface
$(\cdot)^-$, $(\cdot)^+$	Associated with the begin, with the end of the remap step
$(\cdot)_{\mathrm{dev}}$	Deviator of a second-order tensor
$(\cdot)_j$	Associated with the jth control volume
$(\cdot)_n$, $(\cdot)_{n+1}$	Associated with time t_n, t_{n+1}
$(\cdot)_{n+\theta}$	Associated with time $t_{n+\theta}$
$(\cdot)^{\mathrm{T}}$	Transpose of a tensor

D. Aubram (✉) · S.A. Savidis
Chair of Soil Mechanics and Geotechnical Engineering, Technische Universität Berlin, Secr. TIB1-B7, Gustav-Meyer-Allee 25, 13355 Berlin, Germany
e-mail: daniel.aubram@tu-berlin.de

F. Rackwitz
Department of Geotechnical Engineering, Ostbayerische Technische Hochschule Regensburg, Prüfeninger Straße 58, 93049 Regensburg, Germany

Th. Triantafyllidis (ed.), *Holistic Simulation of Geotechnical Installation Processes*, Lecture Notes in Applied and Computational Mechanics 77,
DOI 10.1007/978-3-319-18170-7_5

$\grave{(\cdot)}^{k}$	Material time derivative of a k-phase-related field along $\boldsymbol{v}^k$
$\dot{(\cdot)}^{\beta k}$	Material time derivative of β-species-k-phase-related field along $\boldsymbol{v}^{\beta k}$
$\overset{\triangledown}{(\cdot)}$	Zaremba-Jaumann rate
$\boldsymbol{\nabla}(\cdot)$	Covariant derivative
$\mathfrak{A}$	Corotated set of material state variables
$\boldsymbol{b}$	Microscopic body force per unit mass
$\boldsymbol{b}^{\beta k}$	Average body force per unit mass of the β-species of the k-phase
$\langle \boldsymbol{b} \rangle$	Average body force per unit mass of the mixture
$\boldsymbol{c}$	Convective velocity
$\boldsymbol{\mathsf{c}}$	Generic fourth-order material tangent tensor
$\boldsymbol{\mathsf{c}}^{\mathrm{f}k}$	Material tangent tensor for the fluid species of the k-phase
$\boldsymbol{\mathsf{c}}^{k'}$, $\boldsymbol{\mathsf{c}}^{k''}$	Material tangent tensor for the k-phase effective Cauchy stress
$\boldsymbol{\mathsf{c}}^{\mathrm{s}k}$	Material tangent tensor for the solid species of the k-phase
$\mathbf{d}(\cdot)$	Differential, gradient
$\boldsymbol{d}^{\mathrm{f}k}$	Average spatial rate of deformation of the k-phase fluid species
$\boldsymbol{d}^{k}$	k-phase mean spatial rate of deformation
$\boldsymbol{d}^{\mathrm{s}k}$	Average spatial rate of deformation of the k-phase solid species
$\langle \boldsymbol{d} \rangle$	Mean spatial rate of deformation of the mixture
$\mathrm{div}(\cdot)$	Divergence
$\mathrm{d}\upsilon$	Volume form on the ambient Euclidian space
e	Void ratio
E, E^{p}	Young's modulus, plastic modulus
f^{k}	k-phase volume fraction
$\boldsymbol{f}$	Relative incremental deformation gradient
$\mathfrak{f}$	Constitutive response function
$\mathrm{Fl}(\cdot)$	Averaged convective volume flux
$\boldsymbol{h}$	Generic set of material state variables
$\boldsymbol{h}^{k'}$, $\boldsymbol{h}^{k''}$	Set of material state variables for k-phase effective Cauchy stress
$\boldsymbol{h}^{\mathrm{s}k}$	Set of material state variables for the solid species of the k-phase
$\boldsymbol{I}$	Second-order unit tensor
k	Index for the kth phase of the mixture, $k \in \{1, \ldots, n_{\mathrm{mat}}\}$
K	Microscopic bulk modulus
$K^{\mathrm{f}k}$	Average bulk modulus of the fluid species of the k-phase
K^{k}	Average k-phase bulk modulus (undrained)
K^{k}_{d}	Average k-phase bulk modulus (drained)
K^{k}_{uj}	Average k-phase unjacketed bulk modulus
$K^{\mathrm{s}k}$	Average bulk modulus of the solid species of the k-phase
$\langle K \rangle$	Mean bulk modulus of the mixture
l_{macro}	Characteristic length at the macroscale
l_{meso}	Characteristic length at the mesoscale
l_{micro}	Characteristic length at the microscale
n^{k}	k-phase fluid fraction, k-phase porosity
n_{mat}	Number of material phases in the mixture

$\boldsymbol{n}^*$	Outward normals on an interface
p	Microscopic pressure
$p^{\mathrm{f}k}$	Average pressure in the fluid species of the k-phase
p^k	Average k-phase pressure
$p^{k'}$, $p^{k''}$	k-phase mean effective stress
$p^{\mathrm{s}k}$	Average pressure in the solid species of the k-phase
$p^{\beta k}$	Average pressure in the β-species of the k-phase
P^k	Phase function, indicator function
$P^k S^\beta$	Phase-species function, indicator function
q	Generic scalar-, vector-, or tensor-valued microscopic spatial field
$\hat{q}$	Referential or ALE description of the field q
$\mathfrak{R}$	Rotation related to a corotational rate
$\boldsymbol{s}$	Microscopic extra stress
$\boldsymbol{s}^{\mathrm{f}k}$	Average extra stress in the fluid species of the k-phase
$\boldsymbol{s}^k$	Average k-phase extra stress
$\boldsymbol{s}^{k'}$, $\boldsymbol{s}^{k''}$	k-phase effective extra stress
$\boldsymbol{s}^{\mathrm{s}k}$	Average extra stress in the solid species of the k-phase
$\boldsymbol{s}^{\beta k}$	Average extra stress in the β-species of the k-phase
$\langle \boldsymbol{s} \rangle$	Average extra stress of the mixture
S^β	Species function, indicator function
$\mathcal{S}$	Ambient Euclidian space
$\mathfrak{S}$	Corotated Cauchy stress
t	Time
t_n, t_{n+1}	Time at the beginning, at the end of a time step
$t_{n+\theta}$	Intermediate time in a time step, with $\theta \in [0, 1]$
$\mathrm{tr}(\cdot)$	Trace of a second-order tensor
$\boldsymbol{u}$	Mean spatial displacement field of the mixture
$\boldsymbol{v}$	Microscopic spatial velocity
$\boldsymbol{v}^{\mathrm{f}k}$	Mean spatial velocity of the fluid species of the k-phase
$\boldsymbol{v}_{\mathrm{I}}$	Interface velocity
$\boldsymbol{v}^k$	k-phase mean spatial velocity
$\boldsymbol{v}^{\mathrm{s}k}$	Mean spatial velocity of the solid species of the k-phase
$\boldsymbol{v}^{\beta k}$	Mean spatial velocity of the β-species of the k-phase
$\langle \boldsymbol{v} \rangle$	Mean spatial velocity of the mixture
V'	Volume measure of $\mathcal{V}'$ in the ambient Euclidian space
V_j	Volume measure of the jth control volume
$\mathcal{V}$	Spatial domain of interest
$\mathcal{V}'$	Representative volume element (RVE)
x, x'	Points in the ambient Euclidian space
z	Penetration depth
α^k	k-phase Biot-Willis coefficient
β	index for the βth species of the mixture, $\beta \in \{\mathrm{s}, \mathrm{f}\}$
$\boldsymbol{\Gamma}^{\beta k}$	Momentum transfer onto the k-phase β-species due to drag forces
$\delta_{\mathrm{I}}^{\beta k}$	Dirac delta picking out the β-species-k-phase interface

$\Delta\mathfrak{E}$	Corotated algorithmic finite strain increment
$\Delta\boldsymbol{r}$	Algorithmic finite rotation increment
$\Delta\mathfrak{R}$	Incremental rotation
$\Delta\mathfrak{S}$	Corotated Cauchy stress increment
Δt	Time increment
$\Delta\varepsilon$	Algorithmic finite strain increment
$\dot{\varepsilon}^{\mathrm{s}}_{\mathrm{vol}}$	Average volumetric strain rate of solid grains
$\Pi^{\beta k}$	Mass transfer onto the β-species of the k-phase
ρ	Microscopic spatial mass density
$\rho^{\mathrm{f}k}$	Average mass density of the fluid species of the k-phase
ρ^{k}	Average k-phase mass density
$\rho^{\mathrm{s}k}$	Average mass density of the solid species of the k-phase
$\rho^{\beta k}$	Average mass density of the β-species of the k-phase
$\langle\rho\rangle$	Average spatial mass density of the mixture
σ^{y}, $\sigma^{\mathrm{y}0}$	Yield stress, initial yield stress in uniaxial tension
$\boldsymbol{\sigma}$	Microscopic Cauchy stress
$\boldsymbol{\sigma}^{\mathrm{f}k}$	Average Cauchy stress in the fluid species of the k-phase
$\boldsymbol{\sigma}^{k}$	Average k-phase Cauchy stress
$\boldsymbol{\sigma}^{k'}$, $\boldsymbol{\sigma}^{k''}$	k-phase effective Cauchy stress
$\boldsymbol{\sigma}^{\mathrm{s}k}$	Average Cauchy stress in the solid species of the k-phase
$\boldsymbol{\sigma}^{\beta k}$	Average Cauchy stress in the β-species of the k-phase
$\langle\boldsymbol{\sigma}\rangle$	Average Cauchy stress of the mixture
$\boldsymbol{\Phi}$	Relative motion of the ALE reference domain
$\boldsymbol{\omega}$	Vorticity tensor

Abbreviations

ALE	Arbitrary Lagrangian-Eulerian
CFD	Computational Fluid Dynamics
FEM	Finite Element Method
MMALE	Multi-Material Arbitrary Lagrangian-Eulerian
PIV	Particle Image Velocimetry
RI-pile	Vibro-injection pile ("Rüttelinjektionspfahl")
RVE	Representative Volume Element

1 Introduction

Vibro-injection piles, in German called "Rüttelinjektionspfähle (RI-Pfähle)", are used in sandy soil to tie back the base slab of deep excavations in urban area with high groundwater level. They consist of an H-section steel pile equipped with an injection tube and a welded-on collar located at the pile toe (Figs. 1 and 2). During the installation of the pile into the water-saturated sand by vibration the soil loses its shear strength ("soil liquefaction") and the annular gap generated by the collar is

Fig. 1 Grid of vibro-injection piles to tie back an excavation base slab. Deep excavation of about 20 m in urban area (*left*) and recovered pile (*right*)

continuously injected with grout. The installation process of a vibro-injection pile interacts to a great extent with its neighborhood [32, 34]. This is why the numerical modeling of the installation process is of high practical relevance for the realistic prediction of the deformations and the load bearing behavior of the wall.

Sufficiently realistic computational models must be able to reproduce the basic installation phenomena (Fig. 2, detail A). These include the shearing and liquefaction of the locally undrained saturated soil, the displacement of the liquefied soil by the pile and the grout, as well as the mixing of the grout with the liquefied soil. Stated more generally, the installation of vibro-injection piles can be characterized by a multi-material flow with large material deformations, by free surfaces and non-stationary contact interfaces, and by the complex coupled behavior of the grain skeleton and the pore water. The numerical simulation of such problems is very challenging. Even the penetration of the steel profile alone could not be handled by using the classical Lagrangian formulations of the finite element method (FEM) [44, 47]. It is no surprise, therefore, that up to date there are no FE prediction models for the installation of RI-piles and for the related time-histories of the stress and density states within the soil.

It is the aim of the Subproject 5 as part of the DFG Research Unit FOR 1136 to make a significant contribution to this area through the development of a so-called multi-material arbitrary Lagrangian-Eulerian (MMALE) method. Objectives are, firstly, to predict both qualitatively and quantitatively the stress and density time-histories within the soil in the vicinity of the vibro-injection pile and hence the effects of its installation process on close-by structures and, secondly, the realistic simulation of the load bearing and deformation behavior of the completed pile. Based on a continuum mechanical description, especially the single installation phenomena (driving of the steel profile by vibration, liquefaction and displacement of the undrained soil during vibratory pile driving, grouting of the emerging annular gap

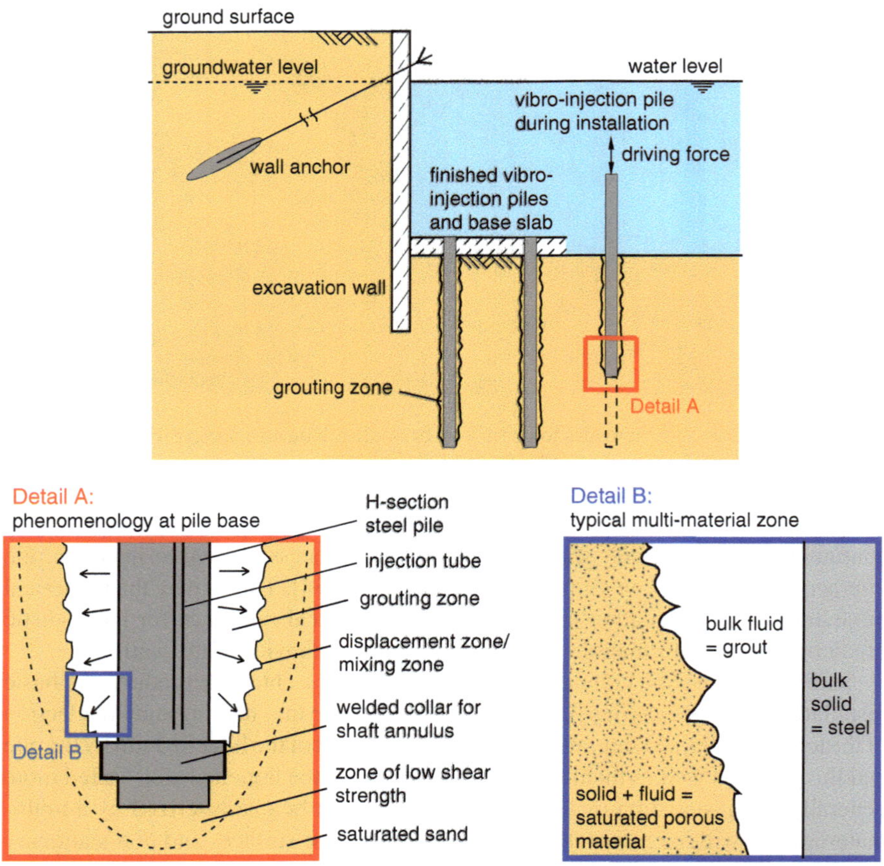

Fig. 2 Illustration of the installation of a vibro-injection pile in saturated sand to tie back an excavation base slab. *Detail A* Phenomenology at the pile base, *Detail B* typical zone of the considered multi-material flow including macroscopic interfaces

between the steel profile and the liquefied soil; see also Fig. 2) should be modeled by using an MMALE finite element method.

MMALE methods fall into the category of arbitrary Lagrangian-Eulerian methods [2, 4, 23, 35] and have no limitations concerning material deformations and the evolution or generation of material interfaces. The mesh can move independently of the material such that material interfaces (boundaries) may flow through the mesh. Elements cut by interfaces contain a mixture of two or more materials and are referred to as multi-material elements. The mixture is treated as an effective single-phase material or homogenized mixture on the element level. Hence, quantities related to each material must be "mixed" in a certain way to yield the corresponding homogenized element quantities. As the latter should be based on physical principles, the development of an MMALE method for vibro-injection pile installation in sand has to start with the mathematical and physical modeling of the associated multi-material flow, which is the content of Part 1 in this series of papers. Details of the new method and of our numerical and experimental investigations concerning vibro-injection pile installation will be presented in Part 2.

Part 1 is structured as follows. In Sect. 2 we introduce the three different spatial scales (microscale, mesoscale, macroscale) of the continuum mechanical problem and formulate the balance principles and jump conditions of the microscopic multi-material flow. This flow is homogenized in Sect. 3 to yield a macroscopic multi-material flow forming the physical model of the numerical method. One of the basic techniques to achieve this homogenization is spatial averaging, a theory which is widely used in the field of multiphase flow and the modeling of porous media. Section 4 then addresses the closure of the resulting system of equations by making reasonable assumptions and establishing interfacial transfer laws, constitutive laws, and topological laws. The final form of the macroscopic model for multi-material flow is derived in Sect. 5 through model reduction based on additional, a priori closure assumptions. The paper closes with concluding remarks and outlook in Sect. 6.

2 Spatial Scales and Balance Principles

2.1 Three Spatial Scales

Consider the process of vibro-injection pile installation into sand as illustrated in Fig. 2. Detail B of that figure can be regarded as a still image of a multi-material flow recorded through a spatially fixed observation window. We take this zone as characteristic of the multi-material (multi-constituent) flow and assume that the basic features of the flow are independent of the specific arrangement resp. distribution of the constituents. The characteristic zone is filled with an immiscible mixture consisting of a bulk solid phase (steel), a bulk fluid phase (grout), and a compound phase consisting of a solid species and a fluid species which represents the fluid-saturated porous material (sand)—the term "species" is used here to distinguish these constituents from the bulk solid and bulk fluid phases. Void (empty space) is considered as a particular bulk fluid phase in our model. Immiscibility of the mixture is characterized by the fact that the constituents are separated by interfaces.

Three spatial scales are introduced in accordance with [6]; see Fig. 3. The porous material is constituted by an assembly of sand grains, whose typical diameter defines the microscale of the problem, l_{micro}. However, in the present research we have to properly reproduce the nonlinear coupled mechanical behavior of the water-saturated sand. In this regard the best models currently available are phenomenological two-phase models that rely on a continuum representation of the soil and not on micro-mechanics. Therefore, we introduce the characteristic length at which the saturated sand can be represented by a continuum as the mesoscale l_{meso}. A similar issue of upscaling, but on a larger scale, has to be faced in multi-material elements of the

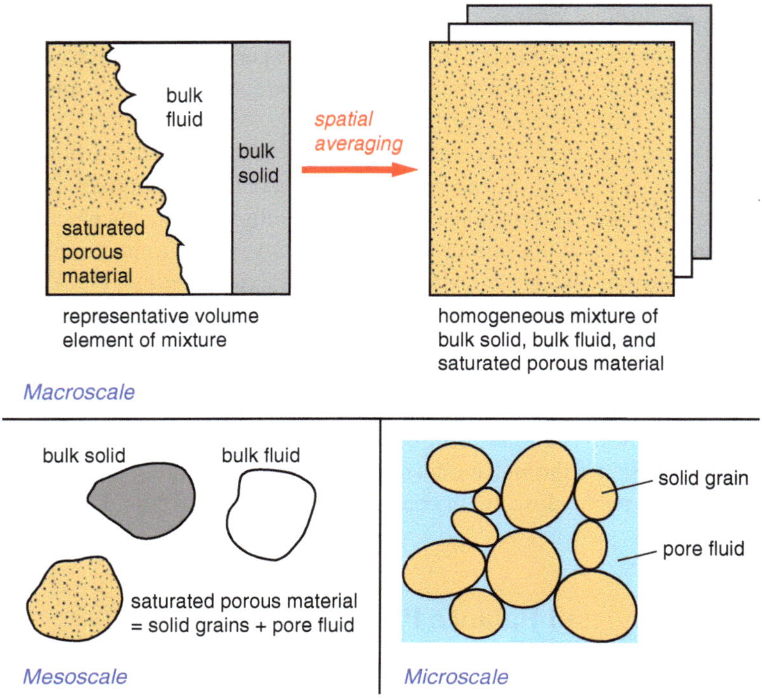

Fig. 3 Three spatial scales to model vibro-injection pile installation in saturated sand

MMALE method. Accordingly, we postulate the existence of a representative volume element (RVE) at some macroscale, l_{macro}, through which the immiscible mixture of mesoscale continua (bulk solid, bulk fluid, and saturated sand) can be equivalently modeled as an effective single-phase material (homogenized immiscible mixture) treated by the MMALE method. The real world is modeled on an even larger scale; in the literature this is sometimes called the megascale.

To make this point clear, we remark that our definitions of a constituent, a phase, and a species is comparable to those generally used in chemistry and thermodynamics. Constituents are called the individual (chemically-independent) materials composing the mixture on the microscale (constituent = micro-continuum). The heterogeneous mixture consists of different material phases separated by macroscopic interfaces. A phase is physically distinct and mechanically separable and, in our case, might be identified with a constituent if there would be no saturated porous medium in the mixture. However, we generally define a phase to be a compound material consisting of solid species and a fluid species, and which can be regarded homogeneous on a mesoscale (phase = meso-continuum). Finally, as indicated by Fig. 3, on the postulated macroscale the mixture of phases can be addressed as a homogeneous mixture (macro-continuum).

2.2 Microscopic Balance Principles

On the microscale all constituents of the mixture can be regarded as micro-continua (Fig. 3), governed by the equations of continuum mechanics [1, 20, 27, 28, 38, 39]. Generally these include conservation of mass

$$\frac{\partial \rho}{\partial t} + \operatorname{div}(\rho \boldsymbol{v}) = 0 \tag{1}$$

and balance of momentum

$$\frac{\partial \rho \boldsymbol{v}}{\partial t} + \operatorname{div}(\rho \boldsymbol{v} \otimes \boldsymbol{v}) = \rho \boldsymbol{b} + \operatorname{div} \boldsymbol{\sigma}, \tag{2}$$

assumed to hold in a domain of interest $\mathcal{V} \subset \mathcal{S}$ in the ambient Euclidian space and time interval $[0, T] \subset \mathbb{R}$. In the equations, $\boldsymbol{v}$ denotes the material velocity in the spatial description, ρ is the mass density, $\boldsymbol{b}$ is a prescribed body force per unit mass (e.g. gravity), $\boldsymbol{\sigma} = \boldsymbol{\sigma}^{\mathrm{T}}$ is the symmetric Cauchy stress, div is the spatial divergence operator, and $\otimes$ is the tensor product. The notation is found at the beginning of this paper.

In the process of vibro-injection pile installation, the velocities are moderate and thermal effects can be ignored, so that Eqs. (1) and (2) alone serve as the balance principles of that initial boundary value problem. At significantly larger velocities, on the other hand, thermal and strain rate effects may constitute an important part of the solution. In this case balance of energy must be added to the set of governing equations.

While the balance principles hold in the interior of each constituent, the jump conditions for mass and momentum given by

$$[\![\rho(\boldsymbol{v} - \boldsymbol{v}_{\mathrm{I}})]\!] \cdot \boldsymbol{n}^{*} = 0 \qquad \text{and} \qquad [\![\rho \boldsymbol{v} \otimes (\boldsymbol{v} - \boldsymbol{v}_{\mathrm{I}}) - \boldsymbol{\sigma}]\!] \cdot \boldsymbol{n}^{*} = \boldsymbol{0}, \tag{3}$$

respectively, must hold at the interfaces; the jump conditions express the balance principles at an interface. Here $\cdot$ denotes the single contraction of tensors and $[\![q]\!] \stackrel{\text{def}}{=} q^{[l]} - q^{[k]}$ the jump of q across the interface between material l and material k, with $q^{[k]}$ being the limit value of q on the k-side of the interface, $\boldsymbol{v}_{\mathrm{I}}$ is the interface velocity, and $\boldsymbol{n}^{*}$ is the field of normals on the interface pointing outward of the kth micro-continuum. In the jump condition for momentum $(3)_2$, no surface tension is taken into account.

We remark that the balance principles (1) and (2) are in the so-called Eulerian conservation form, and that all equations in this section refer to a spatial reference volume instantaneously occupied by the material on the microscale. The quantities $\rho(x, t)$, $\boldsymbol{v}(x, t)$, etc., represent microscopic (non-averaged) time-dependent spatial fields, with $x \in \mathcal{V}$ and $t \in [0, T]$. This means that a spatial point is viewed as being currently occupied by a single constituent.

3 Macroscopic Multi-material Flow

3.1 Fundamentals of Spatial Averaging

In our approach the effective multi-material dynamics on the macroscale is derived from physical principles by making use of spatial averaging [6, 20, 26, 42]. Spatial averaging results in continuum mechanical equations which are similar to those that can be derived using the continuum theory of mixtures and the theory of porous media [16, 18, 22, 38]; see also Subproject 2 of this research unit. However, the advantage of the averaging approach is that information available at a smaller scale is transferred to the larger scale with respect to which averaging is carried out.

Let $\mathcal{V}' \subset \mathcal{V}$ be a time-independent RVE having the characteristic length l_{macro}, and $q(x, t)$ be an arbitrary time-dependent spatial microscopic field for all $x \in \mathcal{V}$ and $t \in [0, T]$. Then, the volume average of q is defined through

$$\langle q \rangle(x, t) \overset{\text{def}}{=} \frac{1}{V'} \int_{\mathcal{V}'} q(x + \boldsymbol{\xi}, t)\, \mathrm{d}v. \tag{4}$$

Here $\mathrm{d}v$ is the volume form on $\mathcal{S}$ and $V' \overset{\text{def}}{=} \int_{\mathcal{V}'} \mathrm{d}v$ is the volume measure of $\mathcal{V}'$. Moreover, $\boldsymbol{\xi} = x' - x$ and $x' \in \mathcal{V}'$. By (4) and the definition of the RVE, averaged quantities always refer to the macroscale.

A phase function (or indicator function) is defined by

$$P^k(x, t) \overset{\text{def}}{=} \begin{cases} 1 & \text{if } x \text{ is in phase } k \text{ at time } t, \\ 0 & \text{else} \end{cases} \tag{5}$$

for all $x \in \mathcal{V}$ and $t \in [0, T]$, where $k \in \{1, \ldots, n_{\text{mat}}\}$ and n_{mat} being the number of phases in the mixture. The phase function is a so-called generalized function resp. distribution and picks out the generally time-dependent k-phase volume $\mathcal{V}^k \subset \mathcal{V}'$. By the definition of a volume average (4), then, the volume fraction of the kth phase with respect to $\mathcal{V}'$ is defined through

$$f^k \overset{\text{def}}{=} \langle P^k \rangle = \frac{1}{V'} \int_{\mathcal{V}'} P^k\, \mathrm{d}v = \frac{1}{V'} \int_{\mathcal{V}^k} \mathrm{d}v = \frac{V^k}{V'} \quad \in [0, 1], \tag{6}$$

where $V^k \overset{\text{def}}{=} \int_{\mathcal{V}^k} \mathrm{d}v$. Equation (6) is a natural, that is, derived definition of volume fraction. By contrast, volume fraction is postulated in the continuum theory of mixtures and the theory of porous media.

We now assume that all material phases of the considered mixture are composed of a solid species (denoted by $\beta = \text{s}$) and a fluid species (denoted by $\beta = \text{f}$). In fact the portion of one of these species in a particular phase might be zero. For all $x \in \mathcal{V}$ and $t \in [0, T]$, we then define another indicator function S^β, called species function, which equals zero everywhere except on the spatial region occupied by the β-species at time t, where it is equal to one. Concerning the RVE, we denote this region by $\mathcal{V}^\beta \subset \mathcal{V}'$. Therefore, by the property of indicator functions, the product $P^k S^\beta$ is the phase-species function picking out the partial volume occupied by the β-species in the k-phase:

$$(P^k S^\beta)(x, t) = \begin{cases} 1 & \text{if } x \in \mathcal{V} \text{ is in species } \beta \text{ of phase } k \text{ at } t \in [0, T], \\ 0 & \text{else.} \end{cases} \tag{7}$$

The volume fraction of the β-species in the k-phase is then obtained from

$$\pi^{\beta k} \overset{\text{def}}{=} \frac{\langle P^k S^\beta \rangle}{f^k} = \frac{1}{f^k V'} \int_{\mathcal{V}'} P^k S^\beta \, \mathrm{d}v = \frac{1}{V^k} \int_{\mathcal{V}^{\beta k}} \mathrm{d}v = \frac{V^{\beta k}}{V^k} \qquad \in [0, 1], \tag{8}$$

where $\mathcal{V}^{\beta k} \overset{\text{def}}{=} \mathcal{V}^k \cap \mathcal{V}^\beta$ and $V^{\beta k} \overset{\text{def}}{=} \int_{\mathcal{V}^{\beta k}} \mathrm{d}v$ is the volume of the β-species in the k-phase in the RVE. The macroscopic β-species fraction in the mixture is $\pi^\beta \overset{\text{def}}{=} \sum_k f^k \pi^{\beta k}$, and

$$\sum_k f^k = 1 \quad \text{and} \quad \sum_\beta \pi^{\beta k} = 1. \tag{9}$$

As we are concerned with phases solely composed of a solid species and a fluid species, we simply define the fluid fraction (or porosity) of the k-phase through

$$n^k \overset{\text{def}}{=} \pi^{\text{f}k} \tag{10}$$

so that the solid fraction within the k-phase becomes $\pi^{\text{s}k} = 1 - n^k$ by using $(9)_2$. If the k-phase consists of a solid without significant porosity (e.g. steel), then $n^k = 0$. If on the other hand the k-phase is a fluid, then $n^k = 1$ applies. Pure solid mixtures are characterized by $n^k = 0$ and pure fluid mixtures by $n^k = 1$ for all $k \in \{1, \ldots, n_{\text{mat}}\}$, that is, $n \overset{\text{def}}{=} \sum_k n^k = 0$ or 1 respectively. The mixture represented by a single fluid-saturated porous medium is characterized by $f^k = 1$ ($n_{\text{mat}} = 1$) and $0 < n^k < 1$.

Spatial averaging of microscopic physical fields on the constituents is done in a similar way as in (8) in conjunction with (4) and (6). For example, the averaged mass density and velocity of the solid species in the k-phase are given by

$$\rho^{\text{s}k} \overset{\text{def}}{=} \frac{\langle P^k S^{\text{s}} \rho \rangle}{f^k (1 - n^k)} \quad \text{and} \quad \boldsymbol{v}^{\text{s}k} \overset{\text{def}}{=} \frac{\langle P^k S^{\text{s}} \rho \boldsymbol{v} \rangle}{f^k (1 - n^k) \rho^{\text{s}k}}, \tag{11}$$

indicating that $\boldsymbol{v}^{\text{s}k}$ in fact is a mass-weighted volume average. While $\rho^{\text{s}k}$ is the intrinsic (or material) mass density, the bulk mass density of the k-phase solid species with respect to the RVE is obtained from $f^k(1-n^k)\rho^{sk}$. The mass density of the mixture, by the properties (9), can be computed from

$$\langle \rho \rangle = \sum_k \sum_\beta f^k \pi^{\beta k} \rho^{\beta k} = \sum_k f^k \rho^k = \sum_k f^k \left((1-n^k)\rho^{sk} + n^k \rho^{fk} \right), \tag{12}$$

where ρ^k is the (intrinsic) mass density of the k-phase in the mixture. As an example, consider a specimen of dry sand in which the pores are filled with gas of negligible density such that $\rho^{fk} \approx 0$ and $n_{\text{mat}} = 1$. Then, $\rho^{sk} = \rho^{s}$ represents the grain mass density and $(1-n)\rho^{s} \approx \langle \rho \rangle$ is the bulk mass density, which is approximately equal to the mass density $\langle \rho \rangle$ of the solid-gas mixture.

We emphasize that k, β, etc., are labels and not coordinate indices. No summation on repeated labels in a term is enforced unless the sigma notation is employed.

Based on (11) and (12) the mean spatial velocity of the mixture is related to the momentum and mass densities of the constituents by

$$\langle \boldsymbol{v} \rangle = \frac{\sum_k \sum_\beta f^k \pi^{\beta k} \rho^{\beta k} \boldsymbol{v}^{\beta k}}{\sum_k \sum_\beta f^k \pi^{\beta k} \rho^{\beta k}} = \frac{\sum_k f^k \rho^k \boldsymbol{v}^k}{\sum_k f^k \rho^k} = \langle \rho \rangle^{-1} \sum_k f^k \rho^k \boldsymbol{v}^k, \tag{13}$$

where

$$\boldsymbol{v}^k \overset{\text{def}}{=} \frac{\langle P^k \rho \boldsymbol{v} \rangle}{f^k \rho^k} = \sum_\beta \frac{f^k \pi^{\beta k} \rho^{\beta k}}{f^k \rho^k} \boldsymbol{v}^{\beta k} = \frac{1}{\rho^k} \left((1-n^k)\rho^{sk} \boldsymbol{v}^{sk} + n^k \rho^{fk} \boldsymbol{v}^{fk} \right) \tag{14}$$

is the mean spatial velocity of the k-phase. It should be pointed out here that averaged quantities are defined for all $x \in \mathcal{V}$ and $t \in [0, T]$. In other words, the mixture after averaging is represented by a superposition of continuous bodies with independent motions (i.e. overlapping and interpenetrating continua), as in the continuum theory of mixtures according to [38].

Averaging of the stress field is not straightforward; see [6], or [26, Chap. 2] for the case where $S^\beta \equiv 1$. For example, the averaged microscopic Cauchy stress within the solid species of the k-phase is given by

$$\boldsymbol{\sigma}^{sk} \overset{\text{def}}{=} \frac{\langle P^k S^s \boldsymbol{\sigma} \rangle}{f^k(1-n^k)} - \frac{\langle P^k S^s \rho \bar{\boldsymbol{v}}^{sk} \otimes \bar{\boldsymbol{v}}^{sk} \rangle}{f^k(1-n^k)}, \tag{15}$$

in which $\bar{\boldsymbol{v}}^{sk}$ denotes the deviation of the microscopic velocity of the solid species in the k-phase from its spatial average $\boldsymbol{v}^{sk}$. If, as in the case considered here, the solid species of the k-phase is formed by a grain skeleton, then $\boldsymbol{\sigma}^{sk}$ represents the microscopic Cauchy stress within the grains averaged with respect to the RVE. The second term on the right hand side of (15) is a residual stress, called Reynolds stress, due to mass flux relative to the averaging volume. The averaged stress in the grain skeleton, $\boldsymbol{\sigma}^{\mathrm{s}k}$, should not be confused with Terzaghi's effective stress whose proper definition is given below.

Our three-scale approach to obtain the macroscopic fields differs significantly from that in [6]. Consider a representative volume element $\mathcal{V}''$ smaller than $\mathcal{V}'$ for each phase k having the characteristic length l_{meso} at which the mixture of the two species $\beta \in \{\mathrm{s}, \mathrm{f}\}$ in the k-phase can be represented by a continuum. Then, spatial averaging over $\mathcal{V}'$ after averaging over $\mathcal{V}''$ will result in the same averaged (i.e. macroscopic) physical field as spatial averaging over $\mathcal{V}'$ alone. Therefore, we do not introduce a separate operator for averaging over $\mathcal{V}''$, as done in [6], so that stressing the term "mesoscale" is somewhat superfluous; the averaged equations of a binary mixture are obtained from those presented herein simply by setting $P^k \equiv 1$ (resp. $f^k \equiv 1$). In our approach we use only one macroscopic averaging operator in conjunction with a composition of indicator functions, called the phase-species function, which picks out a particular species in a particular phase. A detailed comparison our approach and that in [6] is left to future work.

3.2 Macroscopic Balance Principles for Each Species

By using spatially averaged fields and taking into account basic averaging theorems, the macroscopic (averaged) balance principles for each species in the mixture can be derived. In particular, conservation of mass and balance of momentum of the βth species of the kth phase with respect to the whole mixture in the representative volume element read

$$
\begin{aligned}
\frac{\partial f^k \pi^{\beta k} \rho^{\beta k}}{\partial t} + \operatorname{div}(f^k \pi^{\beta k} \rho^{\beta k} \boldsymbol{v}^{\beta k}) &= \Pi^{\beta k} \\
\frac{\partial f^k \pi^{\beta k} \rho^{\beta k} \boldsymbol{v}^{\beta k}}{\partial t} + \operatorname{div}(f^k \pi^{\beta k} \rho^{\beta k} \boldsymbol{v}^{\beta k} \otimes \boldsymbol{v}^{\beta k}) &= f^k \pi^{\beta k} \rho^{\beta k} \boldsymbol{b}^{\beta k} + \operatorname{div}(f^k \pi^{\beta k} \boldsymbol{\sigma}^{\beta k}) \\
&\quad + \Pi^{\beta k} \bar{\boldsymbol{v}}_{\mathrm{I}}^{\beta k} + \boldsymbol{\Gamma}^{\beta k}
\end{aligned}
\tag{16}
$$

in which $\beta \in \{\mathrm{s}, \mathrm{f}\}$, $k \in \{1, \ldots, n_{\mathrm{mat}}\}$, and

$$
\begin{aligned}
\Pi^{\beta k} &\stackrel{\text{def}}{=} \left\langle \delta_{\mathrm{I}}^{\beta k} \rho^{[\beta k]} (\boldsymbol{v} - \boldsymbol{v}_{\mathrm{I}})^{[\beta k]} \cdot \boldsymbol{n}^{\beta k *} \right\rangle, \\
\Pi^{\beta k} \bar{\boldsymbol{v}}_{\mathrm{I}}^{\beta k} &\stackrel{\text{def}}{=} \left\langle \delta_{\mathrm{I}}^{\beta k} \rho^{[\beta k]} \boldsymbol{v}^{[\beta k]} \otimes (\boldsymbol{v} - \boldsymbol{v}_{\mathrm{I}})^{[\beta k]} \cdot \boldsymbol{n}^{\beta k *} \right\rangle, \\
\text{and} \quad \boldsymbol{\Gamma}^{\beta k} &\stackrel{\text{def}}{=} -\langle \delta_{\mathrm{I}}^{\beta k} \boldsymbol{\sigma}^{[\beta k]} \cdot \boldsymbol{n}^{\beta k *} \rangle.
\end{aligned}
\tag{17}
$$

$\langle \cdot \rangle$ is the spatial average as defined by (4), $\delta_{\mathrm{I}}^{\beta k}$ is a Dirac delta function which picks out the interface of the βth species of the kth phase, and $\boldsymbol{n}^{\beta k *}$ is the field of outward normals on that interface. Moreover, for the microscopic field q, $q^{[\beta k]}$ is the restriction to the side of the interface lying in the β-species of the k-phase.

The terms $\Pi^{\beta k}$, $\Pi^{\beta k}\bar{\boldsymbol{v}}_{\mathrm{I}}^{\beta k}$, and $\boldsymbol{\Gamma}^{\beta k}$ are due to the interaction between the species and phases and can be interpreted as follows. $\Pi^{\beta k}$ describes the (average) mass transfer onto the β-species of the k-phase by all other constituents of the mixture through chemical reactions or erosion, for example. The momentum transfer term $\boldsymbol{\Gamma}^{\beta k}$ includes drag force densities per unit volume generated by the relative motion of the constituents. It accounts for surface forces, but not for momentum exchange owing to transfer of inertial mass (i.e. diffusion) which is described by the term $\Pi^{\beta k}\bar{\boldsymbol{v}}_{\mathrm{I}}^{\beta k}$. Concerning locally drained fluid-saturated porous media a specific constitutive equation for $\boldsymbol{\Gamma}^{\beta k}$ will result in Darcy's law for the fluid flow. Further details about the derivation of (16) will be given in a future paper [3].

3.3 Mixture Balance Principles

Summation of conservation of mass $(16)_1$ and balance of momentum $(16)_2$ over all species and phases in consideration of (9) yields the corresponding macroscopic balance principles of the homogenized mixture:

$$\begin{aligned}\frac{\partial\langle\rho\rangle}{\partial t}+\operatorname{div}\langle\rho\boldsymbol{v}\rangle&=0\\ \frac{\partial\langle\rho\boldsymbol{v}\rangle}{\partial t}+\operatorname{div}\langle\rho\boldsymbol{v}\otimes\boldsymbol{v}\rangle&=\langle\rho\boldsymbol{b}\rangle+\operatorname{div}\langle\boldsymbol{\sigma}\rangle.\end{aligned}\tag{18}$$

Here we have used the fact that the sum of the transfer terms over all constituents vanishes, that is,

$$\sum_k\sum_\beta \Pi^{\beta k}=0 \quad \text{and} \quad \sum_k\sum_\beta\left(\Pi^{\beta k}\bar{\boldsymbol{v}}_{\mathrm{I}}^{\beta k}+\boldsymbol{\Gamma}^{\beta k}\right)=\boldsymbol{0}\tag{19}$$

in accordance with the microscopic jump conditions (3). Only averaged quantities will be considered in the remainder of this paper.

It proofs convenient to display the balance principles derived so far in a different form. Note that each of (16) and (18) provides the so-called Eulerian conservation form of the balance principle with respect to fixed spatial points. An equivalent representation more common in solid mechanics is the (updated) Lagrangian form referring to the current configuration of the mixture in the ambient space. Conservation of mass and balance of momentum of the mixture then read

$$\langle\dot{\rho}\rangle+\langle\rho\rangle\operatorname{div}\langle\boldsymbol{v}\rangle=0 \quad \text{and} \quad \langle\rho\dot{\boldsymbol{v}}\rangle=\langle\rho\boldsymbol{b}\rangle+\operatorname{div}\langle\boldsymbol{\sigma}\rangle,\tag{20}$$

respectively. Here $\langle\dot{q}\rangle$, in which q is an arbitrary time-dependent spatial field, is used as an abbreviation for

$$\langle\dot{q}\rangle\overset{\text{def}}{=}\sum_k\sum_\beta f^k\pi^{\beta k}\dot{q}^{\beta k}=\sum_k f^k\left((1-n^k)\dot{q}^{\mathrm{s}k}+n^k\dot{q}^{\mathrm{f}k}\right),\tag{21}$$

where

$$\dot{q}^{\beta k} \stackrel{\text{def}}{=} \frac{\partial q^{\beta k}}{\partial t} + \boldsymbol{v}^{\beta k} \cdot \nabla q^{\beta k} \tag{22}$$

denotes the material time derivative of a β-species-k-phase-related quantity following the individual motion with velocity $\boldsymbol{v}^{\beta k}$. The term $\langle\rho\rangle\,\mathrm{div}\langle\boldsymbol{v}\rangle$ in $(21)_1$ is defined as the difference between $(18)_1$ and $\langle\dot{\rho}\rangle$. We note that $(20)_2$ is equivalent to $(18)_2$ if and only if conservation of mass $(20)_1$ resp. $(18)_1$ is satisfied. We also remark that (22) is generally different form the material time derivative of a k-phase-related quantity q^k along the k-phase mean velocity, which is denoted by

$$\grave{q}^{k} \stackrel{\text{def}}{=} \frac{\partial q^{k}}{\partial t} + \boldsymbol{v}^{k} \cdot \nabla q^{k}. \tag{23}$$

4 Closure of the Model

The two equations (18) in conjunction with (19), (16), and (17) are the balance principles governing the flow of a mixture of multiple materials with several species, by including the flow of a single-phase material as a special case. Modeling the specific multi-material flow associated with vibro-injection pile installation, however, requires closure of this set of equations, which is otherwise underdetermined. Generally the following closure laws (or closure models) have to be specified [14, 15]:

1. Transfer laws expressing the physics at the material interfaces.
2. Constitutive laws characterizing the physical behavior of each material.
3. Topological laws accounting for the evolution of variables characterizing the interfacial structure.

Restrictions on the form of the closure laws result from the principles of constitutive theory (e.g. objectivity) and from the fact that a material phase containing fractions of both a solid species and a fluid species must represent a fluid-saturated porous medium.

4.1 Interfacial Transfer Closure Laws

With regard to the transfer laws for the mixture it is assumed that no momentum transfer occurs, i.e. $\boldsymbol{\Gamma}^{\beta k} = \mathbf{0}$ for all $\beta \in \{\mathrm{s}, \mathrm{f}\}$ and $k \in \{1, \ldots, n_{\mathrm{mat}}\}$, resulting in uncoupled constituents at this stage of the derivation. Microscopic or molecular shear resistance within a constituent (e.g. grain contact forces, fluid viscosity) has to be modeled by the associated constitutive equation. Furthermore, any mass transfer, no matter between which constituents, phases, or species of the mixture, remains unconsidered such that $\Pi^{\beta k} = 0$ for all $\beta \in \{\mathrm{s}, \mathrm{f}\}$ and $k \in \{1, \dots, n_{\mathrm{mat}}\}$. That is to say, constituents do not chemically react, no diffusion and dispersion occurs, and interfaces are impermeable. As a result the injection of grout into the pore space of the soil is not described by the model and fluid-saturated porous media in the mixture are regarded locally undrained; the drained case including consolidation effects will be considered in future research. We are aware that this is a drastic simplification, but is deemed necessary concerning the development from scratch of the new MMALE method outlined in Part 2.

Notwithstanding this, inflow and outflow of the homogenized mixture across non-Lagrangian boundaries of the computational domain are generally allowed. Moreover, no limitations whatsoever exist with regard to the miscibility of the water-saturated soil with grout material on the macro level, in which the material phases maintain their original properties. The underlying averaged description of the multi-material flow likewise captures separated mixtures (analogy: oil on water) as well as disperse mixtures (analogy: emulsion of oil and water).

4.2 Constitutive Closure Laws

The constitutive closure laws summarized next characterize the mechanical behavior of the material phases during vibro-injection pile installation (cf. Fig. 2, detail B), namely of the bulk solid phase, the bulk fluid phase, and the phase forming a fluid-saturated porous medium (sand). The objective is to determine for each phase k the Cauchy stress given by

$$\boldsymbol{\sigma}^k = \sum_{\beta} \pi^{\beta k} \boldsymbol{\sigma}^{\beta k} = (1 - n^k)\boldsymbol{\sigma}^{\mathrm{s}k} + n^k \boldsymbol{\sigma}^{\mathrm{f}k}, \tag{24}$$

with the fluid fraction being either $\pi^{\mathrm{f}k} = n^k = 0$ (bulk solid, $\boldsymbol{\sigma}^k = \boldsymbol{\sigma}^{\mathrm{s}k}$), $n^k = 1$ (bulk fluid, $\boldsymbol{\sigma}^k = \boldsymbol{\sigma}^{\mathrm{f}k}$), or $0 < n^k < 1$ (fluid-saturated porous medium). In order to treat the mechanics of all materials of the problem (cf. Fig. 2, detail B) in a unified fashion, we recall from [27, 39] that the Cauchy stress tensor of any material can be decomposed into a pressure stress $-p^{\beta k}\boldsymbol{I}$ and an extra stress $\boldsymbol{s}^{\beta k}$, but also into a spherical part and the stress deviator $\boldsymbol{\sigma}^{\beta k}_{\mathrm{dev}} \overset{\mathrm{def}}{=} \boldsymbol{\sigma}^{\beta k} - \frac{1}{3}(\mathrm{tr}\, \boldsymbol{\sigma}^{\beta k})\boldsymbol{I}$ according to

$$\boldsymbol{\sigma}^{\beta k} = -p^{\beta k}\boldsymbol{I} + \boldsymbol{s}^{\beta k} = -\bar{p}^{\beta k}\boldsymbol{I} + \boldsymbol{\sigma}^{\beta k}_{\mathrm{dev}}, \tag{25}$$

where $\boldsymbol{I}$ is the second-order unit tensor and $\bar{p}^{\beta k} \overset{\mathrm{def}}{=} -\frac{1}{3}\,\mathrm{tr}\,\boldsymbol{\sigma}^{\beta k}$ is referred to as the (negative) mean stress. Generally one has $p^{\beta k} \neq \bar{p}^{\beta k}$ unless $\boldsymbol{s}^{\beta k} = \boldsymbol{\sigma}^{\beta k}_{\mathrm{dev}}$ resp. $\mathrm{tr}\,\boldsymbol{s}^{\beta k} = 0$, which is usually assumed for pure solids. Fluids may possess a non-deviatoric $\boldsymbol{s}^{\beta k}$ through volume viscosity but this is not considered here; cf. $(27)_2$.

The decomposition of stress (25) is useful to model both compressible and nearly incompressibile materials. We then assume for phases composed of a single species (i.e. $\pi^{\beta k} \equiv 1$), like bulk fluid or bulk solid, that there is a compression model of the form

$$-\left.\frac{1}{V^{\beta k}}\frac{\partial V^{\beta k}}{\partial p^{\beta k}}\right|_{M^{\beta k}} = \frac{1}{\rho^{\beta k}}\frac{\mathrm{d}\rho^{\beta k}}{\mathrm{d}p^{\beta k}} \overset{\text{def}}{=} \frac{1}{K^{\beta k}} \quad \text{resp.} \quad \dot{p}^{\beta k} \overset{\text{def}}{=} \frac{K^{\beta k}}{\rho^{\beta k}}\dot{\rho}^{\beta k} \tag{26}$$

for each constituent, relating the rate of pressure to the rate of mass density through a finite bulk modulus $K^{\beta k}$. In (26), $V^{\beta k}$ and $M^{\beta k} \overset{\text{def}}{=} \rho^{\beta k} V^{\beta k}$ are the volume and intrinsic mass of the β-species in the k-phase in the RVE, respectively, and $|_{M^{\beta k}}$ means that mass is kept constant along with differentiation.

First, let us consider the bulk solid and bulk fluid phases, whose behavior is assumed to be the same at all length scales defined (micro, meso, or macro). The following constitutive assumptions are made. Effects of turbulence in the bulk fluid are currently neglected. Bulk solid is either rigid or hypoelasto-plastic, and bulk fluid is a Newtonian fluid with deviatoric viscous stress. The latter two assumptions can be formalized as

$$\begin{aligned} \overset{\nabla}{\boldsymbol{\sigma}}{}^{sk} &\overset{\text{def}}{=} \mathbf{c}^{sk}(\boldsymbol{\sigma}^{sk}, \boldsymbol{h}^{sk}) : \boldsymbol{d}^{sk} \\ \boldsymbol{\sigma}^{fk} &\overset{\text{def}}{=} -p^{fk}\boldsymbol{I} + \mathbf{c}^{fk} : \boldsymbol{d}^{fk} \quad \text{with} \quad \mathrm{tr}(\mathbf{c}^{fk} : \boldsymbol{d}^{fk}) = 0, \end{aligned} \tag{27}$$

respectively, where $\boldsymbol{h} \overset{\text{def}}{=} \{h_1, \ldots, h_m\}$ is a set of material state variables, $\overset{\nabla}{\boldsymbol{\sigma}} \overset{\text{def}}{=} \dot{\boldsymbol{\sigma}} + \boldsymbol{\sigma}\cdot\boldsymbol{\omega} - \boldsymbol{\omega}\cdot\boldsymbol{\sigma}$ denotes the Zaremba-Jaumann rate of the considered second order tensor, $\boldsymbol{\omega} \overset{\text{def}}{=} \frac{1}{2}(\nabla\boldsymbol{v} - (\nabla\boldsymbol{v})^{\mathrm{T}})$ is the vorticity tensor, $\boldsymbol{d} \overset{\text{def}}{=} \frac{1}{2}(\nabla\boldsymbol{v} + (\nabla\boldsymbol{v})^{\mathrm{T}})$ is the spatial rate of deformation tensor, $\mathrm{tr}(\cdot)$ returns the trace of a second-order tensor, and : indicates double contraction. The bulk fluid phase representing macroscopic void zones (e.g. ambient atmosphere) is modeled by

$$\boldsymbol{\sigma}^{fk} \overset{\text{def}}{=} -p^{fk}\boldsymbol{I}, \qquad \text{with} \quad \dot{p}^{fk} = -K^{fk}\,\mathrm{div}\,\boldsymbol{d}^{fk} \quad \text{and} \quad K^{fk} \approx 0,\ \rho^{fk} \approx 0. \tag{28}$$

The application of (25) in conjunction with (26) to a rate constitutive equation of the form $(27)_1$ generally results in

$$\overset{\nabla}{\boldsymbol{s}}{}^{sk} = \overset{\nabla}{\boldsymbol{\sigma}}{}^{sk}_{\mathrm{dev}} \overset{\text{def}}{=} \mathbf{c}^{sk}(\boldsymbol{\sigma}^{sk}, \boldsymbol{h}^{sk}) : \boldsymbol{d}^{sk}_{\mathrm{dev}} \quad \text{and} \quad \dot{p}^{sk} \overset{\text{def}}{=} \frac{\dot{\rho}^{sk}}{3\rho^{sk}}\boldsymbol{I} : \mathbf{c}^{sk} : \boldsymbol{I}. \tag{29}$$

Clearly, the rate of the extra stress (rate of stress deviator) in the bulk solid can be calculated from the usual rate constitutive equation by using the deviatoric rate of deformation, and the corresponding bulk modulus can be calculated from the material tangent tensor as $3K^{sk} = \boldsymbol{I} : \mathbf{c}^{sk} : \boldsymbol{I}$.

It should be noted that in a mixture of materials the material time derivative and the velocity $\boldsymbol{v}$ entering the previous relations are those related to the macroscopic motion of the actual constituent, in accordance with (22). We also remark that compressive stress is taken with negative sign, but pressure has positive sign whenever stress is compressive. This corresponds to the sign convention of general mechanics.

Two different approaches can be employed to model the mechanical behavior of sand, or porous media in general, on the mesoscale. The first approach describes the behavior on the microscale of grain size and then applies spatial averaging to obtain the behavior on the mesoscale [43]. In the second approach, which is followed here, each species is regarded as a continuum at the mesoscale defined over all space. Concerning fluid-saturated porous media in the mixture, we first notice that $\boldsymbol{v}^{\mathrm{f}k} = \boldsymbol{v}^{\mathrm{s}k} = \boldsymbol{v}^k$ due to the locally undrained conditions assumed. The pore fluid is taken ideal (non-viscous) such that $\boldsymbol{\sigma}^{\mathrm{f}k} \stackrel{\mathrm{def}}{=} -p^{\mathrm{f}k}\boldsymbol{I}$, in contrast to $(27)_2$. As a consequence, the solid species velocity and pore fluid pressure are the only degrees of freedom of the undrained fluid-saturated porous medium.

Terzaghi's principle of effective stress [45, 46] is introduced as

$$\boldsymbol{\sigma}^{k'} = \boldsymbol{\sigma}^k + p^{\mathrm{f}k}\boldsymbol{I}, \tag{30}$$

where $\boldsymbol{\sigma}^{k'}$ is Terzaghi's effective stress and $\boldsymbol{\sigma}^k$ is the total Cauchy stress acting on the saturated porous medium representing the k-phase. Using (30) in conjunction with (25), the k-phase mean effective stress $p^{k'} \stackrel{\mathrm{def}}{=} -\frac{1}{3}\operatorname{tr}\boldsymbol{\sigma}^{k'}$ can be obtained from

$$p^{k'} = (1-n^k)(p^{\mathrm{s}k} - p^{\mathrm{f}k}) = p^k - p^{\mathrm{f}k}, \qquad \text{so that} \qquad \dot{p}^{k'} = \dot{p}^k - \dot{p}^{\mathrm{f}k}. \tag{31}$$

Hence, the mean effective stress divided by the solid fraction equals the averaged excess pressure in the solid species, that is, $p^{k'}/(1-n^k) = p^{\mathrm{s}k} - p^{\mathrm{f}k}$; in a suspension each sand grain is completely surrounded by water such that $p^{\mathrm{s}k} = p^{\mathrm{f}k}$ and $p^{k'} = 0$. The second equation in (31) is the rate form of the first equation. The superposed dot indicates the material time derivative along the velocity $\boldsymbol{v}^{\mathrm{s}k}$ of the solid species in the k-phase (which equals the velocity of the fluid species at locally undrained conditions, see above).

The solid and fluid species of a fluid-saturated porous medium are generally compressible. This particularly means that the density of both the water and the grain material of water-saturated sand can change due to pressure loading. The mass density of the pore fluid can be considered as a function of the pore fluid pressure alone, that is, $\rho^{\mathrm{f}k} = \tilde{\rho}^{\mathrm{f}k}(p^{\mathrm{f}k})$ such that (26) holds. The compressibility resp. the bulk modulus of pore water generally depends not only on pressure and temperature, but also on the gas content indicating partial saturation. A small air content reduces the bulk modulus of an air-water mixture considerably, hence should be considered in the calculation of the pore fluid pressure. In the present research we use the relation [25]

$$K^{\mathrm{f}k} = \left(\frac{S^k}{K_0^{\mathrm{f}k}} + \frac{1-S^k}{p^{\mathrm{f}k}}\right)^{-1}, \tag{32}$$

where K^{fk} is the bulk modulus of an air-water mixture, S^k is the degree of water saturation, K_0^{fk} is the bulk modulus of pure water ($S^k = 1.0$; $K_0^{fk} = 4.78 \times 10^{-10}\,\mathrm{Pa}^{-1}$ under atmospheric pressure at 10 °C), und p^{fk} is the absolute fluid pressure.

As opposed to the pore fluid the mass density of the solid species (grain material) in saturated porous media not only depends on the pressure but also on the porosity [7–9, 21]:

$$\rho^{sk} = \tilde{\rho}^{sk}(p^{sk}, n^k), \qquad \text{or, equivalently,} \qquad p^{sk} = \tilde{p}^{sk}(\rho^{sk}, n^k). \tag{33}$$

Therefore, the total change in pressure within the solid species in the k-phase,

$$\mathrm{d}p^{sk} = \left.\frac{\partial p^{sk}}{\partial \rho^{sk}}\right|_{n^k} \mathrm{d}\rho^{sk} + \left.\frac{\partial p^{sk}}{\partial n^k}\right|_{\rho^{sk}} \mathrm{d}n^k, \tag{34}$$

consists of a pressure change due to compression of the grains leaving the porosity unchanged and a pressure change due to rearrangement of the grains (configurational compression) at constant mass density of the grain material. In the partial derivatives the superposed ˜ is omitted for notational brevity.

Because the porosity is not directly measurable, we choose a different set of independent variables in (33) which can be combined to back out the original variables. In consideration of (31), we replace in $(33)_1$ the set of independent variables (p^{sk}, n^k) by the equivalent set $(p^{k'}, p^{fk})$, so that $\rho^{sk} = \tilde{\rho}^{sk}(p^{k'}, p^{fk}) = \tilde{\rho}^{sk}(p^k - p^{fk}, p^{fk})$. Then, the total change of volume of the k-phase solid species can be calculated from

$$\begin{aligned} -\left.\frac{\mathrm{d}V^{sk}}{V^{sk}}\right|_{M^{sk}} &= \frac{\mathrm{d}p^{sk}}{K^{sk}} = \frac{\mathrm{d}\rho^{sk}}{\rho^{sk}} = \frac{1}{\rho^{sk}} \left.\frac{\partial \rho^{sk}}{\partial (p^k - p^{fk})}\right|_{p^{fk}} \mathrm{d}p^{k'} + \frac{1}{\rho^{sk}} \left.\frac{\partial \rho^{sk}}{\partial p^{fk}}\right|_{p^{k'}} \mathrm{d}p^{fk} \\ &= \frac{1}{\rho^{sk}} \left.\frac{\partial \rho^{sk}}{\partial p^k}\right|_{p^{fk}} \mathrm{d}p^{k'} + \frac{1}{\rho^{sk}} \left.\frac{\partial \rho^{sk}}{\partial p^{fk}}\right|_{p^{k'}} \mathrm{d}p^{fk} = \frac{\mathrm{d}p^{k'}}{K_\mathrm{d}^k} + \frac{\mathrm{d}p^{fk}}{K_\mathrm{uj}^k}. \end{aligned} \tag{35}$$

K^{sk} is the material bulk modulus of the solid species, which is assumed to be a function of the mass density ρ^{sk} alone so that (26) applies. Note that in general, the solid species bulk modulus is also a function of the porosity. The bulk moduli K_d^k and K_uj^k have likewise been introduced in accordance with (26) and are referred to as the drained bulk modulus and unjacketed bulk modulus of the k-phase porous medium, respectively [8]. K_uj^k is approximately equal to K^{sk} under the assumption that the solid species volume changes only little during an unjacketed test. K_d^k is the bulk modulus of the drained porous medium as measured in a jacketed test.

By replacing in (35) the total differential with the material time derivative along the velocity $\boldsymbol{v}^{sk}$, and noting that $\dot{V}^{sk}/V^{sk}|_{M^{sk}} = \operatorname{div} \boldsymbol{v}^{sk}$, one obtains

$$\dot{p}^{k'} = -K_\mathrm{d}^k \left(\operatorname{div} \boldsymbol{v}^{sk} + \frac{\dot{p}^{fk}}{K_\mathrm{uj}^k} \right) = \dot{p}^{k''} - \frac{K_\mathrm{d}^k}{K_\mathrm{uj}^k} \dot{p}^{fk}. \tag{36}$$

where $\dot{p}^{k''} \overset{\text{def}}{=} -K^k_{\text{d}} \operatorname{div} \boldsymbol{v}^{\text{s}k}$. Using $(31)_2$ and $K^k_{\text{uj}} \approx K^{\text{s}k}$ one arrives at

$$\dot{p}^k = \dot{p}^{k''} + \alpha^k \dot{p}^{\text{f}k}, \qquad \text{where} \quad \alpha^k \overset{\text{def}}{=} 1 - \frac{K^k_{\text{d}}}{K^{\text{s}k}}. \tag{37}$$

The coefficient α^k is due to [12, 13] and accounts for the compressibility of the solid material forming the porous medium. A common approximation for sand is $\alpha^k = 1$, meaning that the grain material is incompressible. A suspension of sand and water is characterized by $K^k_{\text{d}} = 0$, leading to $\alpha^k = 1$ likewise. From (36) together with $\dot{p}^{\text{s}k}/K^{\text{s}k} = -\operatorname{div} \boldsymbol{v}^{\text{s}k}$ we also have

$$\dot{p}^{\text{s}k} = \dot{p}^{k'} \frac{K^{\text{s}k}}{K^k_{\text{d}}} + \dot{p}^{\text{f}k}. \tag{38}$$

We now turn to the mechanical behavior of sand under general deformations. In the present work, sand is modeled as a hypoplastic porous medium, in accordance with Subproject 1 of the DFG Research Unit FOR 1136. A widely-used hypoplastic model for sand neglecting viscous effects is due to [41] and has been extended by [33]. Its general form,

$$\overset{\nabla}{\boldsymbol{\sigma}}{}^{k'} \overset{\text{def}}{=} \boldsymbol{\mathsf{c}}^{k'}(\boldsymbol{\sigma}^{k'}, \boldsymbol{h}^{k'}) : (\boldsymbol{d}^{\text{s}k} - \dot{\varepsilon}^{\text{s}}_{\text{vol}} \boldsymbol{I}) \overset{\text{def}}{=} \overset{\nabla}{\boldsymbol{\sigma}}{}^{k''} - \dot{\varepsilon}^{\text{s}}_{\text{vol}} \boldsymbol{\mathsf{c}}^k : \boldsymbol{I}, \tag{39}$$

provides a spatial rate constitutive equation for the effective Cauchy stress, but here we subtracted from $\boldsymbol{d}^{\text{s}k}$ the average volumetric strain rate $\dot{\varepsilon}^{\text{s}}_{\text{vol}} \boldsymbol{I}$ of the solid grains under the assumption of a compressible solid species. The stress tensor $\boldsymbol{\sigma}^{k''}$ whose Zaremba-Jaumann rate is defined in (39) is responsible for all deformation of the solid species in the k-phase. The particular form of $\boldsymbol{\mathsf{c}}^{k'}(\boldsymbol{\sigma}^{k'}, \boldsymbol{h}^{k'})$ can be found in the references and is assumed to be valid for dry, fully-saturated drained, and locally undrained sand. Moreover, we postulate $\boldsymbol{\mathsf{c}}^{k'}(\boldsymbol{\sigma}^{k'}, \boldsymbol{h}^{k'}) \equiv \boldsymbol{\mathsf{c}}^{k''}(\boldsymbol{\sigma}^{k''}, \boldsymbol{h}^{k''})$ because in soil mechanics the grain material is usually taken incompressible.

A basic property of the Zaremba-Jaumann rate is $\operatorname{tr} \overset{\nabla}{\boldsymbol{\sigma}} = \boldsymbol{I} : \overset{\nabla}{\boldsymbol{\sigma}} = \boldsymbol{I} : \dot{\boldsymbol{\sigma}}$ [5, p. 139], so that the trace of (39) yields

$$\dot{p}^{k'} = \dot{p}^{k''} + \frac{\dot{\varepsilon}^{\text{s}}_{\text{vol}}}{3} \boldsymbol{I} : \boldsymbol{\mathsf{c}}^k : \boldsymbol{I}, \tag{40}$$

where $\dot{p}^{k''} \overset{\text{def}}{=} -\frac{1}{3} \operatorname{tr} \dot{\boldsymbol{\sigma}}^{k''} = -\frac{1}{3} \operatorname{tr}(\boldsymbol{\mathsf{c}}^k : \boldsymbol{d}^{\text{s}k})$. Comparison with (36) shows that

$$\dot{\varepsilon}^{\text{s}}_{\text{vol}} = \frac{\dot{p}^{\text{f}k}}{K^k_{\text{uj}}} \approx \frac{\dot{p}^{\text{f}k}}{K^{\text{s}k}} \qquad \text{and} \qquad K^k_{\text{d}} = \frac{\boldsymbol{I} : \boldsymbol{\mathsf{c}}^k : \boldsymbol{I}}{3}. \tag{41}$$

Here $\mathbf{c}^k$ is the hypoplastic material tangent of (39), but it can be any other material tangent representing generally non-isotropic porous media.

In (39) the generally unsymmetric tensor $\mathbf{c}^k : \boldsymbol{I}$ is replaced with its average $K_{\mathrm{d}}^k \boldsymbol{I}$ derived from $(41)_2$ [45]. Then, by taking the Zaremba-Jaumann rate of (30),

$$\overset{\nabla}{\boldsymbol{\sigma}}{}^k = \overset{\nabla}{\boldsymbol{\sigma}}{}^{k''} - \alpha^k \dot{p}^{\mathrm{f}k} \boldsymbol{I}. \tag{42}$$

Integrating this equation by starting from a zero initial state ($\boldsymbol{\sigma}^{k''}|_{t=0} = \mathbf{0}$ and $p^{\mathrm{f}k}|_{t=0} = 0$) yields the modified principle of effective stress [12, 45, 46]

$$\boldsymbol{\sigma}^{k''} = \boldsymbol{\sigma}^k + \alpha^k p^{\mathrm{f}k} \boldsymbol{I}. \tag{43}$$

We will use this principle instead of (30) in what follows despite the fact that, strictly speaking (43) in contrast to (30) holds only if the porous medium is characterized by linear isotropic behavior. However, for sand the coefficient α^k is indeed very close to unity.

Based on (25) and (43), it can be shown that the effective stress is related to the pressures and stresses in the constituents of the saturated porous media through

$$\begin{aligned} \boldsymbol{\sigma}^{k''} &= (1 - n^k)\boldsymbol{\sigma}^{\mathrm{s}k} + (\alpha^k - n^k) p^{\mathrm{f}k} \boldsymbol{I} \\ &= -((1 - n^k) p^{\mathrm{s}k} - (\alpha^k - n^k) p^{\mathrm{f}k})\boldsymbol{I} + (1 - n^k)\boldsymbol{s}^{\mathrm{s}k}, \end{aligned} \tag{44}$$

with $\boldsymbol{s}^{\mathrm{s}k} = \boldsymbol{\sigma}^{\mathrm{s}k}_{\mathrm{dev}} = \boldsymbol{\sigma}^{k''}_{\mathrm{dev}}/(1 - n) = \boldsymbol{s}^{k''}/(1 - n) = \boldsymbol{s}^k/(1 - n)$, and

$$\overset{\nabla}{\boldsymbol{s}}{}^{k''} = \overset{\nabla}{\boldsymbol{\sigma}}{}^{k''}_{\mathrm{dev}} = \mathbf{c}^k(\boldsymbol{\sigma}^{k''}, \boldsymbol{h}^{k''}) : \boldsymbol{d}^{\mathrm{s}k}_{\mathrm{dev}} \tag{45}$$

by (39). Since locally undrained conditions have been assumed, the rate of pore fluid pressure can be determined from the deformation of the porous medium as [45, 46]

$$\dot{p}^{\mathrm{f}k} = -\alpha^k Q^k \operatorname{div} \boldsymbol{v}^{\mathrm{s}k}, \qquad \text{with} \quad \frac{1}{Q^k} \overset{\mathrm{def}}{=} \frac{\alpha^k - n^k}{K^{\mathrm{s}k}} + \frac{n^k}{K^{\mathrm{f}k}}, \tag{46}$$

This relation can be derived using conservation of mass for each species. Therefore,

$$\dot{p}^k = -K^k \operatorname{div} \boldsymbol{v}^{\mathrm{s}k} \tag{47}$$

by (37), in which

$$K^k = K_{\mathrm{d}}^k \left(1 + \frac{(\alpha^k)^2}{\alpha^k K_{\mathrm{d}}^k / K^{\mathrm{s}k} + n^k \left(K_{\mathrm{d}}^k / K^{\mathrm{f}k} - K_{\mathrm{d}}^k / K^{\mathrm{s}k} \right)} \right) \tag{48}$$

represents, in the most general case, the bulk modulus of an undrained saturated porous material with compressible constituents [19]. The remaining cases are:

- Bulk solid phase ($n^k = 0$, $K_{\mathrm{d}}^k = K^{\mathrm{s}k}$, $\alpha^k = 0$), for which $K^k = K^{\mathrm{s}k}$.
- Bulk fluid phase ($n^k = 1$, $K_{\mathrm{d}}^k = 0$, $\alpha^k = 1$), for which $K^k = K^{\mathrm{f}k}$.
- Dry porous medium ($0 < n^k < 1$, $K^{\mathrm{f}k} \approx 0$), for which $K^k = K_{\mathrm{d}}^k$.
- Suspension of sand and water ($K_{\mathrm{d}}^k = 0$, $\alpha^k = 1$), for which $p^{\mathrm{s}k} = p^{\mathrm{f}k}$ and $K^k = \big((1 - n^k)/K^{\mathrm{s}k} + n^k/K^{\mathrm{f}k}\big)^{-1}$.

In concluding this section, we remark that the velocity $\boldsymbol{v}^k$ and pressure p^k of each phase k represent its degrees of freedom (primary unknowns). Each phase can be a solid, a fluid, or a fluid-saturated porous medium depending on the fluid fraction n^k assigned at the outset. Accordingly, the developed model is able to describe the flow and large deformation of mixtures of solids, fluids, and porous media in a unified fashion.

4.3 Topological Closure Laws

The application of spatial averaging to multi-material flows entails a loss of information as it smoothes out details of the flow structure, like the geometry of the material interfaces [14, 15]. The topological closure laws should restore the lost information. Because the flow structure results from quantities related to the problem as a whole and not from intrinsic material properties alone, topological closure laws are not closure laws in a strict sense. Concerning the multi-material flow associated with vibro-injection pile installation and the assumptions and restrictions made so far, the only topological laws required are those that account for the evolution of the k-phase volume fraction f^k and fluid fraction n^k. To show this, we analyze the governing equations (16) subject to the constraints (9).

The external loads $\boldsymbol{b}^{\mathrm{s}k}$ and $\boldsymbol{b}^{\mathrm{f}k}$ are assumed to be given, and the unknowns for each phase k are the mass densities $\rho^{\mathrm{s}k}$ and $\rho^{\mathrm{f}k}$, the velocities $\boldsymbol{v}^{\mathrm{s}k}$ and $\boldsymbol{v}^{\mathrm{f}k}$, the stresses $\boldsymbol{\sigma}^{\mathrm{s}k}$ and $\boldsymbol{\sigma}^{\mathrm{f}k}$, the volume fraction f^k, the fractions of the solid and fluid species $\pi^{\mathrm{s}k}$ and $\pi^{\mathrm{f}k}$, respectively, the mass transfer terms $\Pi^{\mathrm{s}k}$ and $\Pi^{\mathrm{f}k}$, as well as the momentum transfer terms $\boldsymbol{\Gamma}^{\mathrm{s}k}$ and $\boldsymbol{\Gamma}^{\mathrm{f}k}$. Mass and momentum transfer was assumed zero, and the constraint $(9)_2$ yields $\pi^{\mathrm{s}k} = 1 - \pi^{\mathrm{f}k} = 1 - n^k$. Since the stress tensors are symmetric one is left with a total of $22n_{\mathrm{mat}}$ unknowns in three-dimensional space: $2n_{\mathrm{mat}}$ mass densities (resp. pressures, by (26)), $6n_{\mathrm{mat}}$ velocity components, $12n_{\mathrm{mat}}$ stress components, n_{mat} volume fractions, and n_{mat} fluid fractions (porosities). These unknowns are accompanied by the $2n_{\mathrm{mat}}$ equations of conservation of mass and $6n_{\mathrm{mat}}$ equations of balance of momentum for both species in all material phases, and by the constraint $(9)_1$. The balance principles of the mixture (18) do not provide additional information, but for the $12n_{\mathrm{mat}}$ stress components of $\boldsymbol{\sigma}^{\mathrm{s}k}$ and $\boldsymbol{\sigma}^{\mathrm{f}k}$, respectively of combinations of these, the constitutive equations of the previous section are substituted. Therefore, the number of unknowns is finally reduced to $2n_{\mathrm{mat}} - 1$, namely $n_{\mathrm{mat}} - 1$ volume fractions and n_{mat} porosities.

The basic equations to tackle this problem of closure are the constraints (9) as well as the mass conservation equation of the β-species of the k-phase $(16)_1$, with

$\beta \in \{\mathrm{s}, \mathrm{f}\}$ and $k \in \{1, \ldots, n_{\mathrm{mat}}\}$, under the assumption of zero mass transfer between the constituents. First, we notice that time derivation of the constraints (9) and the application of a basic averaging rule [20] results in

$$\sum_k \frac{\partial f^k}{\partial t} = 0 \quad \text{and} \quad \sum_\beta \frac{\partial \pi^{\beta k}}{\partial t} = 0. \tag{49}$$

The second condition is automatically satisfied since $\pi^{\mathrm{s}k} = 1 - \pi^{\mathrm{f}k}$. Moreover, due to the fact that $\boldsymbol{v}^{\mathrm{f}k} = \boldsymbol{v}^{\mathrm{s}k} = \boldsymbol{v}^k$ under the assumption of locally undrained conditions the n_{mat} unknown porosities can be determined. This is done by expanding $(16)_1$ for the case where $n_{\mathrm{mat}} = 1$, so that $f^k = 1$:

$$\frac{\partial \pi^{\beta k} \rho^{\beta k}}{\partial t} + \operatorname{div}(\pi^{\beta k} \rho^{\beta k} \boldsymbol{v}^{\beta k}) = \rho^{\beta k} \dot{\pi}^{\beta k} + \pi^{\beta k} \dot{\rho}^{\beta k} + \pi^{\beta k} \rho^{\beta k} \operatorname{div} \boldsymbol{v}^{\beta k} = 0. \tag{50}$$

Hence, the porosity $\pi^{\mathrm{f}k} = n^k$ is updated by

$$\dot{n}^k = (1 - n^k) \left(\frac{\dot{\rho}^{\mathrm{s}k}}{\rho^{\mathrm{s}k}} + \operatorname{div} \boldsymbol{v}^{\mathrm{s}k} \right) = (1 - n^k) \left(\frac{\dot{p}^{\mathrm{s}k}}{K^{\mathrm{s}k}} + \operatorname{div} \boldsymbol{v}^{\mathrm{s}k} \right). \tag{51}$$

Next, we consider $(16)_1$ for arbitrary $n_{\mathrm{mat}} > 1$ but we take the sum over all species in consideration of $(9)_2$:

$$\begin{aligned}
0 &= \sum_\beta \frac{\partial f^k \pi^{\beta k} \rho^{\beta k}}{\partial t} + \sum_\beta \operatorname{div}(f^k \pi^{\beta k} \rho^{\beta k} \boldsymbol{v}^{\beta k}) \\
&= \frac{\partial}{\partial t} \sum_\beta f^k \pi^{\beta k} \rho^{\beta k} + \operatorname{div} \sum_\beta f^k \pi^{\beta k} \rho^{\beta k} \boldsymbol{v}^{\beta k} = \frac{\partial f^k \rho^k}{\partial t} + \operatorname{div}(f^k \rho^k \boldsymbol{v}^k) \\
&= \rho^k \left(\frac{\partial f^k}{\partial t} + \mathbf{d} f^k \cdot \boldsymbol{v}^k \right) + f^k \left(\frac{\partial \rho^k}{\partial t} + \operatorname{div}(\rho^k \boldsymbol{v}^k) \right) \\
&= \rho^k \grave{f}^k + f^k \grave{\rho}^k + f^k \rho^k \operatorname{div} \boldsymbol{v}^k,
\end{aligned} \tag{52}$$

where $\mathbf{d}(\cdot)$ returns the differential (or gradient) of a scalar-valued function. That is,

$$\grave{f}^k = -f^k \left(\frac{\grave{\rho}^k}{\rho^k} + \operatorname{div} \boldsymbol{v}^k \right). \tag{53}$$

The equality of the first and second lines in (52) can again be shown by basic averaging rules [20], and the material time derivative $\grave{q}^k$ of a k-phase-related quantity q^k has been defined through (23).

It is crucial to note that (53) neither yields additional information nor defines a topological closure law for the volume fraction. Equation (53) is just a rearranged form of conservation of mass with regard to the k-phase of the mixture. A proper closure law for volume fraction instead has to account for the physics of the problem and particularly has to specify how the volumetric distribution of the bulk solid, the bulk fluid, and the saturated porous medium evolves during vibro-injection pile installation. Further research is needed to establish such a topological closure law.

4.4 Resulting Model for Multi-material Flow

The derivations presented so far result in a macroscopic model for the mechanics of the multi-material flow associated with vibro-injection pile installation in saturated sand:

$$\begin{aligned} \operatorname{div}\langle \boldsymbol{s} - p\boldsymbol{I}\rangle + \langle \rho \boldsymbol{b}\rangle - \langle \rho \dot{\boldsymbol{v}}\rangle &= \boldsymbol{0} \\ \left\langle \frac{\dot{p}}{K} \right\rangle + \operatorname{div}\langle \boldsymbol{v}\rangle &= 0, \end{aligned} \tag{54}$$

where

$$\begin{aligned} \langle \boldsymbol{s}\rangle &= \sum_k f^k \boldsymbol{s}^k = \sum_k f^k \left((1-n^k)\boldsymbol{s}^{\mathrm{s}k} + n^k \boldsymbol{s}^{\mathrm{f}k}\right) = \sum_k f^k \boldsymbol{s}^{k''}, \\ \langle p\rangle &= \sum_k f^k p^k = \sum_k f^k \left((1-n^k)p^{\mathrm{s}k} + n^k p^{\mathrm{f}k}\right) = \sum_k f^k \left(p^{k''} + \alpha^k p^{\mathrm{f}k}\right), \\ \langle \rho\rangle &= \sum_k f^k \rho^k = \sum_k f^k \left((1-n^k)\rho^{\mathrm{s}k} + n^k \rho^{\mathrm{f}k}\right), \qquad \langle \rho \dot{\boldsymbol{v}}\rangle = \sum_k f^k \rho^k \grave{\boldsymbol{v}}^k, \\ \left\langle \frac{\dot{p}}{K}\right\rangle &= \sum_k \frac{f^k}{K^k}\grave{p}^k, \qquad \operatorname{div}\langle \boldsymbol{v}\rangle = \sum_k \left(\grave{f}^k + f^k \operatorname{div} \boldsymbol{v}^k\right), \\ K^k &= K_{\mathrm{d}}^k \left(1 + \frac{(\alpha^k)^2}{\alpha^k K_{\mathrm{d}}^k / K^{\mathrm{s}k} + n^k \left(K_{\mathrm{d}}^k/K^{\mathrm{f}k} - K_{\mathrm{d}}^k/K^{\mathrm{s}k}\right)}\right), \qquad \alpha^k = 1 - \frac{K_{\mathrm{d}}^k}{K^{\mathrm{s}k}}. \end{aligned} \tag{55}$$

This model, in conjunction with (19), (16), and (17), is closed by the constitutive equations for the bulk solid, bulk fluid, porous medium, pore fluid, and void, and by the evolution equations (topological closure laws) for the porosities n^k and volume fractions f^k. The necessary topological closure laws for the $n_{\mathrm{mat}} - 1$ volume fractions are yet unspecified. It will be shown in the next section, however, that a priori closure respecting the assumptions underlying the applicability of an MMALE method give rise to proper relations.

5 Two-Equation Reduced Model

5.1 A Priori Closure (Subcell Model)

In addition to the closure laws related to the continuous problem addressed in Sect. 4, the application of an MMALE numerical method calls for a special closure model for multi-material elements, referred to as the subcell model, in order to render the discretized problem well-posed [10, 11, 17, 29, 30, 36]. This subcell model solves the problem of relating the evolution of the individual materials in multi-material elements to the macroscopic degrees of freedom of the element. In developing the subcell model we keep things as simple as possible and a priori assume homogeneous distributions of pressure and velocity in the mixture at each instant of time:

$$p^k = \langle p \rangle \quad \text{and} \quad \boldsymbol{v}^k = \langle \boldsymbol{v} \rangle \qquad \text{for all } k \in \{1, \ldots, n_{\text{mat}}\} \text{ and } t \in [0, T]. \tag{56}$$

A direct consequence of these two assumptions is that each element also has a single deviatoric strain rate, that is, $\boldsymbol{d}^k_{\text{dev}} = \langle \boldsymbol{d}_{\text{dev}} \rangle$ for all $k \in \{1, \ldots, n_{\text{mat}}\}$.

From a physical viewpoint (56) means that everything is in homogeneous thermodynamic equilibrium [17, 29, 37]. The assumption of pressure equilibrium is reasonable because pressure is continuous across a material interface. Pressure disequilibration is associated with highly-dynamic compaction or other processes not considered here. If $(56)_1$ holds, then the adjustment of volume fractions and the transfer of pressure and pressure changes is infinitely fast. The requirement that each element has a single velocity, on the other hand, is not appropriate because equilibration of velocity differences is driven by drag forces on material interfaces (momentum transfer). Assumption $(56)_2$ results in fully-bonded material phases without a contact mechanism. However, it is commonly accepted because tangential contact with or without friction is difficult to model in MMALE and multi-material Eulerian methods [40]. As a consequence, shear resistance is accounted for only by the constitutive equation inside of the phases next to the interface.

Based on the assumptions (56), the continuum mechanical two-equation model summarized in Sect. 4.4 can now be reduced; see also [17, 29]. In particular (53) under the assumption $(56)_2$ becomes

$$\dot{f}^k + f^k \operatorname{div}\langle \boldsymbol{v} \rangle = -\frac{\dot{\rho}^k}{\rho^k} = -f^k \frac{\dot{p}^k}{K^k}, \tag{57}$$

in which the superposed dot now represents the material time derivative of any spatial field q along the mean velocity $\langle \boldsymbol{v} \rangle$ of the mixture:

$$\dot{q} \overset{\text{def}}{=} \frac{\partial q}{\partial t} + \langle \boldsymbol{v} \rangle \cdot \nabla q. \tag{58}$$

The assumption $(56)_1$ yields

$$\left\langle \frac{\dot{p}}{K} \right\rangle = \sum_k \frac{f^k \dot{p}^k}{K^k} = \frac{\langle \dot{p} \rangle}{\langle K \rangle}, \quad \text{with} \quad \frac{1}{\langle K \rangle} = \sum_k \frac{f^k}{K^k}. \tag{59}$$

Moreover,

$$\frac{K^k}{\rho^k} \dot{\rho}^k = \dot{p}^k = \langle \dot{p} \rangle = \left\langle \frac{K}{\rho} \dot{\rho} \right\rangle = \langle K \rangle \left\langle \frac{\dot{\rho}}{\rho} \right\rangle, \tag{60}$$

so that (57) in conjunction with conservation of mass of the mixture, $\langle \dot{\rho} \rangle + \langle \rho \rangle \operatorname{div}\langle \boldsymbol{v} \rangle = 0$, results in the self-consistent balance equation

$$\dot{f}^k + f^k \operatorname{div}\langle \boldsymbol{v} \rangle = -f^k \frac{\langle K \rangle}{K^k} \left\langle \frac{\dot{\rho}}{\rho} \right\rangle = f^k \frac{\langle K \rangle}{K^k} \operatorname{div}\langle \boldsymbol{v} \rangle, \tag{61}$$

that is,

$$\dot{f}^k = f^k \left(\frac{\langle K \rangle}{K^k} - 1 \right) \operatorname{div}\langle \boldsymbol{v} \rangle. \tag{62}$$

This is the remaining topological closure law for the volume fraction. Note that it naturally provides for a void collapse mechanism because the material with the smallest bulk modulus contributes most to the total volume change. This feature is of particular importance in cases where the compressibilities of the materials phases are widely different, as for example in a mixture of void and steel.

Because of (60) and the basic constraint $\sum_k f^k = 1$, summation of (62) over the n_{mat} phases in the mixture results in (59). Therefore, the topological closure law also satisfies the constraint $\sum_k \dot{f}^k = 0$.

5.2 Application to the Developed Model

The macroscopic model for the multi-material flow associated with vibro-injection pile installation summarized in Sect. 4.4 and the subcell model developed in the previous section result in following two-equation (two-field) reduced model:

$$\begin{aligned} \operatorname{div}\langle \boldsymbol{s} - p\boldsymbol{I} \rangle + \langle \rho \boldsymbol{b} \rangle - \langle \rho \rangle \langle \dot{\boldsymbol{v}} \rangle &= \boldsymbol{0} \\ \langle \dot{p} \rangle / \langle K \rangle + \operatorname{div}\langle \boldsymbol{v} \rangle &= 0, \end{aligned} \tag{63}$$

where

$$\langle \boldsymbol{s}\rangle = \sum_k f^k \boldsymbol{s}^k = \sum_k f^k((1-n^k)\boldsymbol{s}^{\mathrm{s}k} + n^k \boldsymbol{s}^{\mathrm{f}k}) = \sum_k f^k \boldsymbol{s}^{k''},$$
$$\langle p\rangle = \sum_k f^k p^k = \sum_k f^k((1-n^k)p^{\mathrm{s}k} + n^k p^{\mathrm{f}k}) = \sum_k f^k \left(p^{k''} + \alpha^k p^{\mathrm{f}k}\right),$$
$$\langle \rho\rangle = \sum_k f^k \rho^k = \sum_k f^k((1-n^k)\rho^{\mathrm{s}k} + n^k \rho^{\mathrm{f}k}), \tag{64}$$
$$\langle K\rangle^{-1} = \sum_k f^k/K^k, \qquad \alpha^k = 1 - K_{\mathrm{d}}^k/K^{\mathrm{s}k},$$
$$\text{and} \quad K^k = K_{\mathrm{d}}^k\left(1 + \frac{(\alpha^k)^2}{\alpha^k K_{\mathrm{d}}^k/K^{\mathrm{s}k} + n^k\left(K_{\mathrm{d}}^k/K^{\mathrm{f}k} - K_{\mathrm{d}}^k/K^{\mathrm{s}k}\right)}\right).$$

The model is closed by the general constitutive equations for the

$$\left.\begin{aligned}
\text{bulk solid:}\quad & \overset{\nabla}{\boldsymbol{\sigma}}{}^{\mathrm{s}k} = \mathbf{c}^{\mathrm{s}k}(\boldsymbol{\sigma}^{\mathrm{s}k}, \boldsymbol{h}^{\mathrm{s}k}) : \langle \boldsymbol{d}\rangle,\\
\text{bulk fluid:}\quad & \boldsymbol{\sigma}^{\mathrm{f}k} = -p^{\mathrm{f}k}\boldsymbol{I} + \mathbf{c}^{\mathrm{f}k} : \langle \boldsymbol{d}\rangle \quad \text{subject to}\\
& \mathrm{tr}(\mathbf{c}^{\mathrm{f}k} : \langle \boldsymbol{d}\rangle) = 0 \quad \text{and} \quad \dot{p}^{\mathrm{f}k} = -K^{\mathrm{f}k}\,\mathrm{div}\langle \boldsymbol{d}\rangle,\\
\text{porous medium:}\quad & \overset{\nabla}{\boldsymbol{\sigma}}{}^{k''} = \mathbf{c}^{k''}(\boldsymbol{\sigma}^{k''}, \boldsymbol{h}^{k''}) : \langle \boldsymbol{d}\rangle,\\
\text{pore fluid:}\quad & \boldsymbol{\sigma}^{\mathrm{f}k} = -p^{\mathrm{f}k}\boldsymbol{I} \quad \text{subject to}\\
& \dot{p}^{\mathrm{f}k} = -\alpha^k Q^k\,\mathrm{div}\langle \boldsymbol{d}\rangle \quad \text{and} \quad \frac{1}{Q^k} = \frac{\alpha^k - n^k}{K^{\mathrm{s}k}} + \frac{n^k}{K^{\mathrm{f}k}},\\
\text{void:}\quad & \boldsymbol{\sigma}^{\mathrm{f}k} = -p^{\mathrm{f}k}\boldsymbol{I} \quad \text{subject to}\\
& \dot{p}^{\mathrm{f}k} = -K^{\mathrm{f}k}\,\mathrm{div}\langle \boldsymbol{d}\rangle \quad \text{and} \quad K^{\mathrm{f}k} \approx 0,\ \rho^{\mathrm{f}k} \approx 0,
\end{aligned}\right\} \tag{65}$$

and by the evolution equations for the porosities,

$$\dot{n}^k = (1-n^k)\left(\frac{\dot{p}^{\mathrm{s}k}}{K^{\mathrm{s}k}} + \mathrm{div}\langle \boldsymbol{v}\rangle\right), \qquad \text{where} \quad \dot{p}^{\mathrm{s}k} = \dot{p}^{k'}\frac{K^{\mathrm{s}k}}{K_{\mathrm{d}}^k} + \dot{p}^{\mathrm{f}k}, \tag{66}$$

and volume fractions,

$$\dot{f}^k = f^k\left(\frac{\langle K\rangle}{K^k} - 1\right)\mathrm{div}\langle \boldsymbol{v}\rangle. \tag{67}$$

Finally, the mass density of the mixture is updated by $\langle\dot{\rho}\rangle = \sum_k f^k \dot{\rho}^k$ using the k-phase compression model

$$\dot{\rho}^k = \frac{\rho^k}{K^k}\langle \dot{p}\rangle. \tag{68}$$

We remark that the reduced model might be derived by an asymptotic analysis of the unreduced model in the limit of zero relaxation (equilibration) times instead of using a priori closure through (56) [24, 31].

6 Conclusions and Outlook

We have derived a continuum mechanical model to describe the multi-material flow associated with vibro-injection pile installation in saturated sand. The model has been derived from microscopic balance principles through spatial averaging and treats the mixture of multiple materials as an effective single-phase material or homogenized mixture on the macroscale. In doing so, we have assumed that each phase of the mixture is composed of a solid species and a fluid species, with the portion of the fluid being zero in a pure solid phase (bulk solid) and the portion of the solid being zero in a pure fluid phase (bulk fluid). In general, each phase represents a solid-fluid compound in which the solid species is constituted of grains of a granular material; sand in the present case. The solid-fluid compound might thus represent a fluid-saturated porous medium or a suspension. Each constituent is assumed compressible.

An important step in the development of the macroscopic continuum mechanical model has been the closure of the underlying set of equations in such a way that the specific multi-material flow associated with vibro-injection pile installation in saturated sand is described. General closure models have been defined in order to account for the physics of each material and at the material interfaces. In particular, the macroscopic mechanical behavior of a porous medium representing sand is described by a hypoplastic rate constitutive equation advanced in the Subproject 1 of this research unit. Closure models are also required for the evolution of variables characterizing the interfacial structure. These latter models, called topological closure laws, had been initially left uncompleted because no evolution equation could be specified for the volume fractions of the phases in the mixture. By assuming a priori homogeneous distributions of pressure and velocity for all phases of the mixture the set of equations have finally been closed, resulting in a two-equation (two-field) reduced model. This model will be employed in Part 2 to develop a new multi-material arbitrary Lagrangian-Eulerian (MMALE) numerical method particularly suitable to simulate pile installation.

Acknowledgments The presented work was carried out under the financial support from the German Research Foundation (DFG; grants SA 310/26-1 and SA 310/26-2) as part of the DFG Research Unit FOR 1136, which is gratefully acknowledged. We thank our colleagues in this research unit for several fruitful discussions about our work.

References

1. Aubram, D.: Differential geometry applied to continuum mechanics. In: Veröffentlichungen des Grundbauinstitutes der Technischen Universität Berlin, vol. 44. Shaker, Aachen (2009)
2. Aubram, D.: An arbitrary Lagrangian-Eulerian method for penetration into sand at finite deformation. In: Veröffentlichungen des Grundbauinstitutes der Technischen Universität Berlin, vol. 62. Shaker, Aachen (2013)
3. Aubram, D.: Three-scale hybrid mixture theory for geomechanical multi-material flow (submitted for publication)
4. Aubram, D., Rackwitz, F., Savidis, S.A.: An ALE finite element method for cohesionless soil at large strains: computational aspects and applications. In: Benz, T., Nordal, S. (eds.) Proceedings 7th European Conference on Numerical Methods in Geotechnical Engineering (NUMGE), pp. 245–250. CRC Press, Boca Raton (2010)
5. Belytschko, T., Liu, W.K., Moran, D.: Nonlinear Finite Elements for Continua and Structures. Wiley, Chichester (2000)
6. Bennethum, L.S., Cushman, J.H.: Multiscale, hybrid mixture theory for swelling systems. 1. Balance laws. Int. J. Eng. Sci. **34**, 125–145 (1996)
7. Bennethum, L.S., Weinstein, T.: Three pressures in porous media. Transp. Porous Media **54**, 1–34 (2004)
8. Bennethum, L.S.: Compressibility moduli for porous materials incorporating volume fraction. J. Eng. Mech. **132**, 1205–1214 (2006)
9. Bennethum, L.S.: Theory of flow and deformation of swelling porous materials at the macroscale. Comput. Geotech. **34**, 267–278 (2007)
10. Benson, D.J.: Computational methods in Lagrangian and Eulerian hydrocodes. Comput. Methods Appl. Mech. Eng. **99**, 235–394 (1992)
11. Benson, D.J.: A mixture theory for contact in multi-material Eulerian formulations. Comput. Methods Appl. Mech. Eng. **140**, 59–86 (1997)
12. Biot, M.A.: General theory of three-dimensional consolidation. J. Appl. Phys. **12**, 155–164 (1941)
13. Biot, M.A., Willis, D.G.: The elastic coefficients of the theory of consolidation. J. Appl. Mech. **24**, 594–601 (1957)
14. Bouré, J.A., Delhaye, J.M.: General equations and two-phase flow modeling, Sect. 1.2. In: Hetsroni, G. (ed.) Handbook of Multiphase Systems. Hemisphere Publishing Corporation, Washington (1982)
15. Bouré, J.A.: Two-phase flow models: the closure issue. Multiph. Sci. Technol. **3**(1–4), 3–30 (1987)
16. Bowen, R.M.: Compressible porous media models by use of the theory of mixtures. Int. J. Eng. Sci. **20**(6), 697–735 (1982)
17. Colella, P., Glaz, H.M., Ferguson, R.E.: Multifluid algorithms for Eulerian finite difference methods (1997) (unpublished manuscript)
18. De Boer, R.: Theory of Porous Media. Springer, Berlin (2000)
19. Steinmann, Paul: Elasticity. In: Steinmann, Paul (ed.) . LAMM, vol. 2, pp. 287–364. Springer, Heidelberg (2015)
20. Drew, D.A.: Mathematical modeling of two-phase flow. Annu. Rev. Fluid Mech. **15**, 261–291 (1983)
21. Drumheller, D.S.: A theory for dynamic compaction of wet porous solids. Int. J. Solids Struct. **23**, 211–237 (1987)
22. Ehlers, W.: Grundlegende Konzepte in der Theorie Poröser Medien. Technische Mechanik **16**, 63–76 (1996)
23. Hirt, C.W., Amsden, A.A., Cook, J.L.: An arbitrary Lagrangian-Eulerian computing method for all flow speeds. J. Comput. Phys. **14**, 227–253 (1974)
24. Kapila, A.K., Menikoff, R., Bdzil, J.B., Son, S.F., Stewart, D.S.: Two-phase modeling of deflagration-to-detonation transition in granular materials: reduced equations. Phys. Fluids **13**, 3002–3024 (2001)

25. Koning, H.L.: Some observations on the modulus of compressibility of water. In: Proceedings European Conference on Soil Mechanics and Foundation Engineering-Problems of Settlement and Compressibility of Soils, Sect. 1, Wiesbaden, Germany, pp. 33–36. Deutsche Gesellschaft für Erdund Grundbau e.V. (1963)
26. Lewis, R.W., Schrefler, B.A.: The Finite Element Method in the Static and Dynamic Deformation and Consolidation of Porous Media, 2nd edn. Wiley, Chichester (1998)
27. Malvern, L.E.: Introduction to the Mechanics of a Continuous Medium. Prentice Hall Inc, New Jersey (1969)
28. Marsden, J.E., Hughes, T.J.R.: Mathematical Foundations of Elasticity. Dover Publications, New York (1994)
29. Miller, G.H., Puckett, E.G.: A high-order Godunov method for multiple condensed phases. J. Comput. Phys. **128**, 134–164 (1996)
30. Miller, D.S., Zimmerman, G.B.: An algorithm for time evolving volume fractions in mixed zones in Lagrangian hydrodynamics calculations. Russian J. Phys. Chem. B **3**, 117–121 (2009)
31. Murrone, A., Guillard, H.: A five equation reduced model for compressible two phase flow problems. J. Comput. Phys. **202**, 664–698 (2005)
32. Nikolinakou, M.A., Whittle, A.J., Savidis, S.A., Schran, U.: Prediction and interpretation of the performance of a deep excavation in Berlin sand. J. Geotech. Geoenviron. Eng. **137**(11), 1047–1061 (2011)
33. Niemunis, A., Herle, I.: Hypoplastic model for cohesionless soils with elastic strain range. Mech. Cohesive-Frictional Mater. **2**, 279–299 (1997)
34. Rackwitz, F., Savidis, S.A.: Numerische Untersuchungen zum Tragverhalten von Zugpfählen in Berliner Sand. Bauingenieur **79**(9), 375–383 (2004)
35. Savidis, S.A., Aubram, D., Rackwitz, F.: Arbitrary Lagrangian-Eulerian finite element formulation for geotechnical construction processes. J. Theor. Appl. Mech. **38**(1–2), 165–194 (2008)
36. Shashkov, M.: Closure models for multimaterial cells in arbitrary Lagrangian-Eulerian hydrocodes. Int. J. Numer. Methods Fluids **56**(8), 1497–1504 (2008)
37. Stewart, H.B., Wendroff, B.: Two-phase flow: models and methods. J. Comput. Phys. **56**, 363–409 (1984)
38. Truesdell, C., Toupin R.A.: Encyclopedia of physics. In: Bd. III/1: The Classical Field Theories, pp. 226–793. Springer, Heidelberg (1960)
39. Truesdell, C., Noll, W.: The Non-Linear Field Theories of Mechanics, 3rd edn. Springer, Berlin (2004)
40. Vitali, E., Benson, D.J.: An extended finite element formulation for contact in multi-material arbitrary Lagrangian-Eulerian calculations. Int. J. Numer. Methods Eng. **67**, 1420–1444 (2006)
41. von Wolffersdorff, P.-A.: A hypoplastic relation for granular materials with a predefined limit state surface. Mech. Cohesive-Frictional Mater. **1**, 251–271 (1996)
42. Whitaker, S.: The transport equations for multi-phase systems. Chem. Eng. Sci. **28**, 139–147 (1973)
43. Whitaker, S.: Flow in porous media iii: deforming media. Transp. Porous Media **1**, 127–154 (1986)
44. Wriggers, P.: Nonlinear Finite Element Methods. Springer, Berlin (2008)
45. Zienkiewicz, O.C., Shiomi, T.: Dynamic behaviour of saturated porous media: the generalized Biot formulation and its numerical solution. Int. J. Numer. Analyt. Methods Geomech. **8**, 71–96 (1984)
46. Zienkiewicz, O.C., Chan, A.H.C., Pastor, M., Schrefler, B.A., Shiomi, T.: Computational Geomechanics-With Special Reference to Earthquake Engineering. Wiley, Chichester (1999)
47. Zienkiewicz, O.C., Taylor, R.L.: The Finite Element Method, vol. 3, 5th edn. Butterworth-Heinemann, Oxford (2000)

Vibro-Injection Pile Installation in Sand: Part II—Numerical and Experimental Investigation

S.A. Savidis, D. Aubram and F. Rackwitz

Abstract In Part 1 of this series of papers a macroscopic two-equation (two-field) reduced model for the mechanics of the multi-material flow associated with vibro-injection pile installation in saturated sand was derived. Here we employ this model to develop a so-called multi-material arbitrary Lagrangian-Eulerian (MMALE) method. MMALE avoids the disadvantages of the classical approaches in computational continuum mechanics concerning large deformations and evolving material interfaces. The numerical implementation of this method will be outlined, and then the experimental investigations will be presented that have been carried out in order to validate the computational model. Among these investigations, small-scale model tests in chambers with observing window have been designed step-by-step to reveal penetration and vibro-injection pile installation phenomena.

Keywords Arbitrary Lagrangian-Eulerian · Multi-material · Large deformations · Finite element method · Operator-split · Soil mechanics · Sand

1 Introduction

Subproject 5 as part of the DFG Research Unit FOR 1136 is concerned with the numerical modeling of vibro-injection pile installation into water-saturated sand. The motivation arises from the fact that there are currently no numerical models to accurately predict the effects of this process on the stress and density states within the soil or on close-by structures. However, such a prediction would be of high practical relevance. In Part 1 of our contribution we interpreted the process of vibro-injection pile

S.A. Savidis (✉) · D. Aubram
Chair of Soil Mechanics and Geotechnical Engineering, Technische Universität Berlin,
Secr. TIB1-B7 Gustav-Meyer-Allee 25, 13355 Berlin, Germany
e-mail: savidis@tu-berlin.de

F. Rackwitz
Department of Geotechnical Engineering, Ostbayerische Technische Hochschule Regensburg,
Prüfeninger Straße 58, 93049 Regensburg, Germany

Th. Triantafyllidis (ed.), *Holistic Simulation of Geotechnical Installation Processes*,
Lecture Notes in Applied and Computational Mechanics 77,
DOI 10.1007/978-3-319-18170-7_6

installation as the flow of an immiscible mixture with interfaces (multi-material flow) consisting of a bulk solid phase (steel pile), a bulk fluid phase (injection grout), and a compound phase consisting of a solid species and a fluid species which represents the fluid-saturated porous material (sand). Application of the technique of spatial averaging to the governing equations led to a macroscopic two-equation (two-field) model. In Part 2 we continue our investigation and employ this continuum mechanical model to develop a multi-material arbitrary Lagrangian-Eulerian (MMALE) method for the numerical simulation of vibro-injection pile installation.

MMALE falls into the category of arbitrary Lagrangian-Eulerian (ALE) numerical methods in which the mesh is not fixed as in the Eulerian methods but can move independently of the material [6, 8, 30, 49]. MMALE methods, including the subset of multi-material Eulerian methods, generalize the classical approaches according to Lagrange and Euler in the context of finite element or finite difference methods and are able to address the problems mentioned [11–14, 19, 20, 24, 37–39, 44, 56, 61]. No limitations concerning material deformations do exist because in these methods material boundaries may flow through the mesh. Therefore, elements may arise which contain two or more materials separated by interfaces.

Both MMALE and the subset of multi-material Eulerian methods were originally developed in the defence sector for the numerical simulation of highly dynamical physical problems in which high strain rates appear and new interfaces are generated (hypervelocity multi-material flow). Information is often hard to access because of the classification restriction imposed on many program codes and the related developments. Typical application areas are underwater explosions with fluid-structure-interaction and impact problems. However, with the application of MMALE methods to water-saturated sand at relatively low velocities this subproject has entered new territory on both national and international level.

The change from the Lagrangian to a non-Lagrangian viewpoint inherent to MMALE methods introduces three problems: (i) tracking material interfaces, (ii) treatment of multi-material elements intersected by interfaces, and (iii) advection of the solution variables across element boundaries. Due to the potential presence of multiple materials in a single element, MMALE requires the solution of a multi-material flow problem in consideration of solid mechanical, fluid dynamical, and interaction phenomena. Consequently, the theoretical basis includes aspects of both continuum mechanics and multiphase flow theory. The numerical treatment, on the other hand, requires FEM technology as well as methods from the field of computational fluid dynamics (CFD). This renders MMALE highly interdisciplinary and complex, as visualized in Fig. 1.

The mixture of multiple materials in multi-material elements is treated as an effective single-phase material or homogenized mixture on the element level; empty space (void) is considered as a type of material having vanishing stiffness and mass density. The degrees of freedom of the homogenized mixture solved for at the element nodes are the same for all individual phases of the mixture. In the present model all phases share a common velocity and pressure. Such an approach calls for appropriate mixing rules which relate quantities associated with each material to the corresponding "mixed" (homogenized) variables, and vice versa. The mixing rules play a crucial

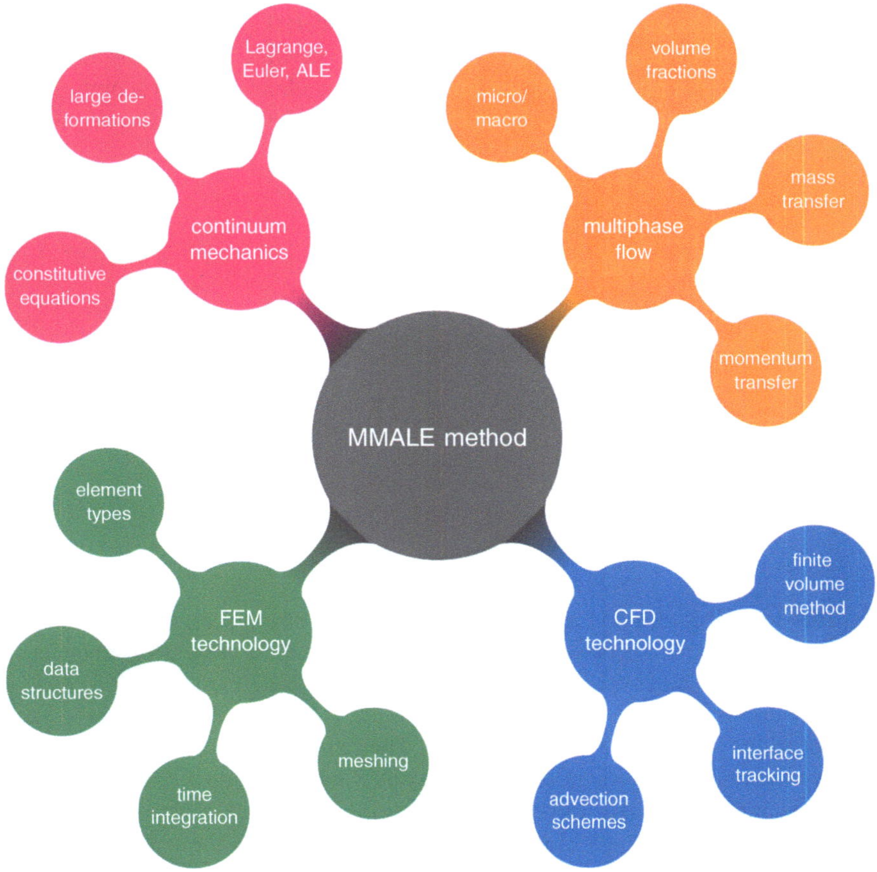

Fig. 1 Mind map of the developed MMALE method to visualize its interdisciplinarity and complexity

role and should be governed by the physics of the problem. In particular, the saturated sand in vibro-injection pile installation must be modeled as a two-phase material in order to account for consolidation and liquefaction phenomena. During the remap step of the method the partial material volumes in multi-material elements must be transported through the mesh. In order to achieve a reasonable accuracy of the procedure, the material volume fluxes across the element boundaries are computed as truncation volumes which requires material interfaces to be tracked along with the flow field.

Part 2 in this series of papers is structured as follows. The development and implementation of the MMALE computational method will be outlined in Sect. 2. It consists of several algorithms specific to multi-material methods which will be briefly addressed. The conception and realization of a model test chamber with observing window and the conducted experimental small-scale tests will be described in Sect. 3. These experimental investigations are undertaken supplementary to the central project of this research unit in order to validate the MMALE method. Section 4 presents example applications of the numerical algorithms as well as preliminary results of back-calculations. The paper closes with concluding remarks and outlook in Sect. 5.

2 Numerical Method and Implementation

2.1 ALE Formulation and Solution Strategy

The governing equations of the two-equation (two-field) reduced model derived in Part 1,

$$\begin{aligned} \operatorname{div}(\boldsymbol{s} - p\boldsymbol{I}) + \rho\boldsymbol{b} - \rho\dot{\boldsymbol{v}} &= \boldsymbol{0} \\ \dot{p}/K + \operatorname{div}\boldsymbol{v} &= 0, \end{aligned} \tag{1}$$

are in the so-called updated Lagrangian form, meaning that the current configuration of the material in space is taken as the reference domain deforming with the material as time elapses. The entire model has been formulated by Eqs. (63)–(68) in that paper. Herein we consider only averaged quantities, so that the angle brackets $\langle\cdot\rangle$ can be dropped for reasons of notational brevity. The basic notation is found in Part 1.

Since the reference domain is represented by the finite element mesh in numerical implementations, large material deformations may cause severe distortion of Lagrangian elements which slows down or even terminates the calculation. The ALE formulation has been developed in order to circumvent these problems by introducing an independently moving reference domain [5, 6]. Accordingly, the spatial description of any scalar-, vector- or tensor-valued physical field q is related to its referential or ALE description $\hat{q}$ by the composition $\hat{q} = q \circ \Phi$, where Φ, called the relative motion, is an embedding that maps the reference points onto the spatial points currently occupied by the material.

The material time derivative of $q = \hat{q} \circ \Phi^{-1}$ leads to the fundamental ALE operator

$$\dot{q} = \frac{\partial \hat{q}}{\partial t} \circ \Phi^{-1} + \boldsymbol{c} \cdot \nabla q\,, \qquad \text{with} \;\; \dot{q} \overset{\text{def}}{=} h(\ldots), \tag{2}$$

and $h(\ldots)$ representing an evolution equation for the field q under consideration. The first term on the right side represents the time derivative of q with respect to fixed reference points. The second term, called the convective term, stems from the relative motion between the material and the reference domain defining the so-called convective velocity $\boldsymbol{c}$ for each material; as we assumed homogeneous distribution of velocity in the mixture, $\boldsymbol{c}$ is the same for all constituents. Note that (2) generalizes the classical Lagrangian ($\boldsymbol{c} = \boldsymbol{0}$; Φ = motion of the body) and Eulerian ($\boldsymbol{c} = \boldsymbol{v}$;

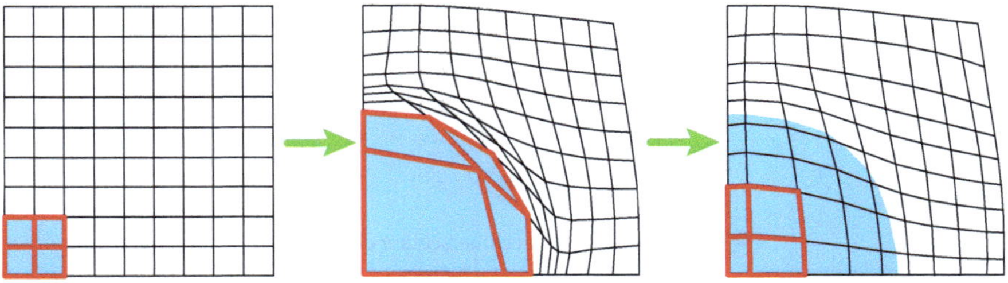

Fig. 2 Schematic diagram of the Lagrange-remap MMALE method. The *blue* area indicates a material zone whose initial configuration is assigned to an element patch highlighted in *red*

$\Phi = \text{id}$) formulations of continuum mechanics. On the other hand, replacing with the ALE operator (2) every material time derivative in the derived model for multi-material flow results in an ALE formulation of that model which can be treated by an MMALE method. The resulting system of equations can be brought into a convenient conservation form [6], but this is not shown here.

The MMALE method developed in this research work is an extension of the simplified or single-material ALE method [6, 8, 49] in which material interfaces are explicitly resolved by element edges. An operator-split or Lagrange-remap strategy [6, 11] is applied to enable the use of simpler and more robust algorithms compared to a monolithic solution approach. It divides the incremental solution of the highly nonlinear problem into a Lagrangian step and remap step, as schematically shown in Fig. 2. Concerning the fundamental ALE operator (2), the Lagrange-remap strategy can be written conceptually as

$$\text{Lagrangian step:} \quad \dot{q} = h(\ldots), \tag{3}$$

$$\text{remap step:} \quad \frac{\partial \hat{q}}{\partial t} \circ \Phi^{-1} + \boldsymbol{c} \cdot \nabla q = 0. \tag{4}$$

Equation (3) represents the updated Lagrangian form of the Eqs. (63)–(68) in Part 1 governing the considered multi-material flow of vibro-injection pile installation into water-saturated sand. During the Lagrangian step the set of equations is solved with common finite element methods for the two-field mixed element formulation by accounting for large deformations [10, 36, 60, 62, 63]. For this purpose the system of Eq. (1) is brought into a weak form, and then this weak form is approximated by discretizing the computational domain. The so-called MINI or P1$^+$/P1 element [3, 18] serves as a basis for the developments. It is a mixed triangle element using linear approximations for the spatial pressure field p and spatial displacement field $\boldsymbol{u}$, with $\dot{\boldsymbol{u}} = \boldsymbol{v}$. An additional bubble function for the displacement field stabilizes the element. The MINI Element is equipped with a multi-material option during the research project for the purposes of MMALE implementation (Fig. 3).

In the Lagrangian step the parametrization of the variables of the problem is chosen such that the element mesh follows the deformation of the mixture and the convective

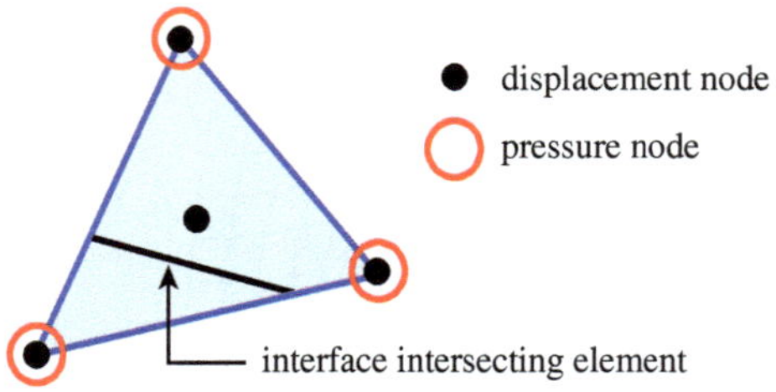

Fig. 3 An essential part of the MMALE method is the stable MINI element [3] which is enriched during this research by a multi-material option in order to account for interfaces intersecting the element

terms disappear. The evolution equations of the variables take the form of (3). That is to say, the Lagrangian step considers the sources, constitutive equations of each material, and topological closure laws for each phase as in standard nonlinear finite element analysis. The remap step accounts for the convective terms by solving (4), which represents the linear advection equation without a source. For this purpose, the nodes are relocated in such a way that mesh distortion is reduced and then the solution variables obtained after the Lagrangian step are transported through the mesh using conservative CFD advection algorithms. If the nodes are simply relocated to their original positions a multi-material Eulerian formulation is obtained. In this case, (4) reduces to

$$\frac{\partial q}{\partial t} + \boldsymbol{v} \cdot \nabla q = 0. \tag{5}$$

A flow chart of the developed MMALE method is shown in Fig. 4. The program steps will be explained in the following.

2.2 Initialization Phase

Before the actual calculation starts the computational domain of the considered initial boundary value problem is discretized with finite elements. Depending on the problem definition, the computational domain has to cover empty space which might be occupied by material during the course of the calculation. Empty space (void) is considered as a type of material within the model; see constitutive equations summarized as Eq. (65) in Part 1. The remap step requires additional information about the mesh connectivity, such as the elements sharing a common node or the adjacent elements of an element. However, this information has to be gathered and stored only once during the initialization phase because the mesh connectivity, by definition, does not change in ALE methods. Furthermore, the material properties must be assigned to the elements. The initialization phase is completed with the specification of the initial conditions and boundary conditions.

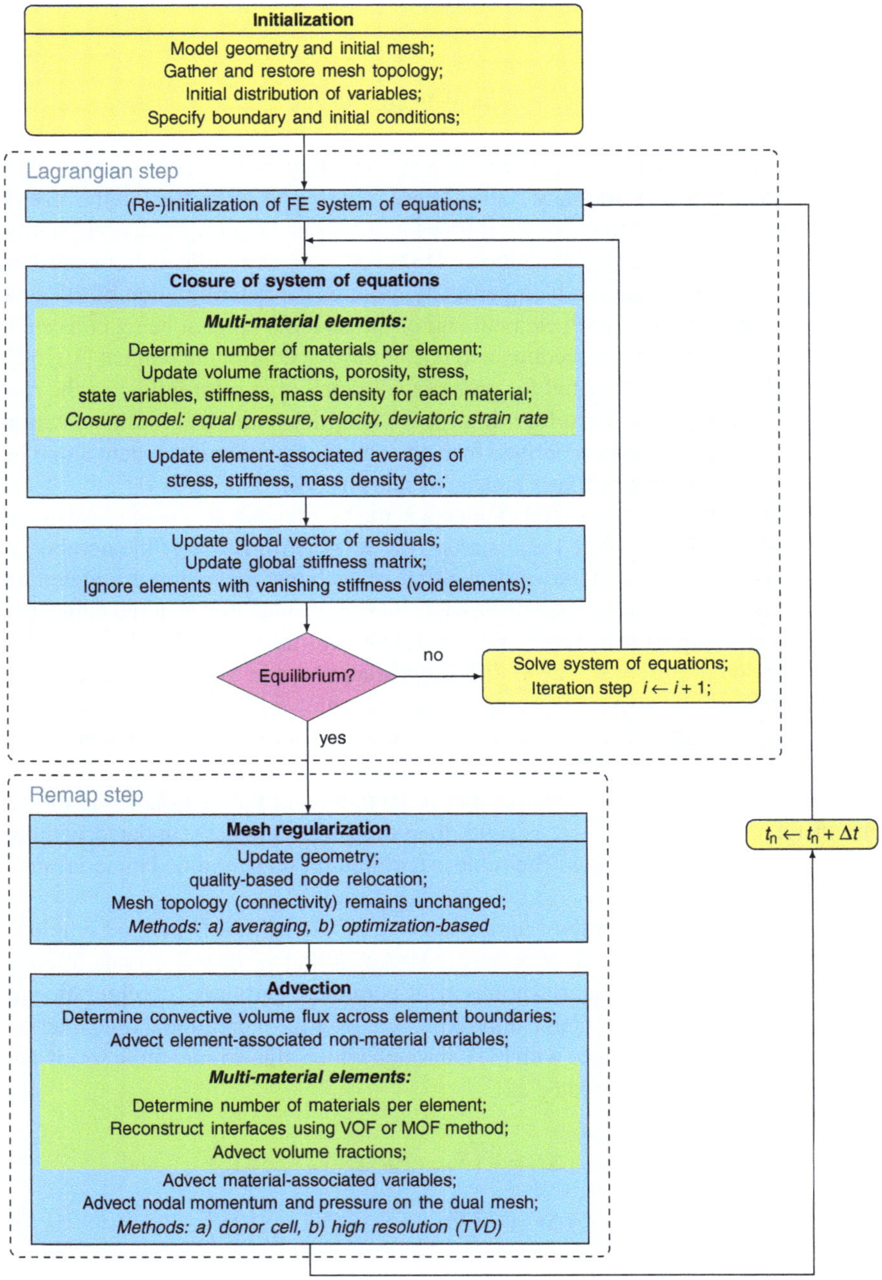

Fig. 4 Flow chart of the developed MMALE method for finite element programs

2.3 Lagrangian Step

A calculational cycle or time increment begins in the Lagrangian step with the re-initialization of the FE system of equations. Solution is advanced implicitly in time by using the Newmark-β method in conjunction with Newton's method. This time integrator is unconditionally stable, hence imposes no restriction on the size of the time step in contrast to the explicit procedures applied in almost all other multi-material Eulerian and MMALE methods. However, the system of equations must be delivered in linearized form which generally requires the determination of the geometric stiffness matrix of each element and of the material tangent being consistent with the stress integration procedure, and not simply the continuum tangent [10, 60]. However, these are not essential for problem solution but at most influence the rate of solution convergence. Therefore, in order to simplify numerical implementation the element stiffness matrix is derived from the continuum material tangent alone by neglecting the geometric stiffness matrix, as this has been done in [14].

The treatment of multi-material elements in the Lagrangian step makes use of the subcell model derived in Part 1 and starts with the determination of the number of materials in each element. Afterwards the state of each material phase is updated. In particular, the volume fraction, porosity, stress, state variables, and material tangent of each material are updated by integrating the related evolution equations in time. The rotational terms of the stress rate at finite deformation render the integration of rate constitutive equations expensive compared to the infinitesimal case. In this context, the restrictions related to the axiom of material frame indifference [54] have led to the notion of incremental objectivity of the integration method over finite time steps [31]. Incremental objectivity requires that if the motion of the material body over a time increment $\Delta t = t_{n+1} - t_n$ is rigid, then the stress is exactly updated without generation of spurious stresses. The same is required for tensor-valued material state variables.

The stress integration in the Lagrangian step of the MMALE method employs the incrementally objective algorithm of [32]. Accordingly, the stress integration is carried out with respect to material points in the corotated (unrotated) configuration of the material body. A time-centered approximation over the incremental time interval $[t_n, t_{n+1}]$ is used in accordance with [31], through which algorithmic finite strain and rotation increments are computed as

$$\Delta\boldsymbol{\varepsilon}_{n+1/2} \overset{\text{def}}{=} \boldsymbol{d}_{n+1/2}\Delta t = \tfrac{1}{2}\left(\nabla_{n+1/2}\boldsymbol{u} + (\nabla_{n+1/2}\boldsymbol{u})^{\mathrm{T}}\right) \quad \text{and}$$
$$\Delta\boldsymbol{r}_{n+1/2} \overset{\text{def}}{=} \boldsymbol{\omega}_{n+1/2}\Delta t = \tfrac{1}{2}\left(\nabla_{n+1/2}\boldsymbol{u} - (\nabla_{n+1/2}\boldsymbol{u})^{\mathrm{T}}\right), \tag{6}$$

respectively, and

$$\nabla_{n+1/2}\boldsymbol{u} = 2(\boldsymbol{f}_{n+1} - \boldsymbol{I})(\boldsymbol{f}_{n+1} + \boldsymbol{I})^{-1}. \tag{7}$$

f_{n+1} is the relative incremental deformation gradient of the configuration at time $t = t_{n+1}$ with respect to the configuration at time $t = t_n$.

Now, consider the general rate constitutive equation

$$\overset{\triangledown}{\boldsymbol{\sigma}} = \mathbf{c}(\boldsymbol{\sigma}, \boldsymbol{\alpha}) : \boldsymbol{d}, \tag{8}$$

where $\boldsymbol{\sigma}^{\triangledown} = \dot{\boldsymbol{\sigma}} + \boldsymbol{\sigma} \cdot \boldsymbol{\omega} - \boldsymbol{\omega} \cdot \boldsymbol{\sigma}$ denotes the corotational Zaremba-Jaumann rate of $\boldsymbol{\sigma}$ defined by the spin $\boldsymbol{\omega}$ representing the vorticity tensor. Since the spin generates a one-parameter group of rotations through $\dot{\mathfrak{R}} = \boldsymbol{\omega} \cdot \mathfrak{R}$, with $\mathfrak{R}|_{t=0} = \boldsymbol{I}$, the Cauchy stress is objectively updated by the general integration algorithm

$$\boldsymbol{\sigma}_{n+1} = \mathfrak{R}_{n+1} \cdot (\mathfrak{S}_n + \Delta\mathfrak{S}_{n+\theta}) \cdot \mathfrak{R}^{\mathrm{T}}_{n+1}, \qquad \text{with} \quad \theta \in [0, 1], \tag{9}$$

and

$$\mathfrak{S}_n \overset{\text{def}}{=} \mathfrak{R}^{\mathrm{T}}_n \cdot \boldsymbol{\sigma}_n \cdot \mathfrak{R}_n, \qquad \Delta\mathfrak{S}_{n+\theta} \overset{\text{def}}{=} \mathfrak{f}_{n+\theta}(\mathfrak{S}_{n+\theta}, \mathfrak{A}_{n+\theta}, \Delta\mathfrak{E}_{n+1/2}),$$
$$\Delta\mathfrak{E}_{n+1/2} \overset{\text{def}}{=} \mathfrak{R}^{\mathrm{T}}_{n+1/2} \cdot \Delta\boldsymbol{\varepsilon}_{n+1/2} \cdot \mathfrak{R}_{n+1/2}, \qquad \mathfrak{S}_{n+\theta} \overset{\text{def}}{=} \mathfrak{R}^{\mathrm{T}}_{n+\theta} \cdot \boldsymbol{\sigma}_{n+\theta} \cdot \mathfrak{R}_{n+\theta}, \tag{10}$$

and $\mathfrak{A}_{n+\theta}$ representing the set $\boldsymbol{\alpha}$ of material state variables under the transformation $\mathfrak{R}_{n+\theta}$. The stress increment $\Delta\mathfrak{S}_{n+\theta}$ is calculated using the response function $\mathfrak{f}_{n+\theta}$. The response function depends on the choice of $\theta \in [0, 1]$ and basically represents an explicit ($\theta = 0$) or implicit ($\theta = 1$) stress-point algorithm for the case of infinitesimal deformations. The rotation and half-step rotation are defined through

$$\mathfrak{R}_{n+1} = \Delta\mathfrak{R} \cdot \mathfrak{R}_n \qquad \text{and} \qquad \mathfrak{R}_{n+1/2} = \Delta\mathfrak{R}^{1/2} \cdot \mathfrak{R}_n, \tag{11}$$

where

$$\Delta\mathfrak{R} = (\boldsymbol{I} - \tfrac{1}{2}\Delta\boldsymbol{r}_{n+1/2})^{-1}(\boldsymbol{I} + \tfrac{1}{2}\Delta\boldsymbol{r}_{n+1/2}) \tag{12}$$

is an approximation to the incremental rotation over the time increment $[t_n, t_{n+1}]$ according to [31]. If the configuration at time $t = t_n$ is taken as the reference configuration, as in an updated Lagrangian description of motion [6, 10], then $\mathfrak{R}_n = \boldsymbol{I}$, otherwise $\mathfrak{R}_{t=0} = \boldsymbol{I}$ is set.

Algorithm (9) is applied to each material phase whose mechanical behavior is characterized by a rate constitutive equation of the form (15). According to (52) in Part 1, this would be the case for bulk solid and porous media. In particular, the behavior of sand is described by an advanced hypoplastic rate constitutive equation [40, 57]. This constitutive equation is advanced in Subproject 1 of the Research Unit FOR 1136. For a reliable numerical simulation of vibro-injection pile installation it is indispensable to carry over new developments to the MMALE method.

After the material-associated variables were updated, their element-associated averaged values can be determined in accordance with (64) in Part 1. The balance between the internal forces and the external loads at the end of the time step, which expresses the identity of the vector of residuals with the null vector, is iterated by means of Newton's method. Elements that partially or completely cover empty space require special treatment in implicit multi-material Eulerian and MMALE calculations [14]. Void elements practically do not have any stiffness or mass density. Therefore, their nodes remain unconsidered when setting up the finite element system of equations. Elements located at material boundaries are partially filled with void. The stiffness of the mixture inside these elements might be low, causing large displacement increments during the equilibrium iterations. Therefore, in order to avoid inverted elements, the incremental nodal displacements of partially filled elements are uniformly scaled.

2.4 Remap Step

When the equilibrium iteration converges, the quality of all elements in the mesh is evaluated. The quality measure employs the radius ratio of the element's incircle and circumcircle. The remap step is initialized if at least one element fails the quality check, and then the nodes of those elements are flagged. Only the flagged nodes and the elements that share these nodes are processed during the remap step for reasons of computational efficiency. Therefore, all algorithms of the remap step are required to work on a local level.

After the initialization of the remap step the geometry of the FE model is updated, so that the totality of nodes in their current position defines the reference domain. In order to increase the quality of the mesh deformed during the Lagrangian step, the flagged nodes are relocated by employing a suitable local mesh smoothing algorithm. Users can choose between different heuristic procedures and an extremely robust optimization-based smoothing algorithm which works on arbitrarily shaped domains [6, 7]. A multi-material Eulerian formulation ($\boldsymbol{c} = \boldsymbol{v}$; $\Phi = \mathrm{id}$) is obtained if the nodes are simply relocated to their original positions.

One of the crucial and at the same time most extensive steps of the MMALE method is the transport (advection) of the variables through the mesh. Technically speaking this means a remap of the solution variables obtained after the Lagrangian step onto the smoothed mesh. Because the mesh topology does not change during the smoothing step, elements have the same neighbors throughout the calculation, so that conservative CFD algorithms can be applied. The remap must be carried out for element-associated variables (e.g. deformation gradient), material-associated variables (e.g. stress and state variables), and nodal variables. The latter are dictated by the primary unknowns of the problem and the underlying balance equations. Concerning the two-equation model considered here (see Part 1), the nodal variables are represented by the total momentum and total volume change of the mixture.

For unstructured meshes, transport algorithms which are based on the finite volume method [9, 35] are best suited. Finite volume (FV) methods solve the integral form of the advection Eq. (1) as an integral conservation law. Hence, they are conservative by definition, leading to mechanically consistent results. In order to solve the integral conservation law for the generic variable q under consideration, a control volume tessellation has to be constructed based on the finite element mesh that is used in the overall MMALE solution procedure. Moreover, a pseudo-time interval $[t^-, t^+] \subset \mathbb{R}$ is introduced because physical time elapses during the Lagrangian step but remains fixed during the remap step. The distribution of the variables at the end of the Lagrangian step ($t = t^-$) is assumed to be given, that is, $q|_{t=t^-} = q^-$. Then, if a first-order accurate integration in time is applied, the variable in the jth control volume after the remap step at $t = t^+$ is obtained by the general formula

$$q_j^+ = \frac{q_j^- V_j^- - \sum_{\text{facets}} \text{Fl}(q)\Delta t}{V_j^+}. \tag{13}$$

Here V_j and q_j denote the volume measure and the average of q associated with the jth control volume, respectively, $\text{Fl}(q)$ is the averaged convective volume flux of q across a facet of the control volume boundary, and $\Delta t \stackrel{\text{def}}{=} t^+ - t^-$. The control volume geometries to compute V_j^- and V_j^+ can be determined from the mesh geometries known at $t = t^-$ and $t = t^+$, respectively.

Inevitably connected with FV methods is the computation of the flux of a solution variable across the boundary of the control volumes; cf. (13). Depending on the accuracy in space of the scheme used for the flux calculation, advection algorithms are divided into first-order, second-order, and higher-order accurate methods. Moreover, the flux of a solution variable can be a linear or nonlinear function of the transported volume, leading to a linear or nonlinear procedure. Despite this, most finite volume advection schemes in ALE methods apply explicit first-order accurate methods to advance solution in time.

A weighted donor-cell linear advection scheme [4, 6, 11, 30, 44, 47] of the Godunov-type is currently implemented. It possesses an accuracy of first order and is stable, conservative, and monotonicity-preserving. The linear procedures of first-order are in fact the most robust and the easiest to implement, but they tend to excessive numerical diffusion through which solution details are smeared. Though with linear schemes of higher order this tendency is less pronounced, the solution can oscillate. Modern nonlinear algorithms can avoid spurious oscillation and at the same time achieve a maximum accuracy of second or higher order in space. The state of the art in the field of MMALE are high-resolution nonlinear schemes, for example total variation diminishing (TVD) algorithms [11, 24, 27, 35]. Such an algorithm will be also implemented into the developed MMALE method.

Concerning multi-material elements cut by one or more non-intersecting interfaces, material-associated variables must be treated separately for each individual material. Accordingly, the field to be used in (2)–(4) is the material-associated variable weighted by the volume fraction, i.e. $q \overset{\text{def}}{=} f^k q^k$. If one would simply use this definition in the general finite volume-based transport algorithm (13), however, the initially coherent material phases would disperse after a few advection steps because the interfacial structure is not accounted for [24].

In order to achieve a reasonable accuracy of the remap procedure and to conserve the volume of the individual phases, the volume fractions and the material-associated fields have to be transported by considering the spatial distribution of the material phases and the locations of the interfaces in every multi-material element. This requires specialized methods to resolve material interfaces during an MMALE or multi-material Eulerian calculation. Such methods, which can be divided into interface tracking and interface reconstruction methods, have been reviewed in [15, 33, 46, 48, 50].

Interface tracking methods update the locations of the material interfaces at each time step. Basic approaches use Lagrangian marker particles [25, 26, 55] or level set functions [42, 43, 52]. Both approaches are extensively used in CFD and computer graphics but they are prone to numerical difficulties when the interfaces experience severe stretching or tearing. In particular, level set methods do not locally conserve volume. Interface reconstruction methods like the volume of fluid (VOF) [29, 45, 61] and moment of fluid (MOF) methods [22, 23], on the other hand, are generally conservative because they track the volume resp. moment (i.e. volume and centroid) of a partial material zone (subcell) in multi-material elements.

Once the interface locations in each multi-material element have been determined by any of the methods mentioned, the material transport volumes across the element facets can be computed as truncation volumes, and then the partial material volumes are integrated to a new time level. The actual calculation is largely geometrical in nature and includes basic algorithms like the point-in-polygon test, intersection tests, area computation, and clipping. Clipping identifies that portion of a material zone which lies inside a mesh element. An example application is shown in Fig. 5; see also [34]. In this example, the TU Berlin logo represents a pseudo-material domain which is clipped against a triangle mesh. The resulting intersection polygons highlighted in different colors correspond to the partial material volumes in each element.

An interface reconstruction procedure will be implemented into the MMALE method developed in this research. Compared to VOF approaches the MOF method is local, that is, it does not require information from neighboring elements or the computation of any terms related to these. Moreover, MOF interface reconstruction works on unstructured meshes and provides an automatic ordering of the materials if an element contains two or more interfaces [2, 23]. The basic equation for both VOF and MOF methods is the volume fraction advection equation

$$\frac{\partial f^k}{\partial t} + \boldsymbol{v} \cdot \nabla f^k = 0 \qquad \text{resp.} \qquad \frac{\partial \hat{f}^k}{\partial t} \circ \Phi^{-1} + \boldsymbol{c} \cdot \nabla f^k = 0, \tag{14}$$

with $\hat{f}^k \overset{\text{def}}{=} f^k \circ \Phi$. Finite volume discretization and first-order approximation in time yields a formula similar to (13),

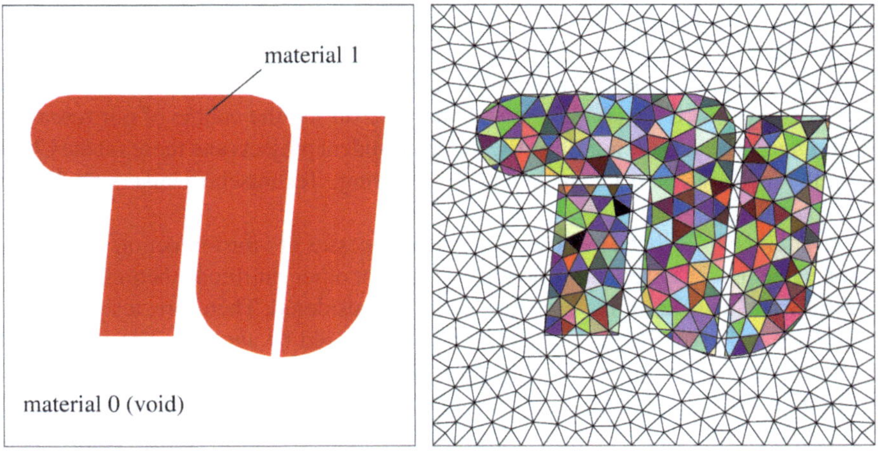

Fig. 5 Clipping of the TU Berlin logo against a triangle mesh. Pseudo-material domains (*left*) and resulting intersection polygons representing partial material volumes (*right*); after [34]

$$f_j^{k+} = \frac{f_j^{k-} V_j^- - \sum_{\text{facets}} \text{Fl}_k(f^k)\Delta t}{V_j^+}, \tag{15}$$

where Fl_k is the convective volume flux across a facet of the element boundary which is aware of the spatial distribution of the kth material, i.e. of the interface locations. The material transport volume across a facet is represented by the term $\text{Fl}_k(f^k)\Delta t$. After the volume fraction has been updated, the material-associated variable at the end of the remap step can be computed from

$$q_j^{k+} = \frac{f_j^{k-} q_j^{k-} V_j^- - \sum_{\text{facets}} \text{Fl}_k(f^k q^k)\Delta t}{f_j^{k+} V_j^+}, \tag{16}$$

which is a particular form of (13).

3 Experimental Model Tests

The following section is concerned with the experimental model tests that have been carried out in order to observe the multi-material flow field during vibro-injection pile installation into sand. The main purposes are (i) the verification of the assumptions underlying the theoretical and numerical investigations and (ii) the validation of the MMALE computational models through back-calculation of the model tests. The latter particularly requires a sufficiently detailed description of the properties of the test sand in such a way that material constants of the hypoplastic rate constitutive equation can be determined. To the best knowledge of the authors, however, there are no qualified experimental tests reported in the literature that could be used. Therefore, a completely new model test equipment had to be designed and manufactured, and series of tests had to be conducted and analyzed during the course of our research. Preliminary work was done within the scope of student projects and theses at the Chair of Soil Mechanics and Geotechnical Engineering, Technische Universität Berlin [16, 17, 41, 51].

No attempts have been made to reproduce processes of vibro-injection pile installation in the field. Consequently no scale models or other similitude theoretical issues to achieve field-scale equivalence needed to be considered. Therefore, tests could be conducted at 1 g and the use of special test sand could be avoided. The absence of scale effects will also facilitate the validation process without placing any further restrictions on the applicability of the MMALE method. MMALE computational models will be built to back-calculate selected experimental tests "as is", that is, without any scaling.

3.1 Test Set-Up and Measurement Concept

The test set-up is shown in Fig. 6. Its main components are a waterproof chamber with glass panel serving as a viewing window, a special model pile, and a device for vibratory pile driving. All components are in-house developments. The vibrator consists of two counter-rotating and synchronously revolving imbalances whose mass and rotational speed are adjustable. The dead weight of the vibrator mounted on the pile head is about 100 kg, so that a counter balance becomes necessary in order to keep the penetration velocity of the driven pile acceptable. The model pile is made up of a 50 mm × 50 mm stainless steel square tube equipped with a welded-on collar at the pile toe as well as a built-in injection tube whose opening is located directly above the collar (Fig. 6c); deflector plates that can be attached to the collar are not shown. The tapering of the toe prevents deviation of the pile from the vertical. Just as with the vibro-injection piles in practice, the shaft annulus created by the collar can be injected with grout material while the pile is being driven into the water-saturated sand. Pressurized injection is enabled by a diaphragm pressure vessel with a maximum operation pressure of 10 bar.

During the tests the pile was guided alongside the glass panel, so that the vibration and grouting process could be digitally filmed through the viewing window of the chamber. A standard Full HD camcorder (Panasonic HDC-SD900) was used to record a series of consecutive still images at 50 Hz with a maximum resolution of 1920 × 1080 pixels. The scene was illuminated by two 500 W halogen floodlights, which were placed in a large enough distance to the camera in order to minimize heating. Measurement of details of the multi-material flow field without on-sample instrumentation then was enabled by analyzing the recorded image sequence using image correlation software.

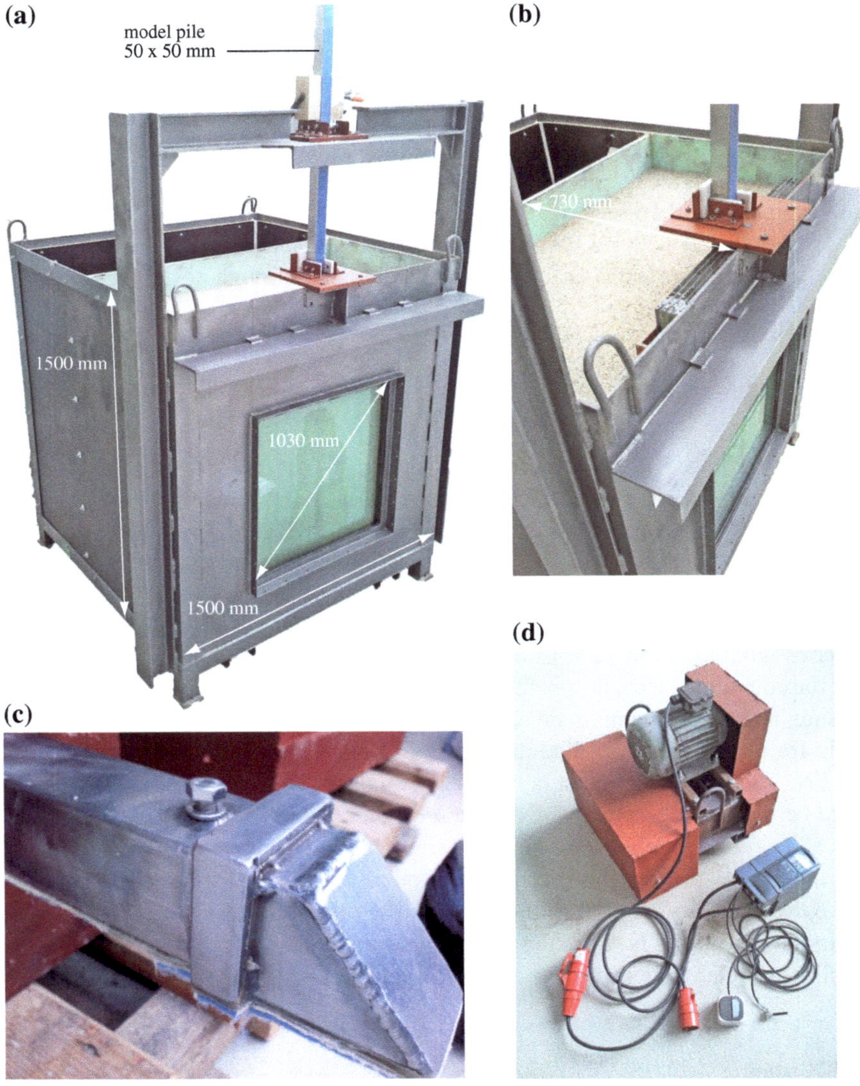

Fig. 6 Experimental investigation of vibro-injection pile installation. **a** Filled test chamber with glass panel and model pile. **b** Detailed view of the glass panel, pile guide, and model pile. **c** Tapered pile toe with welded-on collar and bolt closing the injection tube. **d** Self-made vibrator with controller (frequency converter)

In the present research we use GeoPIV [58] for image correlation, a MATLAB toolbox based on particle image velocimetry (PIV). PIV is a two-dimensional measuring technique for the whole instantaneous spatial velocity field resp. incremental displacement field of a moving and deforming material sample [1, 6, 59]. This is achieved by tracking the texture within areas of an image through a sequence of images. The PIV workflow is as follows. Two images from consecutive configurations of the material sample are subdivided into patches, and then the pixel intensity (luminance) of each pair of patches is cross-correlated in terms of an image displacement vector. The highest correlation occurs when the image displacement vector coincides with the image of the averaged incremental displacement of the material zone captured on the image patch. The totality of local incremental displacement vectors associated with all patches represents the incremental displacement field. From this, the incremental strain field can be obtained through postprocessing.

3.2 Experimental Program

The sand employed in the experimental model tests is a quartz sand with well-rounded to angular grains identified as fine-gravelly coarse Sand (fgrCSa) according to [21]. Its grain size distribution curve is plotted in Fig. 7. The limit void ratios are $e_{\min} = 0.482$ and $e_{\max} = 0.779$, with the void ratio related to the porosity by $e = n/(1-n)$. Further granulometric properties are listed in [6]. In that thesis the same sand was used for quasi-static penetration tests.

Series with a total of 10 tests have been conducted (Table 1). Parameters varied were the counter balance, the degree of saturation, the grouting material, the grouting pressure, and the load amplitude of the vibrator. The vibration frequency was about 20 Hz for all tests and the load amplitude of the vibrator varied between 1.9 and 2.56 kN. The static force, that is, the dead weight of the vibrator and the pile was varied between the tests by using different counter balances.

In all tests the chamber was filled with air dried sand by dry sieve pluviation. Each sand model was prepared in several layers of equal thickness in order to achieve a homogeneous distribution of initial density. The minimum height of pluviation for the first layer of sand is restricted by the inner height of the chamber such that the initial mean relative density was always larger than 85 % (very dense). Two tests

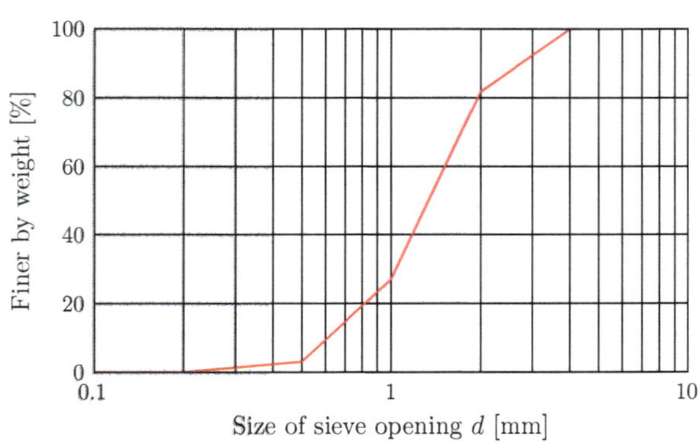

Fig. 7 Grain size distribution curve of the model test sand

Table 1 Details of conducted vibro-injection pile (RI-pile) installation tests

Test ID	Date	Saturation	Grouting material/pressurization (bar)	Frequency (Hz)/load amplitude (kN)[a]
RI-1-D	2011	Air dried	None	20.7/2.56
RI-2-F		Water flooded		20.0/2.39
RI-3-F				20.0/1.90
RI-4-F				
RI-5-B[b]			Bentonite slurry / —[c]	20.0/2.39
RI-6-D	2012	Air dried	None	20.0/2.39
RI-7-B		Water flooded	Bentonite slurry/—[c]	
RI-8-B			Bentonite slurry/1.0	
RI-9-H[b]			Hydraulic binder/3.0	
RI-10-H	2013	Water flooded	Hydraulic binder/3.0	20.0/2.39

D Dry; *F* Flooded; *B* Bentonite; *H* Hydraulic binder
In all tests the initial mean relative density was >85 % (very dense)
[a]With respect to the vibrator; the static force (dead weight) varies between the tests
[b]Analyzed by using particle image velocimetry
[c]Loaded under its own weight

were run in air dried sand, whereas the other were carried out in sand which had been water flooded.

In three tests the pile shaft annulus created by the welded-on collar was injected with pigmented bentonite slurry. In two other experiments we tested a hydraulic binder commonly used in geotechnical engineering for filling, sealing, and solidification.

3.3 Preliminary Test Results

Figure 8 shows a digital photograph of test RI-8-B through the viewing window of the chamber right after the installation of the pile has completed. It can be seen from the figure that the bentonite slurry infiltrated into the pore space of the test sand although the mass concentration of bentonite in the slurry is pretty high. It fails in keeping open the shaft annulus. In contrast to that, the hydraulic binder has a sufficiently high shear strength to stabilize the shaft annulus while possessing excellent flowability during pressurized grouting. Concerning test RI-9-H shown in Fig. 9, this is indicated by the clear soil-grout interface which is almost vertical along the pile shaft. Infiltration of the coarse test sand cannot be completely avoided, that

Fig. 8 Digital photograph of the configuration of test RI-8-B (bentonite slurry injection) through the viewing window of the chamber right after pile installation has completed

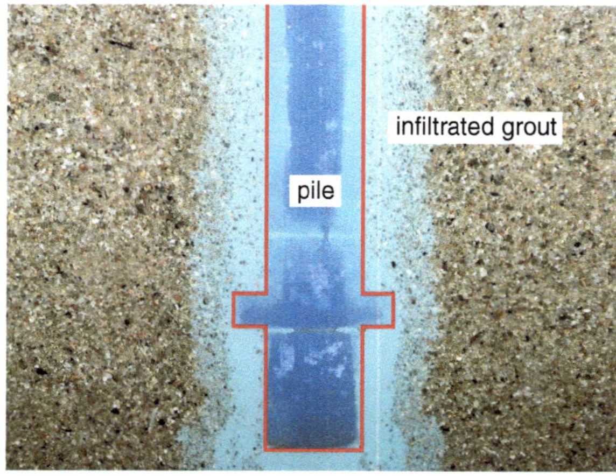

Fig. 9 Digital photograph of the configuration of test RI-9-H (hydraulic binder injection) through the viewing window of the chamber about half an hour after pile installation has completed

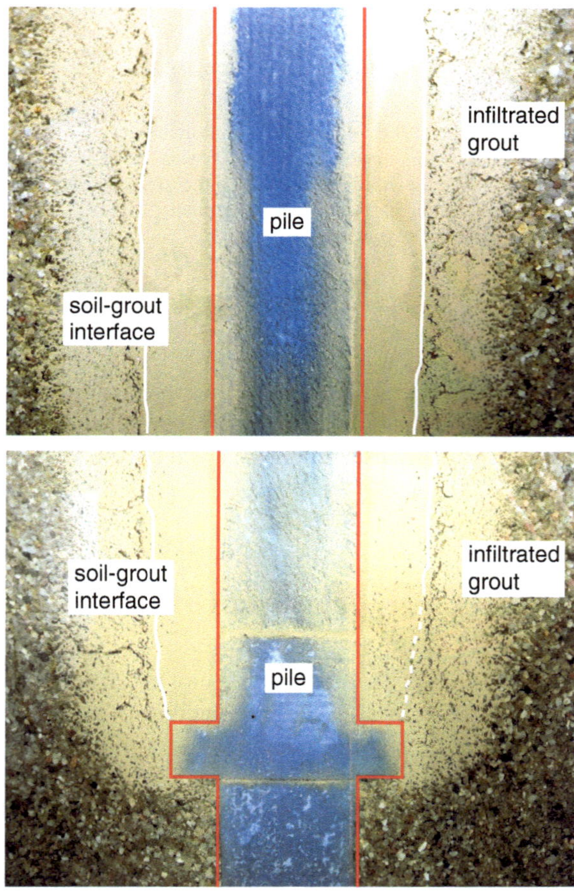

is to say, the model assumption of an impermeable soil-grout interfaces does not appear as a reasonable one. However, it should be noted that the amount of infiltrated hydraulic binder increased with time and that the still images in Fig. 9 were recorded about half an hour after the pile installation has completed.

The results of a PIV analysis of test RI-5-B using bentonite slurry injection is displayed in Fig. 10. Figure 10a shows the time history of the vertical displacement of the pile tip. Those configurations where image capturing took place are marked with black squares. Concerning a time-averaged motion the pile continuously penetrates the soil due to its self-weight. During a vibration cycle, however, the pile moves upward and downward. The displacement increments in the soil which occurred during the downward motion of the pile between image 1 and image 2 are displayed in Fig. 10b using arrows with scaled length. It is clearly visible that the soil is not only displaced below the pile toe and underneath the collar in a predominantly vertical direction but also moves downward above the collar. Figure 10c shows the displacement increments due to upward motion of the pile between image 5 and image 6. Qualitatively the same soil motion can be observed as in Fig. 10b but with reverse signs, indicating that the soil located at the pile shaft is dragged along with the pile motion. The heavings beneath the pile toe and the collar result from the release of the previously compressed soil.

4 Numerical Examples and Validation

The following section presents some numerical examples which belong to a verification and validation (V&V) process controlling and ensuring quality of our new MMALE method. Particular algorithms of the method, like the transport algorithm in the remap step, are verified by running patch tests and basic initial boundary value problems. The validation of the complete MMALE method is enabled through back-analysis of data resulting from the experimental model tests carried out or from example problems found in the literature. Since the numerical implementation of the multi-material option of the method (interface reconstruction, etc.) is still in progress, all the examples shown here consider only a single material per element and time step. Therefore, the computational models used do not utilize all features of MMALE but have to be considered as simplified ALE models.

4.1 Verification of the Transport Algorithm

Example problems towards verification of the advective transport algorithm use a fixed (Eulerian) mesh and prescribe a steady velocity field of the material in conjunction with an initial distribution of a scalar field. The scalar field has no specific physical meaning but represents the quantity being advected through the mesh. In

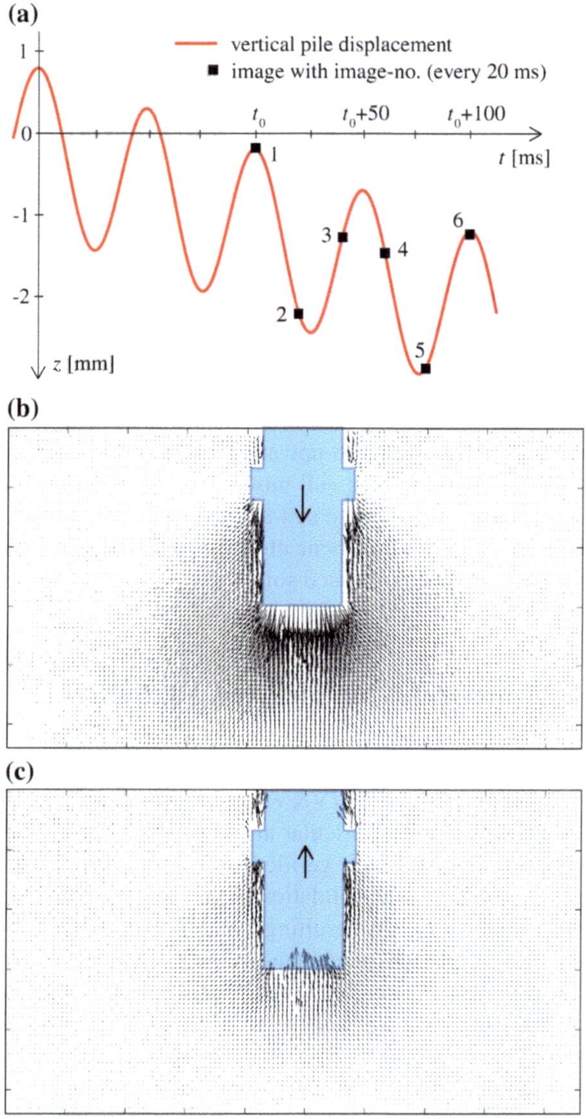

Fig. 10 Results of model test RI-5-B using bentonite slurry injection. **a** Schematic time history of vertical pile displacements. PIV results showing soil displacement increments (**b**) at upward motion of the pile (image 1–image 2), and (**c**) at downward motion of the pile (image 5–image 6)

the present case it can be associated with the solution obtained at the end of the Lagrangian step, e.g. a stress component.

Figure 11a shows the almost uniform unstructured mesh used in the example. The initial signal (Fig. 11b) is a cylindrical pulse implemented as follows. The color function is set to zero in all elements except for those in the circular zone highlighted in Fig. 11a, where the color function is set to 100. For visualization the element-centered values are copied to the nodes and then averaged. In the next step, a steady

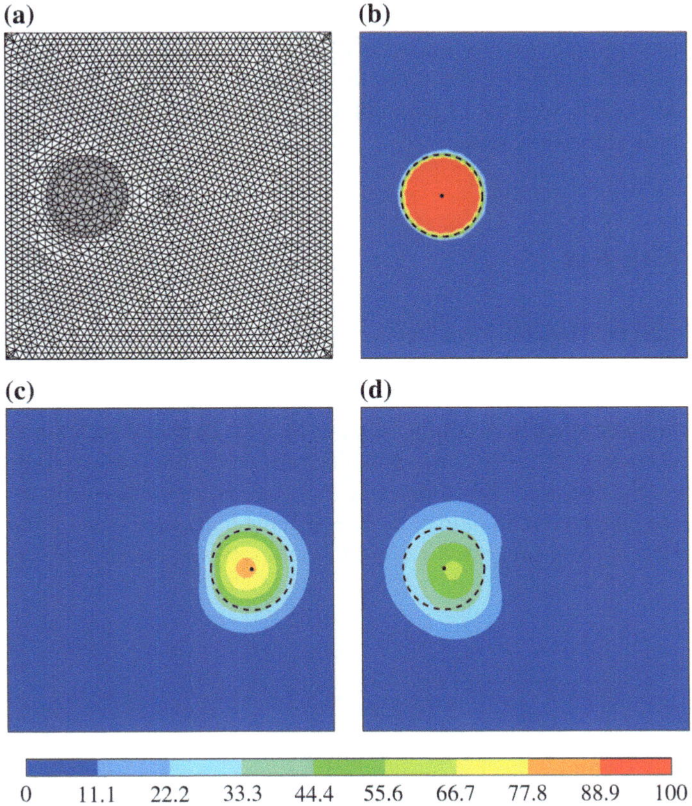

Fig. 11 Rotational advection of a cylindrical pulse by the implemented first-order transport algorithm. The analytical solution is indicated by the *dashed circle*. **a** Mesh, **b** initial scalar distribution, **c** after one half rotation, **d** after a full rotation

velocity field is prescribed in such a way that the material performs a full 360° clockwise rigid rotation about the center of the square domain in 720 advection steps. This means that the analytical solution at the final state and the initial cylindrical pulse are identical. The mesh is kept fixed, meaning that the MMALE method is run in the purely Eulerian mode.

The results of the implemented first-order transport algorithm after one half rotation and after a full rotation are plotted in Fig. 11c, d, respectively. The dashed circle represents the analytical solution. The gradual increase of the area were the color function has values greater than zero indicates numerical diffusion introduced by the finite volume approximation. The plateau erodes and the steep gradients present in the initial signal are getting smeared during the course of rotation. Moreover, the peak moves radially inward, and its maximum value after a full rotation is reduced to only 57 % of its initial value. As already mentioned, numerical diffusion would be less pronounced if a second- or higher-order advective transport algorithm would be used. However, we note that rotation of a cylindrical pulse constitutes an academic extreme example as the solution variables in practical soil mechanical problems often have smaller gradients. Moreover, in the operator-split MMALE method, every transport step is followed by a Lagrangian step that should bring back the solution variables to an admissible state.

4.2 Piercing Test

Piercing is test problem borrowed from the metal forming community which can be employed to validate the MMALE method. In this quasi-static process a billet is held in a heavy walled container and hollowed out by a flat punch (Fig. 12). The example is a plane strain problem and assumes the punch to be rigid and perfectly rough. The vertical and horizontal walls of the container are smooth to ensure sliding contact conditions in the container-billet interface. For the case of plastic-rigid material and a container to punch breadth ratio of $S/B = 0.5$, the maximum penetration pressure, p, in a steady piercing process is related to yield stress in uniaxial tension through

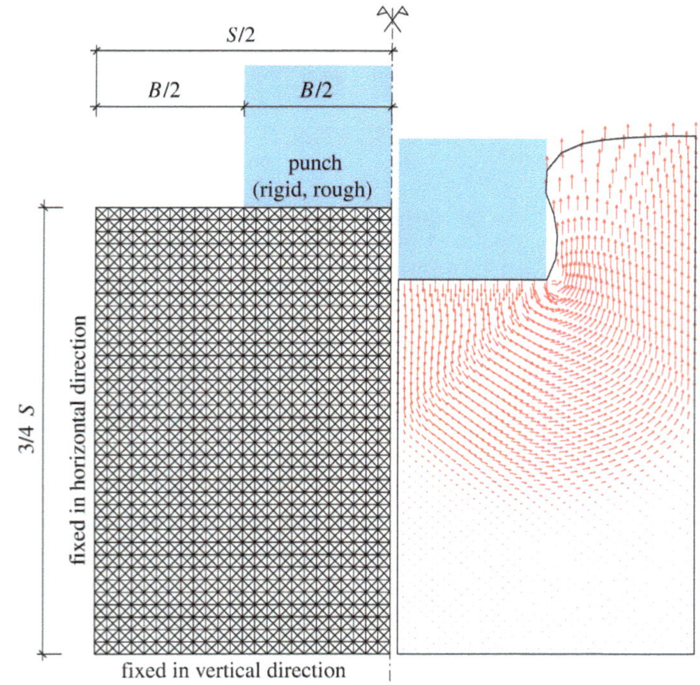

Fig. 12 Simulation of a piercing process. Problem statement and coarse structured mesh (*left*), deformed domain and velocity field at $z/B = 0.25$ (*right*)

$p/\sigma^y = (2+\pi)/\sqrt{3}$ [28]. This is the same relation as for the ultimate bearing capacity of a strip footing on the plane surface of a weightless and frictionless cohesive soil.

The two finite element models used for back-calculation of the analytical solution take advantage of symmetry and differ only in the coarseness of the mesh. The model and the coarse mesh are shown in Fig. 12 left. No contact elements were used. The plastic-rigid material of the billet is modeled by a hypoelasto-J_2-plastic rate constitutive equation (von Mises plasticity). The set of material constants chosen for the simulations consists of Young's modulus $E = 2600$ kPa, Poisson's ratio $\nu = 0.3$, initial yield stress $\sigma^{y0} = 20$ kPa, and plastic modulus $E^p = 0$ kPa (ideal plastic response).

Figure 12 on the right plots the deformed computational domain at a relative penetration depth of $z/B = 0.25$ together with the velocity field on the billet obtained with the coarse mesh. The sudden change in flow direction below the punch is related to a slip line that intersects the axis of symmetry at 45°. A second, less apparent slip line intersects the wall of the container at 45°. The unrealistic deformation of the unconstrained material boundary lateral to the punch (Fig. 12 right) is partly due to its simplified treatment in the ALE method without the multi-material option, but the problem could be resolved in this example by using a finer mesh.

The calculated load-displacement curves and the analytical solution are plotted in Fig. 13. In contrast to the plastic-rigid material behavior assumed in [28], the rate equation governing hypoelastic response in the ALE simulation results in a gradual increase of the punch pressure along with increasing indentation. The relative pressure $p/\sigma^y = 2.97$ of the analytical solution is, however, reached at relative penetration depths of less than $z/B = 0.05$ in all calculations. Beyond that point, the simulated pressure is always larger than the plastic-rigid solution, but the difference decreases with decreasing element size.

Fig. 13 Simulation of a piercing process. Analytical solution $p/\sigma^y = (2+\pi)/\sqrt{3}$ for plastic-rigid material [28] and results of an ALE simulation using the hypoelasto-J_2-plastic model (von Mises plasticity)

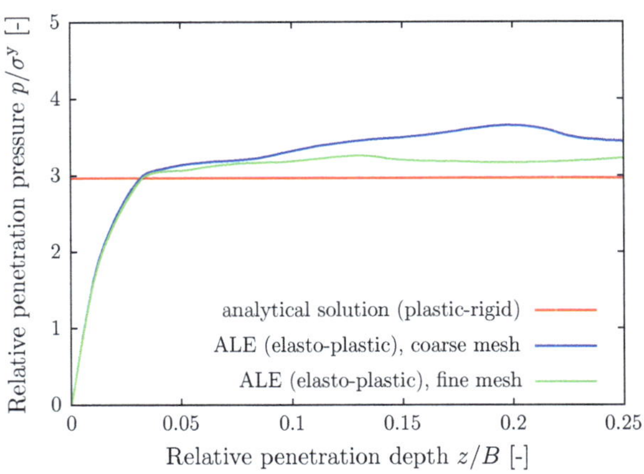

4.3 Pile Penetration into Sand

Until today the realistic simulation of pile penetration into sand is one of the most challenging problems in soil mechanics because of the complex material behavior, large deformations, and contact constraints. Stable and robust simulations are hard to achieve, even when ALE methods are employed. Loading has to be increased very slowly, and adjusting the contact parameters at the beginning of the simulation only is a science of its own.

Recent numerical results of a succeeded ALE simulation of quasi-static pile penetration into sand are shown in Fig. 14. The pile is assumed smooth and rigid, and the initial void ratio of the sand was chosen to $e_0 = 0.678$ (initial relative density $D_{r0} = 0.34$). The initial configuration has a very simple geometry because the penetration was started at the ground surface. Contact elements are attached to the pile and soil surfaces, and a Lagrange multiplier contact algorithm enforces zero penetration of the pile elements when contact is closed. The number of axisymmetric solid elements used for the simulation is approximately 36,000, with the centerline of the pile serving as the axis of radial symmetry. All nodes at the lower boundary of the mesh are fixed in vertical direction, and the nodes of both vertical boundaries are fixed in radial direction. The initial stress state within the sand was prescribed as a K_0-state.

Figure 14 plots the deformed configuration and the predicted void ratio distribution at a relative penetration depth of $z/D = 5.0$, where D is the pile diameter. Recall that void ratio is a material state variable in the formulation of the hypoplastic rate constitutive equation used to model the mechanical behavior of sand. Hence, its

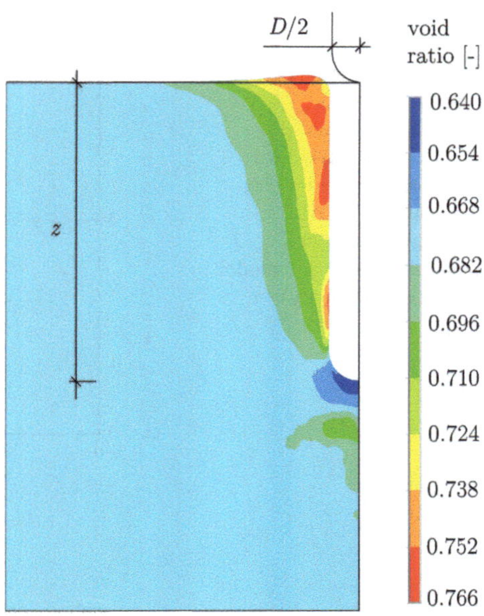

Fig. 14 Quasi-static penetration of a smooth rigid pile into medium dense sand (initial void ratio $e_0 = 0.678$ resp. initial relative density $D_{r0} = 0.34$). Edges of the undeformed configuration, deformed configuration, and predicted void ratio distribution at a relative penetration depth of $z/D = 5.0$

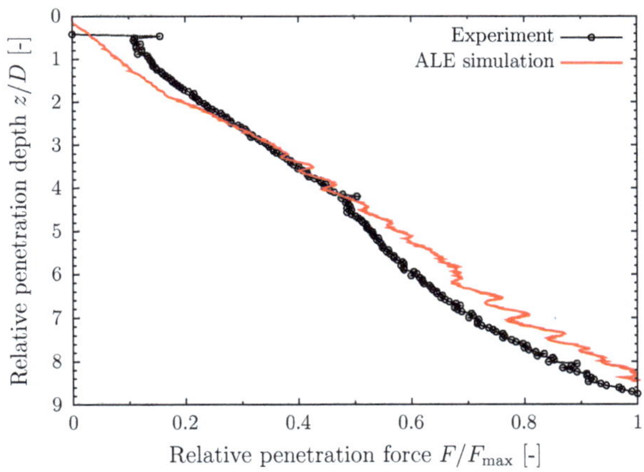

Fig. 15 Quasi-static penetration of a smooth rigid pile into medium dense sand (initial void ratio $e_0 = 0.678$ resp. initial relative density $D_{r0} = 0.34$). Comparison of measured and predicted load-displacement curves

spatial distribution is generally affected by advection during the ALE remap step. The figure indicates a significant densification of the sand immediately underneath the pile base at that stage of penetration. The hemispherical shape and the smooth skin of the pile prevents soil from being trapped in a core zone ahead of the pile base. The sand continuously expands through shearing once the pile base has passed. Consequently, the pile shaft is surrounded by a loosening zone. Since the pile displaces the soil, soil heaving occurs lateral to the pile.

As shown in Fig. 15, the predicted load-displacement curve of the quasi-static pile penetration is in good agreement with experiments [6]. However, the curves considerably differ in relative penetration depths of $z/D < 1$ and $z/D > 5$. The difference at small penetration depths is related to the undesired restraining forces in the experimental set-up not accounted for in the numerical model as well as to the problem of modeling the behavior of sand at very low effective stress levels. In larger depths, the predicted load-displacement curves show a characteristic zig-zag form which stems from the contact interface approximation [53].

5 Conclusions and Outlook

We have developed a multi-material arbitrary Lagrangian-Eulerian (MMALE) method to numerically model and simulate vibro-injection pile installation in water-saturated sand. Because of its applicability on low-velocity problems related to porous media formed by granular material the method is unique on national as well on international level and offers great potential for future research. MMALE describes the motion of the computational mesh independent of the material motion. Material interfaces can flow through the mesh, so that mesh elements may arise which contain two or more materials. Concerning these multi-material elements the two-equation reduced model derived in Part 1 of these papers is applied in order to relate in a mechanically consistent way the element-associated variables with those of the multi-material flow.

The numerical implementation of the method is based on a Lagrange-remap strategy through which each calculational cycle is divided into a Lagrangian step and a remap step. This allows for simpler and more robust algorithms compared to the monolithic problem as well as the implementation into Lagrangian finite element programs. As the remap step contributes most to the overall cost of the calculation, several Lagrangian steps can be performed before the total transport associated with those steps is accounted for in a single remap step.

The Lagrangian step is largely identical to the standard step for advancing solution in time in implicit Lagrangian finite element programs. The mesh follows the material as it deforms, and the rate constitutive equations of the particular materials are integrated over the time step by taking into account the requirement of incremental objectivity. In order to realistically model the behavior of water-saturated sand, especially the tendency to liquefy under cyclic loads, the MMALE method is combined with a highly-developed hypoplastic rate constitutive equation advanced in the Subproject 1 of this DFG Research Unit. Furthermore, in multi-material elements the evolution of the volume fractions must be taken into account because every constituent is generally compressible.

In the remap step the nodes of the element mesh are either moved back to their original positions (Eulerian mode) or relocated such that the overall mesh quality is kept in an acceptable range (ALE mode). The relocation of nodes must be supplemented by a numerical transport of material resp. solution variables relative to the mesh, so as to achieve mechanically consistency. This is implemented by using first-order linear and high-resolution non-linear conservative transport algorithms from the field of computational fluid dynamics. The simple first-order transport algorithm of the Godunov-type currently implemented is, however, sufficient for many applications. In order to precisely determine the amount of lost or gained material volume during the remap step in multi-material elements, interfaces are reconstructed and tracked along with the flow field by using VOF resp. MOF methods.

Parallel to the theoretical and numerical investigations, experimental model tests concerning the installation of vibro-injection piles in sand have been carried out in order to investigate the relevant installation phenomena and to provide data that can be back-calculated using the MMALE method. A test chamber with viewing window together with a special model pile have been designed and manufactured for this purpose. The model pile, like the steel profiles used for vibro-injection piles in practice, allows for the pressurized injection of the pile shaft above a welded-on collar located at the pile toe. During the tests installation process has been digitally filmed through the viewing window, and then the recorded image sequence has been analyzed by using particle image velocimetry. The tests reveal that the grains in a body of sand usually entail adequate image texture for PIV analysis, but not the grouting material unless it would be seeded with marker particles.

The research work reported here is still in progress. Open issues include the completion of the MMALE implementation, the development of MMALE computational models for the simulation of grouting and vibro-injection pile installation in sand, as well as the validation of the computational models through back-calculation of selected experimental model tests.

Acknowledgments The presented work was carried out under the financial support from the German Research Foundation (DFG; grants SA 310/26-1 and SA 310/26-2) as part of the DFG Research Unit FOR 1136, which is gratefully acknowledged. We thank our colleagues in this research unit for several fruitful discussions about our work.

References

1. Adrian, R.J.: Particle-imaging techniques for experimental fluid mechanics. Ann. Rev. Fluid Mech. **23**, 261–304 (1991)
2. Ahn, H.T., Shashkov, M.: Multi-material interface reconstruction on generalized polyhedral meshes. J. Comput. Phys. **226**, 2096–2132 (2007)
3. Arnold, D.N., Brezzi, F., Fortin, M.: A stable finite element for the Stokes equations. Calcolo **21**(4), 337–344 (1984)
4. Askes, H., Bodé, L., Sluys, L.J.: ALE analysis of localization in wave propagation problems. Mech. Cohesive-Frictional Mater. **3**, 105–125 (1998)
5. Aubram, D.: Differential Geometry Applied to Continuum Mechanics. In: Veröffentlichungen des Grundbauinstitutes der Technischen Universität Berlin. vol. 44. Shaker, Aachen (2009)
6. Aubram, D.: An Arbitrary Lagrangian-Eulerian Method for Penetration into Sand at Finite Deformation. In: Veröffentlichungen des Grundbauinstitutes der Technischen Universität Berlin. vol. 62. Shaker, Aachen (2013)
7. Aubram, D.: Optimization-based smoothing algorithm for triangle meshes over arbitrarily shaped domains. (2014). arXiv:1410.5977 [cs.NA]
8. Aubram, D., Rackwitz, F., Savidis, S.A.: An ALE finite element method for cohesionless soil at large strains: computational aspects and applications. In: Benz, T., Nordal, S. (eds.) Proceedings 7th European Conference on Numerical Methods in Geotechnical Engineering (NUMGE), pp. 245–250. CRC Press, Boca Raton (2010)
9. Barth, T., Ohlberger, M.: Finite volume methods: foundation and analysis. In: Stein, E., de Borst, R., Hughes, T.J.R. (eds.) Encyclopedia of Computational Mechanics, vol. 1, Chap. 15. Wiley, Chichester (2004)
10. Belytschko, T., Liu, W.K., Moran, D.: Nonlinear Finite Elements for Continua and Structures. Wiley, New York (2000)
11. Benson, D.J.: Computational methods in Lagrangian and Eulerian hydrocodes. Comput. Meth. Appl. Mech. Eng. **99**, 235–394 (1992)
12. Benson, D.J.: A multi-material Eulerian formulation for the efficient solution of impact and penetration problems. Comput. Mech. **15**, 558–571 (1995)
13. Benson, D.J.: A mixture theory for contact in multi-material Eulerian formulations. Comput. Meth. Appl. Mech. Eng. **140**, 59–86 (1997)
14. Benson, D.J.: An implicit multi-material Eulerian formulation. Int. J. Numer. Meth. Eng. **48**, 475–499 (2000)
15. Benson, D.J.: Volume of fluid interface reconstruction methods for multi-material problems. Appl. Mech. Rev. **55**(2), 151–165 (2002)
16. Beyer, O.: Erweiterung und Optimierung eines Modellversuchsstandes für die Simulation der Rüttelinjektionspfahlherstellung. Bachelor's thesis, Technische Universität Berlin, Germany (2012) (unpublished; in German)
17. Büchler, M.: Qualitative Untersuchung bodenmechanischer Phänomene während der Herstellung von Rüttelinjektionspfählen mittels Particle Image Velocimetry. Bachelor's thesis, Technische Universität Berlin, Germany (2013) (unpublished; in German)

18. Carow, Ch.: Das MINI-Element für Fluid-Kornstruktur-gekoppelte Problemstellungen. Master's Thesis, Technische Universität Berlin, Germany (2012) (unpublished; in German)
19. Colella, P., Glaz, H.M., Ferguson, R.E.: Multifluid algorithms for Eulerian finite difference methods (1997) (unpublished manuscript)
20. DeBar, R.B.: Fundamentals of the KRAKEN Code. Technical Report UCID-17366, Lawrence Livermore Laboratory, Livermore, USA (1974)
21. DIN EN ISO 14688–1. Geotechnische Erkundung und Untersuchung—Benennung, Beschreibung und Klassifizierung von Boden—Teil 1: Benennung und Beschreibung. Beuth Verlag, Berlin (2003) (German Code)
22. Dyadechko, V., Shashkov, M.: Moment-of-fluid interface reconstruction. Technical report LA-UR-05-7571, Los Alamos National Laboratory, New Mexico, USA (2005)
23. Dyadechko, V., Shashkov, M.: Reconstruction of multi-material interfaces from moment data. J. Comput. Phys. **227**, 5361–5384 (2008)
24. Freßmann, D., Wriggers, P.: Advection approaches for single-and multi-material arbitrary Lagrangian-Eulerian finite element procedures. Comput. Mech. **39**, 153–190 (2007)
25. Glimm, J., Isaacson, E., Marchesin, D., McBryan, O.: Front tracking for hyperbolic systems. Computat. Mech. **2**, 91–119 (1981)
26. Glimm, J., Grove, J.W., Li, X.L., Shyue, K.-M., Zeng, Y., Zhang, Q.: Three-dimensional front tracking. SIAM J. Sci. Comput. **19**(3), 703–727 (1998)
27. Harten, A.: High resolution schemes for hyperbolic conservation laws. J. Comput. Phys. **49**(3), 357–393 (1983)
28. Hill, R.: The Mathematical Theory of Plasticity. Oxford University Press, New York (1998) (Reprint; first published 1950)
29. Hirt, C.W., Nichols, B.D.: Volume of fluid (VOF) method for the dynamics of free boundaries. J. Comput. Phys. **39**, 201–225 (1981)
30. Hirt, C.W., Amsden, A.A., Cook, J.L.: An arbitrary Lagrangian-Eulerian computing method for all flow speeds. J. Comput. Phys. **14**, 227–253 (1974)
31. Hughes, T.J.R., Winget, J.: Finite rotation effects in numerical integration of rate constitutive equations arising in large-deformation analysis. Int. J. Numer. Methods Eng. **15**(12), 1862–1867 (1980)
32. Hughes, T.J.R.: Numerical implementation of constitutive models: rate-independent deviatoric plasticity. In: Nemat-Nasser, S., Asaro, R.J., Hegemier, G.A. (eds.) Theoretical Foundation for Large-Scale Computations for Nonlinear Material Behavior, pp. 29–63. Martinus Nijhoff Publishers, Dordrecht (1984)
33. Hyman, J.M.: Numerical methods for tracking interfaces. Physica D **12**, 396–407 (1984)
34. Kirchner, J.: Bestimmung der Schnittpolygone für die Interface-Rekonstruktion in unstrukturierten Dreiecksnetzen. Seminar Paper, Technische Universität Berlin, Germany (2014) (unpublished; in German)
35. LeVeque, R.J.: Finite volume methods for hyperbolic problems, 3rd edn. Cambridge University Press, Cambridge (2002)
36. Lewis, R.W., Schrefler, B.A.: The Finite Element Method in the Static and Dynamic Deformation and Consolidation of Porous Media, 2nd edn. Wiley, Chichester (1998)
37. Luttwak, G., Rabie, R.L.: The multi material arbitrary Lagrangian Eulerian code MMALE and its application to some problems of penetration and impact. Technical report LA-UR-85-2311, Los Alamos National Laboratory, Los Alamos, New Mexico, USA (1985)
38. McGlaun, J.M., Thompson, S.L.: CTH: a three-dimensional shock wave physics code. Int. J. Impact Eng. **10**, 351–360 (1990)
39. Miller, G.H., Puckett, E.G.: A high-order Godunov method for multiple condensed phases. J. Comput. Phys. **128**, 134–164 (1996)
40. Niemunis, A., Herle, I.: Hypoplastic model for cohesionless soils with elastic strain range. Mech. Cohesive-Frictional Mater. **2**, 279–299 (1997)
41. Noack, D.: Anwendung der digitalen Bildkorrelation auf Modellversuche zur Herstellung von Rüttelinjektionspfählen. Diploma thesis, Technische Universität Berlin, Germany (2011) (unpublished; in German)

42. Osher, S., Sethian, J.A.: Fronts propagating with curvature-dependent speed: algorithms based on Hamilton-Jacobi formulations. J. Comput. Phys. **79**, 12–49 (1988)
43. Osher, S., Fedkiw, R.P.: Level set methods: an overview and some recent results. J. Comput. Phys. **169**, 463–502 (2001)
44. Peery, J.S., Carroll, D.E.: Multi-material ALE methods in unstructured grids. Comput. Meth. Appl. Mech. Eng. **187**, 591–619 (2000)
45. Pilliod, J.E., Puckett, E.G.: Second-order accurate volume-of-fluid algorithms for tracking material interfaces. J. Comput. Phys. **199**, 465–502 (2004)
46. Rider, W.J., Kothe, D.B.: Reconstructing volume tracking. J. Comput. Phys. **141**, 112–152 (1998)
47. Rodríguez-Ferran, A., Casadei, F., Huerta, A.: ALE stress update for transient and quasistatic processes. Int. J. Numer. Meth. Eng. **43**, 241–262 (1998)
48. Rudman, M.: Volume-tracking methods for interfacial flow calculations. Int. J. Numer. Meth. Fluids **24**(7), 671–691 (1997)
49. Savidis, S.A., Aubram, D., Rackwitz, F.: Arbitrary Lagrangian-Eulerian finite element formulation for geotechnical construction processes. J. Theor. Appl. Mech. **38**(1–2), 165–194 (2008)
50. Scardovelli, R., Zaleski, S.: Direct numerical simulation of free-surface and interfacial flow. Ann. Rev. Fluid Mech. **31**, 567–603 (1999)
51. Schulz, T.: Konzeption und Realisierung eines kleinmaßstäblichen Modellversuchsstandes für die Herstellung von Rüttelinjektionspfählen. Diploma thesis, Technische Universität Berlin, Germany (2011) (unpublished; in German)
52. Sethian, J.A.: Level Set Methods: Evolving Interfaces in Geometry, Fluid Mechanics, Computer Vision and Material Science. Cambridge University Press, Cambridge (1996)
53. Simo, J.C., Meschke, G.: A new class of algorithms for classical plasticity extended to finite strains. Application to geomaterials. Comput. Mech. **11**, 253–278 (1993)
54. Truesdell, C., Noll, W.: In: Antman, S.S. (ed.) The Non-linear Field Theories of Mechanics, 3rd edn. Springer, Berlin (2004)
55. Tryggvason, G., Bunner, B., Esmaeeli, A., Juric, D., Al-Rawahi, N., Tauber, W., Han, J., Nas, S., Janz, Y.-J.: A front-tracking method for the computations of multiphase flow. J. Comput. Phys. **169**, 708–759 (2001)
56. Vitali, E., Benson, D.J.: An extended finite element formulation for contact in multi-material arbitrary Lagrangian-Eulerian calculations. Int. J. Numer. Meth. Eng. **67**, 1420–1444 (2006)
57. von Wolffersdorff, P.-A.: A hypoplastic relation for granular materials with a predefined limit state surface. Mech. Cohesive-Frictional Mater. **1**, 251–271 (1996)
58. White, D.J., Take, W.A.: GeoPIV: Particle image velocimetry (PIV) software for use in geotechnical testing. Technical report CUED/D-SOILS/TR322, Geotechnical and Environmental Research Group, University of Cambridge, UK (2002)
59. White, D.J., Take, W.A., Bolton, M.D.: Soil deformation measurement using particle image velocimetry (PIV) and photogrammetry. Géotechnique **53**(7), 619–631 (2003)
60. Wriggers, P.: Nonlinear Finite Element Methods. Springer, Berlin (2008)
61. Youngs, D.L.: Time-dependent multi-material flow with large fluid distortion. In: Morton, K.W., Baines, M.J. (eds.) Numerical Methods for Fluid Dynamics, pp. 273–285. Academic Press, London (1982)
62. Zienkiewicz, O.C., Chan, A.H.C., Pastor, M., Schrefler, B.A., Shiomi, T.: Computational Geomechanics: With Special Reference to Earthquake Engineering. Wiley, Chichester (1999)
63. Zienkiewicz, O.C., Taylor, R.L.: The Finite Element Method, vol. 3, 5th edn. Butterworth-Heinemann, Oxford (2000)

Numerical Modelling of the Effective-Stress Evolution in Saturated Soil Around a Vibrating Pile Toe

V.A. Osinov

Abstract The paper presents results of the numerical modelling of the effective-stress evolution in saturated granular soil around the toe of a vertically vibrating pile. The problem is solved in a spherically symmetric formulation using two different types of constitutive models. An incremental hypoplasticity model is used to calculate the stress state after a limited number of cycle at the beginning of the vibration. Further changes in stresses for a large number of cycles are calculated with an explicit cyclic model. The influence of soil permeability and relative density is investigated. It is shown that the cyclic soil deformation results in the reduction of the effective stress around the pile in spite of the pore pressure dissipation in the case of high soil permeability.

Keywords Saturated soil · Pile · Vibration · Liquefaction

1 Introduction

Vibratory pile driving produces cyclic deformation in the surrounding soil and thus brings about permanent changes in the soil stresses and density. These changes may influence neighbouring piles and structures, the bearing capacity of the driven pile and the pile installation process. Although the modelling of pile installation processes is not a novel issue, a survey of the literature reveals a lack of reliable numerical investigations of the soil deformation around a vibratory driven pile. This issue is complicated not only because of numerical difficulties (mesh distortion resulting from large deformations, contact conditions at the pile-soil interface, thousands of loading cycles to be simulated), but also because of stringent requirements to be satisfied by the constitutive model used. On the one hand, the model has to correctly describe cyclic soil deformation with large strain amplitudes (up to several per cent) in the immediate vicinity of the pile, where the constitutive soil behaviour determines

V.A. Osinov (✉)
Institute of Soil Mechanics and Rock Mechanics, Karlsruhe Institute of Technology, Karlsruhe, Germany

Th. Triantafyllidis (ed.), *Holistic Simulation of Geotechnical Installation Processes*, Lecture Notes in Applied and Computational Mechanics 77,
DOI 10.1007/978-3-319-18170-7_7

the soil reaction and the pile penetration rate. On the other hand, the model has to correctly reproduce weak accumulation effects such as compaction under drained conditions and effective-stress reduction under undrained conditions at smaller strain amplitudes (below 10^{-3}). The weak accumulation effects become important for a large number of cycles and are responsible for permanent changes in the soil stresses and density at distances of up to few metres around the pile.

Cyclic soil deformation can be simulated with the use of either conventional plasticity models of incremental type (usually written in rate form) or so-called explicit cyclic models (e.g. [1]). In the latter case, stresses and strains are represented as the sum of an average value and a small oscillating part. The model establishes a rate-type relation between the average values of stresses and strains and also involves the current strain amplitude. The accumulation rates are defined with respect to the number of cycles.

In regard to the pile vibration problem, both types of models—incremental and explicit cyclic—have certain limitations. A drawback of incremental models is the impossibility of calibration for the weak accumulation effects. There is no definite algorithm of how to choose the constitutive parameters in such a way as to obtain the correct accumulation effects and, at the same time, to keep the correct behaviour for large deformation amplitudes. The difficulty of calibration concerns, in particular, the strong dependence of the accumulation effects on the relative soil density for a given strain amplitude. Another drawback of incremental models is a possible loss of accuracy in the calculation of stresses or deformations due to the accumulation of numerical errors during incremental integration. This happens when the numerical error is comparable with the residual stress or deformation accumulated in one cycle. Moreover, incremental models entail high computational costs because of the necessity to integrate each cycle incrementally. This drawback may be important for parametric studies with a large number of calculations.

In view of the limitations of incremental models, explicit cyclic models may be more reliable and convenient for the calculation of a large number of cycles with small strain amplitudes. Such models allow us to calculate tens of thousands of cycles in a reasonable computing time and thus to cover the whole pile installation process. Since the constitutive parameters control the accumulation rates rather than incremental stiffness, it is easier to take into account the influence of various factors (strain amplitude, relative density, principal-stress ratio) on the accumulation effects and to calibrate the model. The main limitation of explicit cyclic models is that they are inapplicable to large strain amplitudes over 10^{-3}. For this reason they cannot be applied to the immediate vicinity of a pile where strain amplitudes are large.

In the present study, in order to better understand how the effective stresses in the soil change during vibratory pile driving, we perform the numerical modelling of the deformation of saturated granular soil around a pile toe caused by the vertical vibration of the pile (without penetration). Following an approach proposed in [2, 3], the problem is solved in two steps using both types of constitutive models mentioned above, Fig. 1. In the first step, which will be subsequently referred to as low-cycle problem, the boundary-value problem is solved for a limited number of cycles (<100) with an incremental plasticity model. The boundary of the computational domain

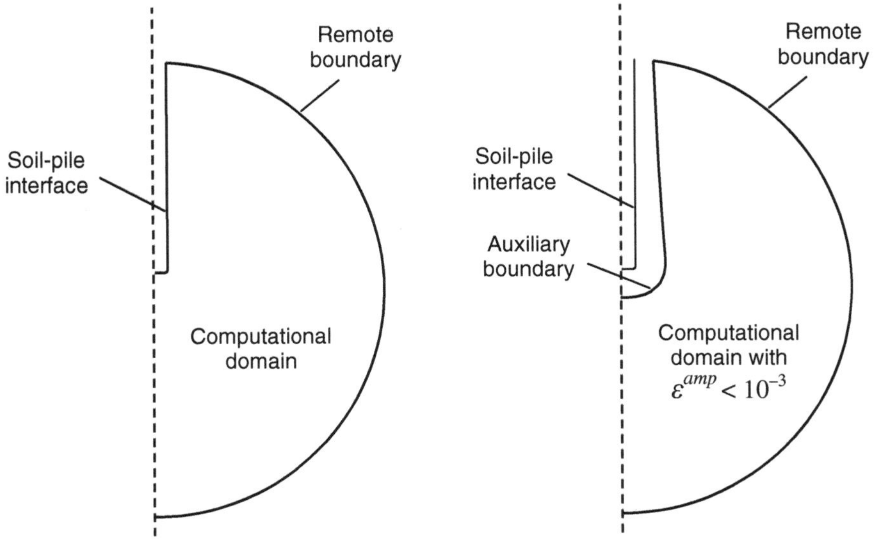

Fig. 1 Low-cycle problem (*left*) and high-cycle problem (*right*)

consists of the soil-pile interface with appropriate boundary conditions, and a remote boundary with non-reflecting boundary conditions for outgoing waves. In the second step, which will be referred to as high-cycle problem, the boundary-value problem is solved for a large number of cycles with an explicit cyclic model and an auxiliary boundary introduced at a certain distance from the pile. The auxiliary boundary must be chosen in such a way that the strain amplitude in the computational domain does not exceed 10^{-3} and the explicit model can therefore be used. The main question in the formulation of the high-cycle problem is how to quantify boundary conditions at the auxiliary surface, as they cannot be directly deduced from the physical problem under study. This issue is discussed in Sect. 3.

The low-cycle problem for a vibrating pile in saturated sand was investigated in [3] under the assumption of locally undrained conditions with the use of a hypoplastic constitutive model. Figure 2 shows the distribution of the mean effective stress around the toe of a cylindrical closed-ended pile after 30 cycles of vibration (compressive stresses are negative). The darkest area in the figure can be considered as a liquefaction zone. The mean effective stress in this zone changes only slightly during a cycle and does not exceed 2 % of the initial effective stress. The permanent liquefaction zone develops around the pile after several cycles of vibration and then grows slowly with the increasing number of cycles. The further evolution of the liquefaction zone for a large number of cycles was studied numerically in [2] with an explicit cyclic model. The problem was solved with locally undrained conditions under the simplifying assumption of spherical symmetry. As seen in Fig. 2, a spherically symmetric problem can serve as an approximation for the lower part of the domain around the pile toe.

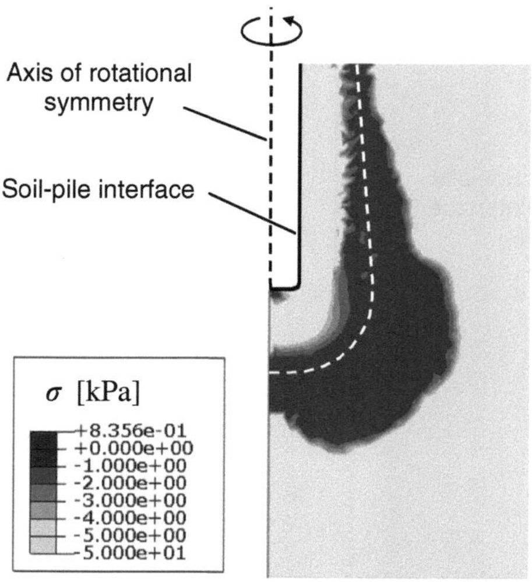

Fig. 2 Mean effective stress in saturated sand around a pile after 30 cycles of vibration calculated with locally undrained conditions for a cylindrical pile with a diameter of 30 cm, a pile displacement amplitude of 2 mm, a hydrostatic initial effective stress of −50 kPa and a frequency of 34 Hz [3]

The constitutive models used in [2, 3] correspond to granular soil whose permeability is rather high as compared, for instance, with that of clay. The high pore pressure gradients obtained in the solutions suggest that the pore pressure would quickly dissipate in a real granular soil. This could presumably lead to soil densification rather than the effective-stress reduction. The question remained open as to whether taking the real soil permeability into account would give a similar effective-stress distribution or would change the solution substantially. In the present paper, both the low-cycle and the high-cycle problems for a vibrating pile are solved in a spherically symmetric formulation without the assumption of locally undrained conditions. Emphasis is placed on the influence of soil permeability and relative density on the effective stresses around the pile toe.

2 Low-Cycle Problem

The spherically symmetric low-cycle problem is formulated for a computation domain bounded by an inner radius R_P and an outer radius R_B, Fig. 3 (left). The inner radius is taken to be equal to the pile radius, whereas the outer radius represents a remote boundary. The equations of motion for a two-phase medium [4] in the spherical coordinates (r, φ, θ) reduce to

$$\frac{\partial \sigma_r}{\partial r} + (n-1)\frac{\partial p_f}{\partial r} + \frac{2}{r}\left(\sigma_r - \sigma_\varphi\right) + \frac{1}{k}\,\varrho_f g n^2 (v_f - v_s) = (1-n)\varrho_s \dot{v}_s, \quad (1)$$

$$-\frac{\partial p_f}{\partial r} - \frac{1}{k}\,\varrho_f g n (v_f - v_s) = \varrho_f \dot{v}_f, \quad (2)$$

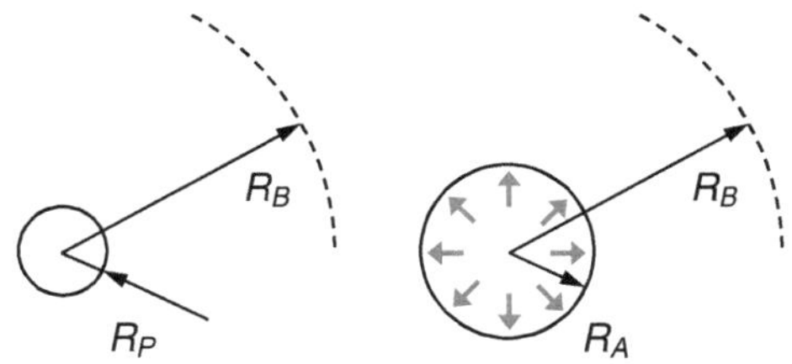

Fig. 3 Spherical low-cycle problem (*left*) and high-cycle problem (*right*)

where σ_r, σ_φ are the effective-stress components (negative for compression), p_f is the fluid pressure (positive for compression), v_s, v_f are the radial velocities of the skeleton and the fluid, the dot stands for the material time derivative, ϱ_s, ϱ_f are the densities of the solid and fluid phases, n is the porosity, k [m/s] is the soil permeability, and g is the acceleration due to gravity.

The constitutive equation for the fluid phase is (neglecting $\partial n/\partial r$)

$$\dot{p}_f = -K_f\left(\frac{\partial v_f}{\partial r} + 2\,\frac{v_f}{r}\right) - K_f\left(\frac{1-n}{n}\right)\left(\frac{\partial v_s}{\partial r} + 2\,\frac{v_s}{r}\right), \tag{3}$$

where K_f is the compression modulus of the pore fluid.

The constitutive behaviour of the skeleton is described within the framework of the hypoplasticity theory. As is well-known, the versions of hypoplastic relations which contain only stresses and density as state variables (e.g. [5]) are unable to correctly describe cyclic deformation. An extended version with so-called intergranular strain as an additional state variable [6] exhibits much more adequacy in the description of the cyclic deformation of granular soil. This version was extensively used for the numerical modelling of dynamic processes in soil with multi-cycle loading (see e.g. references in [3]). Numerical calculations performed in this study with the intergranular-strain extension of hypoplasticity have revealed a drawback of the model which attracted no attention until now—either because this drawback did not manifest itself or because it was overlooked. Hypoplasticity with intergranular strain [6] reduces to hypoelasticity (linearly elastic behaviour with stress-dependent stiffness) when the intergranular-strain tensor vanishes. This occurs under small-amplitude cyclic deformation and also holds true in the presence of additional monotonic deformation superimposed on the cyclic deformation. If the monotonic part of the deformation has a deviatoric component, the hypoelastic constitutive response may produce an unrealistically high principal-stress ratio or even a tensile principal stress. The eventual stress ratio depends not only on the number of cycles but also on the additional intergranular-strain parameters in the extended model. A steadily increasing principal-stress ratio was observed in the numerical solutions to the spherically symmetric problem with finite soil permeability.

The unlimited growth of the principal-stress ratio indicates that hypoplasticity needs further revision. For the time being, rather than resort to another constitutive theory, we propose a simple modification within the existing framework of hypoplasticity by writing the constitutive relation in the form

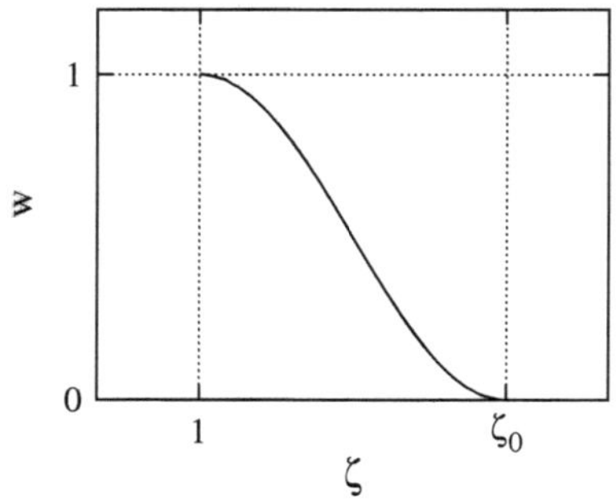

Fig. 4 Function $w(\zeta)$

Table 1 Constitutive parameters of Karlsruhe sand [7]

$\varphi_c(°)$	h_s (MPa)	e_{c0}	e_{d0}	e_{i0}	α	β	n
30	5800	0.84	0.53	1.0	0.13	1.0	0.28

Table 2 Additional constitutive parameters for intergranular strain [8]

R	m_R	m_T	β_r	χ
4×10^{-5}	5.0	5.0	0.05	1.5

$$\dot{\boldsymbol{\sigma}} = w\,\mathbf{F}\,(\dot{\boldsymbol{\varepsilon}}) + (1 - w)\,\mathbf{H}\,(\dot{\boldsymbol{\varepsilon}}), \tag{4}$$

where the tensor-valued functions **F** and **H** represent the stress rates obtained, respectively, with the extended intergranular-strain model [6] and the original hypoplastic model [5], and w is a weight function which depends on the current stress ratio. For the spherical symmetry considered here, we define a stress ratio ζ as the maximum of σ_r/σ_φ, σ_φ/σ_r. The function $w(\zeta)$ used in the present study is shown in Fig. 4, where we put $\zeta_0 = 4$. The constitutive equation (4) corresponds to the intergranular-strain version of hypoplasticity if the stress state is close to a hydrostatic state, and turns into the original version of hypoplasticity without intergranular strain if $\zeta \geq \zeta_0$. The constitutive parameters used in the calculations are given in Tables 1 and 2.

The inner boundary R_P, Fig. 3, is associated with the soil-pile interface and is assumed to be impermeable with the prescribed displacement

$$u_s(R_P, t) = 0.5\,u_{amp}\,(1 - \cos\omega t)\,, \tag{5}$$

where ω is angular frequency, and u_{amp} is taken to be equal to the displacement amplitude of the pile.

The outer boundary R_B should be transparent for outgoing waves. A non-reflecting boundary condition for spherical waves in elastic medium is known [9]. It can be directly used for a two-phase medium with zero permeability substituting $\sigma_r - p_f$ for the radial stress:

$$\dot{\sigma}_r - \dot{p}_f + \frac{c}{r}\left(\sigma_r - p_f\right) + c\varrho\dot{v} + \frac{4\mu}{r}v + \frac{4\mu c}{r^2}u = 0, \tag{6}$$

where λ, μ are the Lamé constants of the skeleton, ϱ is the soil density, and

$$c = \sqrt{\frac{1}{\varrho}\left(\lambda + 2\mu + \frac{K_f}{n}\right)}. \tag{7}$$

A two-phase medium with nonzero permeability requires two boundary conditions at R_B. Considering outgoing spherical waves in a one-phase medium and treating σ_r and p_f as independent stress components with their own constitutive equations, we can obtain two conditions instead of (6):

$$\dot{\sigma}_r + \frac{c}{r}\sigma_r + \frac{\lambda + 2\mu}{c}\dot{v} + \frac{4\mu}{r}v + \frac{4\mu c}{r^2}u = 0, \tag{8}$$

$$-\dot{p}_f - \frac{c}{r}p_f + \frac{K_f}{nc}\dot{v} = 0. \tag{9}$$

Equations (8), (9) can be adapted for the case of nonzero permeability by putting $v = v_s$, $u = u_s$ in (8) and $v = v_f$ in (9). Note that the non-reflecting boundary condition (6) for zero permeability is exact, whereas the non-reflecting boundary conditions (8), (9) for nonzero permeability are approximate. The elastic constants λ, μ can be calculated from the small-strain stiffness of the hypoplastic medium in the far field.

The low-cycle problem was solved with the finite-difference method. Parameters which were not varied in the numerical calculations are given in Table 3. Figure 5 (left) shows an example with zero permeability. Shown in the figure are the maximum and minimum values of the mean effective stress within one cycle after 100 cycles of vibration. Similar to the two-dimensional low-cycle solution in Fig. 2, a liquefaction zone with vanishing effective stresses forms at a certain distance from the inner boundary. Figure 5 (right) shows the solution to the same problem with finite permeability. The distribution of the mean effective stress is similar, although the liquefaction zone is not as wide as in the previous case. It can be therefore concluded that finite permeability does not prevent the effective-stress reduction and the formation of a liquefaction zone in the low-cycle problem.

Table 3 Parameters of the low-cycle problem

R_P (m)	R_B (m)	ϱ_s (kg/m^3)	ϱ_f (kg/m^3)	K_f (GPa)	$\omega/2\pi$ (Hz)
0.15	5.0	2650	1000	2.2	34

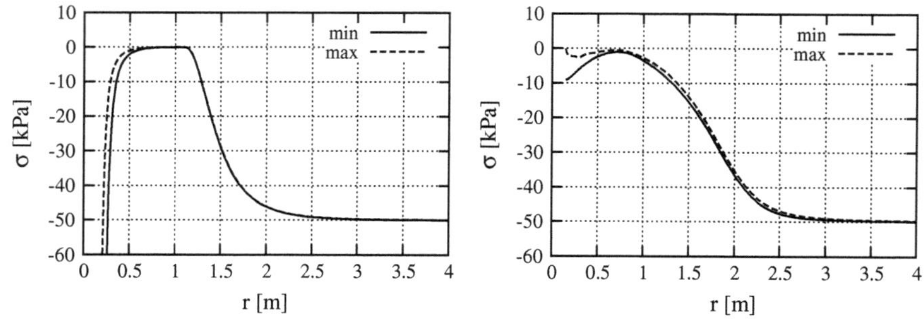

Fig. 5 Mean effective stress variation after 100 cycles in the spherical low-cycle problem for dense sand with $u_{amp} = 4\,\text{mm}$. *Left* $k = 0$, *right* $k = 10^{-3}$ m/s

3 Auxiliary Boundary

The solution of the high-cycle problem requires an auxiliary boundary with appropriate boundary conditions. The main difficulty in the formulation of the high-cycle problem is that we cannot immediately obtain boundary conditions on the auxiliary boundary from the parameters of the physical problem, e.g. from the pile vibration amplitude, because the auxiliary boundary is to be located in the soil. A way to quantify boundary conditions is to use the low-cycle solution obtained with the incremental model. In principle, we could take any surface in the soil as an auxiliary boundary if we know from the low-cycle solution that the strain amplitude in the computational domain does not exceed the threshold of the cyclic model (10^{-3}). However, as the low-cycle solutions show, the stresses and displacements in the soil are in general rather complicated functions of time. The question is how to find an auxiliary boundary with simple and yet correct boundary conditions.

Consider the spherical low-cycle problem from Sect. 2 with $k = 10^{-3}$ m/s. Figure 6 shows the displacements of the solid and fluid phases as functions of time at a radius of 0.69 m. It is seen that, after a certain time, there is no relative motion between the solid and fluid phases, and the average displacement is zero. Assuming that this holds true for later times as well and taking this radius as an auxiliary boundary, we obtain simple boundary conditions, namely impermeability and zero average displacement. An additional boundary condition required for the high-cycle problem is a stress or displacement amplitude obtained from the low-cycle problem.

The radius which corresponds to impermeability and zero average displacement in the low-cycle solution depends on the parameters of the problem. Therefore, the location of the auxiliary boundary with these boundary conditions also depends on the parameters. Figure 7 shows the radius of the auxiliary boundary, denoted by R_A, as a function of the void ratio, the permeability and the displacement amplitude at the inner boundary R_P. The radius R_A increases with increasing amplitude and turns out to be independent of the void ratio.

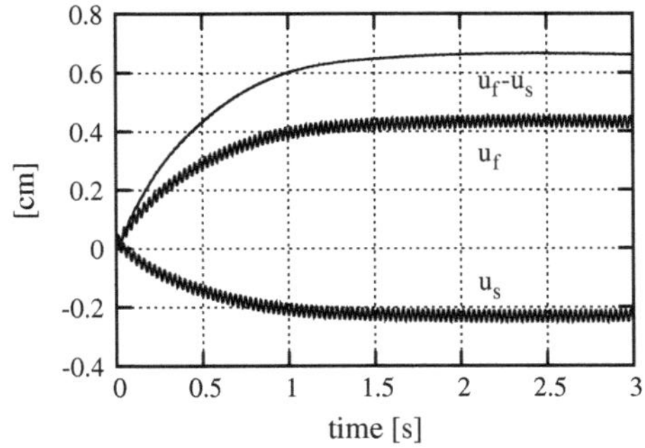

Fig. 6 Displacements of the solid and fluid phases as functions of time at $r = 0.69\,\text{m}$

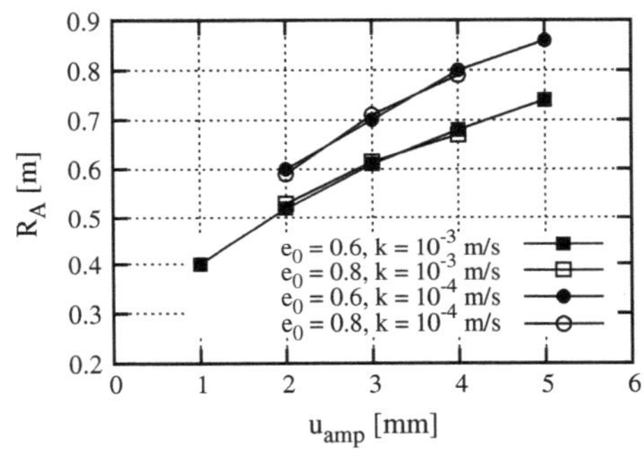

Fig. 7 Radius of the auxiliary boundary, R_A, as a function of the displacement amplitude, u_{amp}, for different values of the initial void ratio, e_0, and permeability, k

4 High-Cycle Problem

The spherical problem for a large number of cycles is posed for the domain between R_A (auxiliary boundary) and R_B (remote boundary), Fig. 3 (right). The problem is solved with the high-cycle accumulation model elaborated in [1, 10]. Displacements, effective stresses and pore pressure represent average values in the sense of the high-cycle model, i.e. values averaged over a cycle. The stress field is quasi-static and must satisfy static equilibrium. The initial distributions of the effective stresses and pore pressure should approximate the eventual values in the low-cycle problem averaged over a cycle.

The determination of stresses and deformations consists in the concurrent solution of two boundary value problems called the first and the second boundary value problems as shown in Fig. 8. Given spatial distributions of the effective stresses and pore pressure at time t, an integration step over a time increment Δt begins with the determination of the current strain amplitude ε^{amp} from the first boundary value problem formulated as a steady-state problem for an elastic medium with time-harmonic functions. The small-strain stiffness of the skeleton is assumed to be a function of the effective stress with the Lamé constants

$$\lambda = \lambda_0 \left(\frac{\sigma}{\sigma_0}\right)^m, \quad \mu = \mu_0 \left(\frac{\sigma}{\sigma_0}\right)^m \tag{10}$$

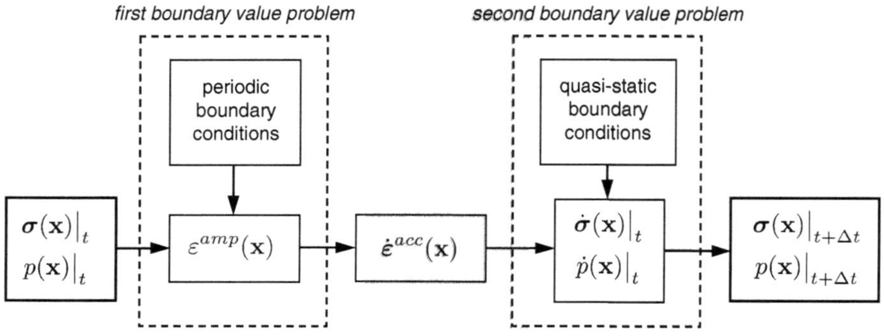

Fig. 8 Solution scheme of the high-cycle problem

(not to be confused with the Lamé constants in Sect. 2), where σ is the mean effective stress, and $\lambda_0, \mu_0, \sigma_0, m$ are parameters. Numerical calculations were performed with $\lambda_0 = 120$ MPa, $\mu_0 = 80$ MPa, $\sigma_0 = -100$ kPa, $m = 0.6$.

According to the choice of the auxiliary boundary R_A as described in Sect. 3, this boundary is considered to be impermeable. The second condition prescribed at this boundary is either a constant displacement amplitude (displacement-controlled case) or a constant radial-stress amplitude (stress-controlled case). These two cases give different solutions and are considered for comparison purposes. The amplitudes are taken from the solution to the low-cycle problem at $r = R_A$. The remote boundary R_B is made transparent for outgoing waves by prescribing non-reflecting boundary conditions.

The spherical steady-state problem with locally undrained conditions was solved in [2]. For the purposes of the present study, the numerical algorithm has been modified to take finite permeability into account. Comparison of the solutions in the drained and undrained cases has shown that the difference in strain amplitudes is insignificant up to permeability of 10^{-2} m/s. In the numerical examples presented below, the first boundary value problem is solved with locally undrained conditions.

The strain amplitude required for the high-cycle model is calculated in the spherically symmetric case as

$$\varepsilon^{amp} = \sqrt{\left(\varepsilon_r^{amp}\right)^2 + 2\left(\varepsilon_\varphi^{amp}\right)^2}, \tag{11}$$

where ε_r^{amp}, $\varepsilon_\varphi^{amp}$ are the radial and circumferential strain amplitudes. The strain amplitude determines a strain accumulation rate $\dot{\varepsilon}^{acc}$ through the tensorial relation

$$\dot{\varepsilon}^{acc} = \mathbf{M}(\varepsilon^{amp}, e, \boldsymbol{\sigma}), \tag{12}$$

which involves the void ratio, the effective-stress tensor and also depends on the current number of cycles. For details, see [1, 10]. Parameters of (12) used in the calculations are given in Table 4.

Table 4 Constitutive parameters of the cyclic model (sand L12 from [11])

C_{amp}	ε_{ref}^{amp}	C_{N1}	C_{N2}	C_{N3}	C_e	e_{ref}	C_p	p_{ref} (kPa)
1.6	10^{-4}	3.6×10^{-3}	0.016	1.05×10^{-4}	0.48	0.829	0.44	100

The second boundary value problem, Fig. 8, is quasi-static and is formulated for the average values of displacements and stresses. The constitutive relation between the effective-stress tensor $\boldsymbol{\sigma}$ and the strain tensor ε of the skeleton is written in rate form as

$$\dot{\boldsymbol{\sigma}} = \mathsf{E}(\boldsymbol{\sigma}) : \left(\dot{\varepsilon} - \dot{\varepsilon}^{acc}\right), \tag{13}$$

where E is a stress-dependent stiffness tensor. It is proposed in [10] to take the tensor E as in an isotropic elastic solid with a Poisson ratio of 0.2 and a pressure-dependent bulk modulus

$$K(\sigma) = A\, p_{atm}^{1-n}\, (-\sigma)^n, \tag{14}$$

where σ is the mean effective stress, $A = 467$, $n = 0.46$ and $p_{atm} = 100\,\text{kPa}$.

A drawback of the constitutive equation (13) is that it allows the unlimited growth of the principal-stress ratio. This drawback is eliminated by introducing a bounding surface in the stress space and projecting the stress state on this surface if the stress state falls outside. The bounding surface is defined by the Matsuoka-Nakai yield condition with a given friction angle.

The stresses and displacements must satisfy the equilibrium equations

$$\frac{\partial \sigma_r}{\partial r} + (n-1)\frac{\partial p_f}{\partial r} + \frac{2}{r}\left(\sigma_r - \sigma_\varphi\right) + \frac{1}{k}\,\varrho_f g n^2 (v_f - v_s) = 0, \tag{15}$$

$$\frac{\partial p_f}{\partial r} + \frac{1}{k}\,\varrho_f g n (v_f - v_s) = 0, \tag{16}$$

the constitutive equation (3) for the pore pressure and the constitutive equation (13) for the effective stress written for the stress components σ_r, σ_φ. Boundary conditions prescribed at R_A and B_R are impermeability and zero displacement. The first and the second boundary value problems are solved with the finite-difference method.

In what follows we will be interested in the dependence of the solutions on the soil permeability and relative density. All numerical solutions below correspond to a pile displacement amplitude of 4 mm (u_{amp} at R_P in the low-cycle problem). The initial effective stresses and pore pressure approximate the time-averaged values in the solution to the low-cycle problem after 100 cycles. The initial mean effective stress and pore pressure are shown in Fig. 9.

The initial relative density is assumed to be spatially homogeneous. We first fix its value at 0.75 and vary the permeability. Solutions in the displacement-controlled case are shown in Fig. 10. Comparison with the initial distributions shows that the mean effective stress is further reduced. The influence of the soil permeability is most noticeable for $r < 1.5\,\text{m}$ where the effective stress is definitely nonzero in distinction

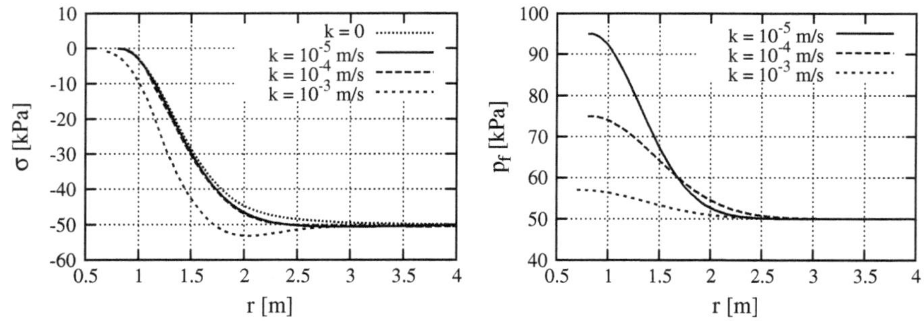

Fig. 9 Initial distributions of the mean effective stress (*left*) and pore pressure (*right*) in the high-cycle problem

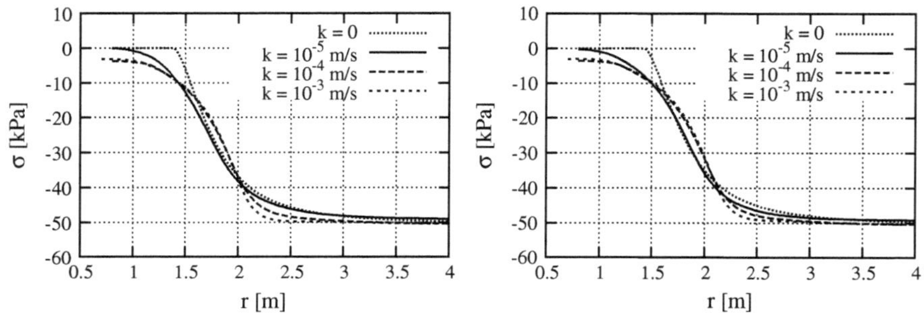

Fig. 10 Mean effective stress distribution in the displacement-controlled case after 1 min (*left*) and 2 min (*right*)

to the locally undrained case. The effective stress at farther distances is reduced to nearly the same extent as in the undrained case. Note also that the difference between the distributions after 1 and 2 min is insignificant. We can conclude that taking permeability into account prevents full liquefaction in the near field but does not prevent the effective-stress reduction in general.

Assuming that the stress amplitude at R_A remains constant, we observe a remarkable difference in the influence of permeability as compared with the displacement-controlled case, Fig. 11. The reduction of the mean effective stress becomes larger as the permeability is changed from 10^{-5} to 10^{-3} m/s. Thus, finite permeability may increase the effective-stress reduction in the stress-controlled case. This is a consequence of larger strain amplitudes in the soil. The strain amplitudes in the displacement-controlled case remain constant because they are controlled by the boundary condition. The strain amplitudes in the stress-controlled case vary with time. They depend on the soil stiffness determined by the current effective stress.

In order to see how the effective-stress reduction depends on the relative soil density, consider the point at which the mean effective stress is equal to 50 % of its value in the far field. The coordinate of this point will be denoted by R_{50}. Figure 12 shows R_{50} after 1 and 2 min of vibration for various values of the relative density, I_D,

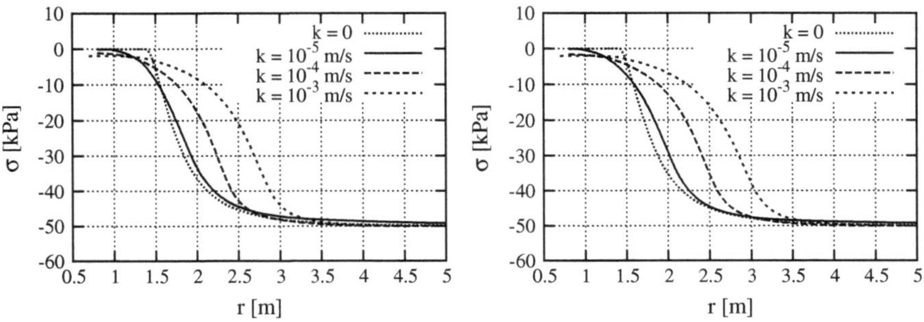

Fig. 11 The same as in Fig. 10 in the stress-controlled case

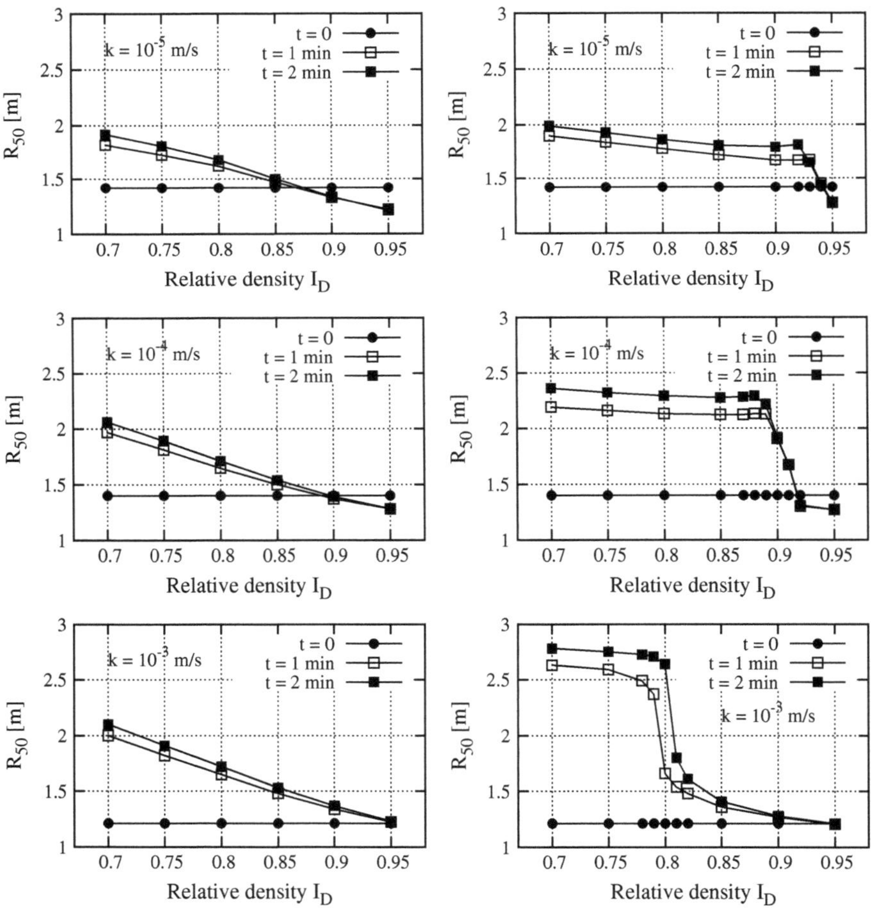

Fig. 12 Dependence of R_{50} on the relative density and permeability. *Left column* displacement-controlled case. *Right column* stress-controlled case

and permeability, k. In the displacement-controlled case shown in the left column of the figure, R_{50} changes linearly with I_D and depends only slightly on k. As expected, the radius R_{50} is greater for looser soil. For each value of permeability there exists a relative density at which there is no increase in R_{50} in the high-cycle problem. The radius R_{50} may even decrease if the soil is dense enough. The curves in the stress-controlled case shown in the right column of Fig. 12 look different: they have a jump at a certain density. The overall dependence on the density and permeability becomes rather complicated. For instance, R_{50} increases monotonically with k if $I_D = 0.75$, and has a maximum at $k = 10^{-4}$ m/s if $I_D = 0.85$.

5 Conclusion

The requirements imposed on the description of the soil behaviour for the modelling of vibratory pile driving can hardly be fully satisfied by the existing constitutive models. To circumvent this difficulty, the problem of the vibration-induced soil deformation is solved in this study using two constitutive models: the incremental hypoplasticity model for a limited number of cycles in the first step, and the explicit high-cycle accumulation model for a large number of cycles in the second step.

Numerical modelling of the effective-stress evolution around a vibrating pile toe was performed recently in [2, 3] assuming locally undrained conditions. The solutions obtained in the present paper show that the influence of permeability on the effective-stress changes is not as strong as it might be expected. Although a zone of full liquefaction near the pile does not develop under locally drained conditions, the effective-stress reduction is not prevented. The extent to which the effective stress is reduced may be the same or even larger. The effective-stress reduction in the high-cycle problem is smaller for denser soil. There is no reduction in the effective stress in the high-cycle problem if the soil is dense enough.

It should be mentioned that the spherical symmetry assumed in this study is a simplification which restricts the soil displacement to the radial direction and thus reduces the possibility of densification—in particular, densification due to vertical settlement. Additional degrees of freedom in the soil displacement will change the effective stresses. Answering the question of whether these changes are essential requires the solution of two- or three-dimensional problems.

Acknowledgments The study was financed by the Deutsche Forschungsgemeinschaft as part of the Research Unit FOR 1136 'Simulation of geotechnical construction processes with holistic consideration of the stress strain soil behaviour', Subproject 6 'Soil deformations close to retaining walls due to vibration excitations'.

References

1. Niemunis, A., Wichtmann, T., Triantafyllidis, T.: A high-cycle accumulation model for sand. Comput. Geotech. **32**, 245–263 (2005)
2. Osinov, V.A.: Application of a high-cycle accumulation model to the analysis of soil liquefaction around a vibrating pile toe. Acta Geotechnica **8**, 675–684 (2013)
3. Osinov, V.A., Chrisopoulos, S., Triantafyllidis, T.: Numerical study of the deformation of saturated soil in the vicinity of a vibrating pile. Acta Geotechnica **8**, 439–446 (2013)
4. Zienkiewicz, O.C., Chan, A.H.C., Pastor, M., Schrefler, B.A., Shiomi, T.: Computational Geomechanics with Special Reference to Earthquake Engineering. John Wiley, Chichester (1999)
5. von Wolffersdorff, P.A.: A hypoplastic relation for granular materials with a predefined limit state surface. Mech. Cohesive-frictional Mater. **1**(3), 251–271 (1996)
6. Niemunis, A., Herle, I.: Hypoplastic model for cohesionless soils with elastic strain range. Mech. Cohesive-frict. Mater. **2**(4), 279–299 (1997)
7. Herle, I., Gudehus, G.: Determination of parameters of a hypoplastic constitutive model from properties of grain assemblies. Mech. Cohesive-frictional Mater. **4**, 461–486 (1999)
8. Osinov, V.A., Gudehus, G.: Dynamics of hypoplastic materials: theory and numerical implementation. In: Hutter K., Kirchner N. (eds.) Dynamic Response of Granular and Porous Materials Under Large and Catastrophic Deformations, pp. 265–284. Springer, Berlin (2003)
9. Wolf, J.P.: Soil-Structure-Interaction Analysis in Time Domain. Prentice Hall, New Jersey (1988)
10. Wichtmann, T., Niemunis, A., Triantafyllidis, T.: On the 'elastic' stiffness in a high-cycle accumulation model for sand: a comparison of drained and undrained cyclic triaxial tests. Canadian Geotech. J. **47**(7), 791–805 (2010)
11. Wichtmann, T., Niemunis, A., Triantafyllidis, T.: Simplified calibration procedure for a high-cycle accumulation model based on cyclic triaxial tests on 22 sands. In: Gourvenec S., White D. (eds.) Frontiers in Offshore Geotechnics II, pp. 383–388. Taylor & Francis, London (2011)

A Numerical Approach to the Solution of Dynamic Boundary Value Problems for Fluid-Saturated Solids

V.A. Osinov and C. Grandas-Tavera

Abstract A dynamic boundary value problem for a fluid-saturated solid can be represented as two coupled boundary value problems for one-phase media. This allows us to solve the problem with a commercial computer program without a built-in procedure for the solution of dynamic problems with non-zero permeability, provided that the user is able to establish the required coupling between the two problems. This approach has been implemented in the present paper with the computer program Abaqus/Standard using the dynamic analysis for one-phase media as a built-in procedure without the need to construct a user-defined finite element.

Keywords Fluid-saturated solid · Dynamic problem

1 Motivation

Dynamic boundary value problems for fluid-saturated porous solids are dealt with in various branches of mechanics, including soil and rock mechanics. However, most commercial computer codes do not provide a built-in solution procedure for such problems. In particular, this is true for the widely used finite-element program Abaqus. As far as Abaqus is concerned, these limitations can be overcome by the implementation of the subroutine `UEL` which allows us to construct a user-defined finite element with the desired degrees of freedom. The construction of a user-defined element requires a thorough knowledge of the finite-element method and entails careful debugging. In this regard, alternative ways of solving the problem without resort to the `UEL` subroutine may be useful (e.g. [1]).

This paper presents a numerical approach to the solution of the dynamic initial boundary value problem for a two-phase porous medium. The approach itself is not related to a specific discretization method such as the finite-element or finite-difference method. It consists in the representation of the boundary value

V.A. Osinov (✉) · C. Grandas-Tavera
Institute of Soil Mechanics and Rock Mechanics, Karlsruhe Institute of Technology, Karlsruhe, Germany

Th. Triantafyllidis (ed.), *Holistic Simulation of Geotechnical Installation Processes*, Lecture Notes in Applied and Computational Mechanics 77,
DOI 10.1007/978-3-319-18170-7_8

problem for a two-phase medium as two coupled problems for one-phase media. Using this approach, the original problem for a two-phase medium can be solved with a computer program which solves boundary value problems for one-phase media, provided that the user is able to establish the required coupling between the two problems and to solve them concurrently. The proposed method has been implemented with Abaqus/Standard. Since Abaqus performs the dynamic analysis for one-phase media as a built-in procedure, a user-defined finite element is not needed.

2 Governing Equations

Assuming that the solid phase of a two-phase medium is much stiffer than the skeleton, the total stress is represented as the sum of the effective stress $\boldsymbol{\sigma}$ (compressive stresses are negative) and an isotropic stress $-p_f\mathbf{I}$, where p_f is the fluid pressure (positive for compression) and $\mathbf{I}$ is the unit tensor. We write the dynamic equations in the small-strain approximation neglecting the convective terms and replacing the material time derivatives with the partial ones. The equations of motion for the solid and fluid phases without mass forces are [2, 3]

$$\operatorname{div}\boldsymbol{\sigma} + (n-1)\operatorname{grad} p_f + \xi(\mathbf{v}_f - \mathbf{v}_s) = (1-n)\varrho_s \frac{\partial \mathbf{v}_s}{\partial t}, \tag{1}$$

$$-n \operatorname{grad} p_f - \xi(\mathbf{v}_f - \mathbf{v}_s) = n\varrho_f \frac{\partial \mathbf{v}_f}{\partial t}, \tag{2}$$

where $\mathbf{v}_s$, $\mathbf{v}_f$ are the velocities of the skeleton and the fluid, ϱ_s, ϱ_f are the densities of the solid and fluid phases, and n is the porosity of the skeleton. The coefficient ξ is inversely proportional to the permeability of the skeleton: $\xi = \varrho_f g n^2/k$, where k is the permeability (m/s), and g is the acceleration due to gravity.

A constitutive equation for the skeleton can be written in the general form

$$\frac{\partial \boldsymbol{\sigma}}{\partial t} = \mathbf{F}\left(\mathbf{D}_s, \boldsymbol{\sigma}, S\right), \tag{3}$$

where

$$\mathbf{D}_s = \frac{1}{2}\left[\operatorname{grad}\mathbf{v}_s + (\operatorname{grad}\mathbf{v}_s)^T\right] \tag{4}$$

is the stretching tensor of the skeleton, and the tensor-valued function $\mathbf{F}$ corresponds to an elasticity or plasticity model. In the latter case, Eq. (3) may contain a set $S = (S_1, ..., S_m)$ of additional state variables which have their own evolution equations denoted here by a function M:

$$\frac{\partial S}{\partial t} = M\left(\mathbf{D}_s, \boldsymbol{\sigma}, S\right). \tag{5}$$

Pore pressure changes are determined by the constitutive equation (neglecting grad n)

$$\frac{\partial p_f}{\partial t} = -K_f \operatorname{tr} \mathbf{D}_f - K_f \frac{(1-n)}{n} \operatorname{tr} \mathbf{D}_s, \tag{6}$$

where

$$\mathbf{D}_f = \frac{1}{2}\left[\operatorname{grad} \mathbf{v}_f + \left(\operatorname{grad} \mathbf{v}_f\right)^T\right] \tag{7}$$

is the stretching tensor of the fluid phase, and K_f is the compression modulus of the fluid.

Let us introduce new velocities $\mathbf{v}^{(1)}$, $\mathbf{v}^{(2)}$ and stresses $\boldsymbol{\sigma}^{(1)}$, $\boldsymbol{\sigma}^{(2)}$ defined as

$$\mathbf{v}^{(1)} = a_{11}\mathbf{v}_s + a_{12}\mathbf{v}_f, \tag{8}$$

$$\mathbf{v}^{(2)} = a_{21}\mathbf{v}_s + a_{22}\mathbf{v}_f, \tag{9}$$

$$\boldsymbol{\sigma}^{(1)} = a_{11}\boldsymbol{\sigma} + (n-1)\left(a_{11} + \frac{\varrho_s}{\varrho_f} a_{12}\right) p_f \mathbf{I}, \tag{10}$$

$$\boldsymbol{\sigma}^{(2)} = a_{21}\boldsymbol{\sigma} + (n-1)\left(a_{21} + \frac{\varrho_s}{\varrho_f} a_{22}\right) p_f \mathbf{I}, \tag{11}$$

where $a_{11}, a_{12}, a_{21}, a_{22}$ are constant coefficients. Provided that

$$a_{11}a_{22} - a_{12}a_{21} \neq 0, \tag{12}$$

transformation (8)–(11) from $\mathbf{v}_s$, $\mathbf{v}_f$, $\boldsymbol{\sigma}$, $p_f\mathbf{I}$ to $\mathbf{v}^{(1)}$, $\mathbf{v}^{(2)}$, $\boldsymbol{\sigma}^{(1)}$, $\boldsymbol{\sigma}^{(2)}$ can be inverted to give

$$\mathbf{v}_s = \frac{a_{22}}{a_0}\mathbf{v}^{(1)} - \frac{a_{12}}{a_0}\mathbf{v}^{(2)}, \tag{13}$$

$$\mathbf{v}_f = -\frac{a_{21}}{a_0}\mathbf{v}^{(1)} + \frac{a_{11}}{a_0}\mathbf{v}^{(2)}, \tag{14}$$

$$\boldsymbol{\sigma} = \kappa_1\boldsymbol{\sigma}^{(1)} + \kappa_2\boldsymbol{\sigma}^{(2)}, \tag{15}$$

$$p_f\mathbf{I} = \frac{\varrho_f}{(1-n)\varrho_s a_0}\left(a_{21}\boldsymbol{\sigma}^{(1)} - a_{11}\boldsymbol{\sigma}^{(2)}\right), \tag{16}$$

where

$$a_0 = a_{11}a_{22} - a_{12}a_{21}, \tag{17}$$

$$\kappa_1 = \frac{1}{a_0}\left(a_{22} + \frac{\varrho_f}{\varrho_s} a_{21}\right), \tag{18}$$

$$\kappa_2 = -\frac{1}{a_0}\left(a_{12} + \frac{\varrho_f}{\varrho_s} a_{11}\right). \tag{19}$$

In numerical calculations, p_f can be taken as a diagonal component of the right-hand side of (16), or as

$$p_f = \frac{\varrho_f}{3(1-n)\varrho_s a_0}\left(a_{21}\operatorname{tr}\boldsymbol{\sigma}^{(1)} - a_{11}\operatorname{tr}\boldsymbol{\sigma}^{(2)}\right). \tag{20}$$

Substituting (13)–(16) into (1), (2), we obtain

$$\operatorname{div}\boldsymbol{\sigma}^{(1)} + b_{11}\xi\mathbf{v}^{(1)} + b_{12}\xi\mathbf{v}^{(2)} = (1-n)\varrho_s\frac{\partial\mathbf{v}^{(1)}}{\partial t}, \tag{21}$$

$$\operatorname{div}\boldsymbol{\sigma}^{(2)} + b_{21}\xi\mathbf{v}^{(1)} + b_{22}\xi\mathbf{v}^{(2)} = (1-n)\varrho_s\frac{\partial\mathbf{v}^{(2)}}{\partial t}, \tag{22}$$

where

$$b_{11} = \frac{(a_{21}+a_{22})}{a_0}\left(-a_{11} + \frac{(1-n)\varrho_s}{n\varrho_f}a_{12}\right), \tag{23}$$

$$b_{12} = \frac{(a_{11}+a_{12})}{a_0}\left(a_{11} - \frac{(1-n)\varrho_s}{n\varrho_f}a_{12}\right), \tag{24}$$

$$b_{21} = \frac{(a_{21}+a_{22})}{a_0}\left(-a_{21} + \frac{(1-n)\varrho_s}{n\varrho_f}a_{22}\right), \tag{25}$$

$$b_{22} = \frac{(a_{11}+a_{12})}{a_0}\left(a_{21} - \frac{(1-n)\varrho_s}{n\varrho_f}a_{22}\right). \tag{26}$$

Equation (21) can be viewed as the equation of motion of a one-phase medium with the stress tensor $\boldsymbol{\sigma}^{(1)}$, the velocity vector $\mathbf{v}^{(1)}$ and the density $(1-n)\varrho_s$. Similarly, Eq. (22) can be viewed as the equation of motion of a one-phase medium with the stress tensor $\boldsymbol{\sigma}^{(2)}$, the velocity vector $\mathbf{v}^{(2)}$ and the same density $(1-n)\varrho_s$. Each equation contains two additional terms that play the role of a mass force. One term is proportional to the velocity of the same medium, while the other is proportional to the velocity of the other medium (a coupling term).

Defining the stretching tensors

$$\mathbf{D}^{(1)} = \frac{1}{2}\left[\operatorname{grad}\mathbf{v}^{(1)} + \left(\operatorname{grad}\mathbf{v}^{(1)}\right)^T\right], \tag{27}$$

$$\mathbf{D}^{(2)} = \frac{1}{2}\left[\operatorname{grad}\mathbf{v}^{(2)} + \left(\operatorname{grad}\mathbf{v}^{(2)}\right)^T\right], \tag{28}$$

and using the transformation relations, we can write constitutive equations for $\boldsymbol{\sigma}^{(1)}$, $\boldsymbol{\sigma}^{(2)}$ in terms of $\mathbf{D}^{(1)}$, $\mathbf{D}^{(2)}$, $\boldsymbol{\sigma}^{(1)}$, $\boldsymbol{\sigma}^{(2)}$:

$$\frac{\partial\boldsymbol{\sigma}^{(1)}}{\partial t} = a_{11}\frac{\partial\boldsymbol{\sigma}}{\partial t} + (n-1)\left(a_{11} + \frac{\varrho_s}{\varrho_f}a_{12}\right)\frac{\partial p_f}{\partial t}\mathbf{I}, \tag{29}$$

$$\frac{\partial \boldsymbol{\sigma}^{(2)}}{\partial t} = a_{21}\frac{\partial \boldsymbol{\sigma}}{\partial t} + (n-1)\left(a_{21} + \frac{\varrho_s}{\varrho_f} a_{22}\right)\frac{\partial p_f}{\partial t}\mathbf{I}, \tag{30}$$

where

$$\frac{\partial \boldsymbol{\sigma}}{\partial t} = \mathbf{F}\left(\frac{a_{22}}{a_0}\mathbf{D}^{(1)} - \frac{a_{12}}{a_0}\mathbf{D}^{(2)},\ \kappa_1\boldsymbol{\sigma}^{(1)} + \kappa_2\boldsymbol{\sigma}^{(2)},\ S\right), \tag{31}$$

$$\frac{\partial p_f}{\partial t} = \frac{K_f}{a_0}\left(a_{21} - \frac{(1-n)}{n}a_{22}\right)\operatorname{tr}\mathbf{D}^{(1)} - \frac{K_f}{a_0}\left(a_{11} - \frac{(1-n)}{n}a_{12}\right)\operatorname{tr}\mathbf{D}^{(2)}. \tag{32}$$

Equation (29) expresses the rate of $\boldsymbol{\sigma}^{(1)}$ as a function of the stretching tensor $\mathbf{D}^{(1)}$ of the same medium, $\boldsymbol{\sigma}^{(1)}$ itself and the state variables S. As distinct from conventional constitutive equations, it contains external terms with $\mathbf{D}^{(2)}$ and $\boldsymbol{\sigma}^{(2)}$. The same holds for Eq. (30): it expresses the rate of $\boldsymbol{\sigma}^{(2)}$ as a function of $\mathbf{D}^{(2)}$, $\boldsymbol{\sigma}^{(2)}$, S and contains external terms with $\mathbf{D}^{(1)}$ and $\boldsymbol{\sigma}^{(1)}$. In this way the original system (1)–(3), (6) for a two-phase medium is split into two systems, namely (21), (29) and (22), (30), for two one-phase media. The two systems are coupled with each other through the mass-force terms in the equations of motion and the additional terms in the constitutive equations. In the case of a plastic skeleton, the two systems are supplemented by Eq. (5) written as

$$\frac{\partial S}{\partial t} = M\left(\frac{a_{22}}{a_0}\mathbf{D}^{(1)} - \frac{a_{12}}{a_0}\mathbf{D}^{(2)},\ \kappa_1\boldsymbol{\sigma}^{(1)} + \kappa_2\boldsymbol{\sigma}^{(2)},\ S\right). \tag{33}$$

Thus, we have shown that a boundary value problem for a two-phase medium can be represented as two boundary value problems for one-phase media to be solved in parallel. Note that the splitting of the original problem into two coupled one-phase problems and the corresponding numerical implementation could be performed only for the particular trivial case $\mathbf{v}^{(1)} = \mathbf{v}_s$, $\mathbf{v}^{(2)} = \mathbf{v}_f$ with $a_{11} = a_{22} = 1$ and $a_{12} = a_{21} = 0$ in (8), (9). The reason why this may be insufficient is possible numerical instability which may depend not only on the discretization method but also on the choice of the transformation coefficients a_{ij}. For this reason, it is advantageous to develop a numerical algorithm for the general case.

3 Boundary Conditions

The two boundary value problems for one-phase media are coupled not only through the additional terms in the governing equations but also through boundary conditions. In what follows, $\boldsymbol{\tau}$, $\boldsymbol{\tau}^{(1)}$, $\boldsymbol{\tau}^{(2)}$ will denote the traction vectors corresponding to the stress tensors $\boldsymbol{\sigma}$, $\boldsymbol{\sigma}^{(1)}$, $\boldsymbol{\sigma}^{(2)}$. Vectors $\mathbf{n}$, $\mathbf{p}$, $\mathbf{q}$ will denote an orthonormal basis, where $\mathbf{n}$ is the outer normal unit vector and $\mathbf{p}$, $\mathbf{q}$ are tangential unit vectors on the boundary. The projections of a vector on $\mathbf{n}$, $\mathbf{p}$, $\mathbf{q}$ will be written with the subscripts n, p, q. For example, $v_{sn} = \mathbf{v}_s \cdot \mathbf{n}$, $v_{sp} = \mathbf{v}_s \cdot \mathbf{p}$. Known quantities in the boundary conditions will be marked with a tilde.

The following boundary conditions may be prescribed (in combination) at the boundary of a two-phase media: impermeable boundary, velocity of the solid phase, effective-stress vector, fluid pressure. These conditions can be written for the two one-phase media with the help of the transformation formulae of the previous section:

- impermeable boundary, $v_{fn} - v_{sn} = 0$,

$$(a_{21} + a_{22})v_n^{(1)} - (a_{11} + a_{12})v_n^{(2)} = 0, \tag{34}$$

- given velocity of the solid phase, $\mathbf{v}_s = \tilde{\mathbf{v}}$,

$$\frac{a_{22}}{a_0}\,\mathbf{v}^{(1)} - \frac{a_{12}}{a_0}\,\mathbf{v}^{(2)} = \tilde{\mathbf{v}}, \tag{35}$$

- given effective-stress vector, $\boldsymbol{\tau} = \tilde{\boldsymbol{\tau}}$,

$$\kappa_1 \boldsymbol{\tau}^{(1)} + \kappa_2 \boldsymbol{\tau}^{(2)} = \tilde{\boldsymbol{\tau}}, \tag{36}$$

- given fluid pressure, $p_f = \tilde{p}_f$,

$$\frac{\varrho_f}{(1-n)\varrho_s a_0}\left(a_{21}\tau_n^{(1)} - a_{11}\tau_n^{(2)}\right) = \tilde{p}_f. \tag{37}$$

The numerical solution of problems formulated originally for infinite domains may require non-reflecting boundary conditions prescribed on an artificial boundary in order to imitate the infinite domain. In this study we consider non-reflecting boundary conditions for plane normally incident waves in a two-phase medium with an isotropic elastic skeleton. These boundary conditions are (see Eqs. (44), (71), (72) in [4])

$$\tau_n = -\frac{\lambda + 2\mu}{c_1}\, v_{sn}, \tag{38}$$

$$\tau_p = -\frac{\mu}{c_2}\, v_{sp}, \tag{39}$$

$$\tau_q = -\frac{\mu}{c_2}\, v_{sq}, \tag{40}$$

$$p_f = \frac{K_f}{n c_1}\left[(1-n)v_{sn} + n v_{fn}\right], \tag{41}$$

where

$$c_1 = \sqrt{\left(\lambda + 2\mu + \frac{K_f}{n}\right)\frac{1}{\varrho}}, \tag{42}$$

$$c_2 = \sqrt{\frac{\mu}{\varrho}}, \tag{43}$$

$$\varrho = (1-n)\varrho_s + n\varrho_f, \tag{44}$$

and λ, μ are the Lamé constants of the skeleton.

Equations (38)–(41) written for the two one-phase media become

$$\kappa_1 \tau_n^{(1)} + \kappa_2 \tau_n^{(2)} = -\frac{\lambda + 2\mu}{c_1 a_0} \left(a_{22} v_n^{(1)} - a_{12} v_n^{(2)} \right), \tag{45}$$

$$\kappa_1 \tau_p^{(1)} + \kappa_2 \tau_p^{(2)} = -\frac{\mu}{c_2 a_0} \left(a_{22} v_p^{(1)} - a_{12} v_p^{(2)} \right), \tag{46}$$

$$\kappa_1 \tau_q^{(1)} + \kappa_2 \tau_q^{(2)} = -\frac{\mu}{c_2 a_0} \left(a_{22} v_q^{(1)} - a_{12} v_q^{(2)} \right), \tag{47}$$

$$a_{21} \tau_n^{(1)} - a_{11} \tau_n^{(2)} = \frac{K_f (1-n) \varrho_s}{n \varrho_f c_1} \left\{ [(1-n) a_{22} - n a_{21}] \, v_n^{(1)} - [(1-n) a_{12} - n a_{11}] \, v_n^{(2)} \right\}. \tag{48}$$

The correct formulation of a boundary value problem for a fluid-saturated solid in the general three-dimensional case requires four scalar boundary conditions. A boundary value problem for a one-phase solid involves three scalar boundary conditions. Hence, two problems for one-phase media would require six boundary conditions (three for each problem). Nevertheless, the two boundary value problems for one-phase media considered here are well-posed with four boundary conditions, namely those derived from the four original boundary conditions for the two-phase medium. This is because the two problems are not independent but coupled and must be solved concurrently. In this connection, difficulties may arise in numerical implementation when using available software with limited possibilities of modification. The computer program that solves the boundary value problem for a one-phase medium will demand three boundary conditions from the user, so that we will need six boundary conditions altogether to be able to use the software. It might seem that a simple way to overcome this difficulty is to use two of the four boundary conditions in both problems and thus to gain six boundary conditions. However, this would lead to a wrong numerical solution (see Appendix for detail).

The required additional boundary conditions can be obtained from Eq. (16) which shows that the tensor on the right-hand side is spherical. This gives two boundary conditions

$$a_{21} \tau_p^{(1)} - a_{11} \tau_p^{(2)} = 0, \tag{49}$$

$$a_{21} \tau_q^{(1)} - a_{11} \tau_q^{(2)} = 0. \tag{50}$$

If boundary conditions (49), (50) are used in the first boundary value problem, they represent conditions imposed on $\tau_p^{(1)}$, $\tau_q^{(1)}$ with given $\tau_p^{(2)}$, $\tau_q^{(2)}$. If (49), (50) are used in the second boundary value problem, then $\tau_p^{(1)}$, $\tau_q^{(1)}$ are known and the conditions are imposed on $\tau_p^{(2)}$, $\tau_q^{(2)}$.

Plane-strain and axisymmetric problems involve three scalar boundary conditions for a fluid-saturated medium and two scalar boundary conditions for a one-phase medium. In the implementation of the present method, two boundary conditions are required for each one-phase medium, and hence, one additional boundary condition in the form (49), (50) for the tangential stress component is needed.

4 Numerical Implementation with Abaqus

The proposed approach has been implemented with the finite-element program Abaqus/Standard. Each of the two boundary value problems is solved in a separate Abaqus job. The two problems have the same mesh and element type. They have independent input files, user subroutines and do not share variables in memory. The first boundary value problem (BVP1) is described by Eqs. (21), (29), while the second one (BVP2) is described by Eqs. (22), (30). The two BVPs are coupled with each other through the mass-force terms in the equations of motion (21), (22), the additional terms in the constitutive Eqs. (29), (30), and also through boundary conditions.

The coupling algorithm is presented schematically in Fig. 1. Each BVP is solved within a time increment between times t_k and t_{k+1} using the required coupling quantities of the other BVP at time t_k. The calculation cycle of each BVP consists of the numerical integration over the increment to obtain the solution at time t_{k+1}, a waiting phase if the solution of the other BVP has not yet been completed, and a phase for reading the data of the other BVP calculated for time t_{k+1}. This solution procedure can be made possible with the user subroutine `UEXTERNALDB` in each BVP. The subroutine allows the user to introduce waiting and reading phases immediately before proceeding to the next time increment. Quantities necessary for the coupling are read by each BVP from the Abaqus result file (`*.fil`) of the other BVP during the reading phase and saved in a global Fortran module accessible to all user subroutines. The global module also contains quantities of the same BVP necessary for the calculation of the next increment, e.g. velocities.

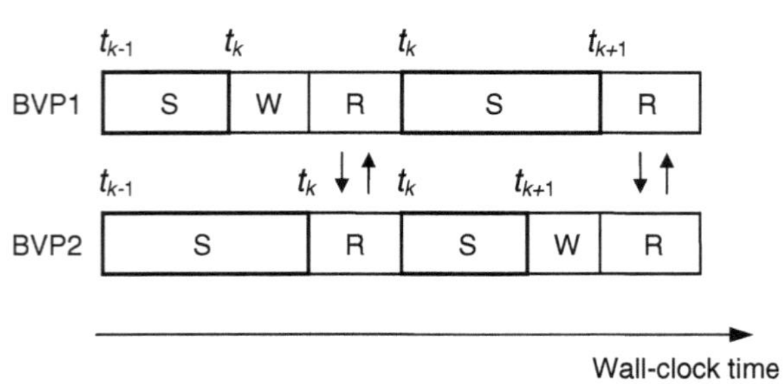

Fig. 1 Concurrent solution of two boundary value problems. Notation: *S* solving for a time increment, *W* waiting for the results of the other problem, *R* reading the results of the other problem

The mass-force terms $b_{ij}\xi\mathbf{v}^{(j)}$ in the equations of motion (21), (22) are treated as an inhomogeneously distributed load. This load is computed at the integration points with the help of the user subroutine DLOAD. Velocities available at the nodal points are interpolated to obtain values at the integration points. Constitutive Eqs. (29), (30) are implemented in the user subroutines UMAT. The global module of BVPj contains the deformation fields of BVPi for the last two times. This allows the UMAT subroutine of BVPj to calculate the external term $\mathbf{D}^{(i)}$ in the constitutive equations. The global module of BVPj also contains the external stress $\boldsymbol{\sigma}^{(i)}$ of BVPi which may appear in the constitutive equation of BVPj. Additional state variables governed by Eq. (33) can be calculated in either of the BVPs and stored in the global module. In the numerical example presented below in Sect. 5, the constitutive equation for the solid skeleton is linearly elastic and has no additional state variables.

Boundary conditions (34), (35) for velocity are implemented with the user subroutine DISP. Boundary conditions (36), (37), (49), (50) for tractions are implemented with the user subroutine UTRACLOAD. This subroutine is also used to implement non-reflecting boundary conditions (45)–(48). According to the scheme shown in Fig. 1, a boundary condition written for velocity or traction for time t_{k+1} contains a given quantity (with a tilde) for time t_{k+1} and known quantities for time t_k taken from the global module.

It should be mentioned that a somewhat similar approach to the solution of dynamic problems for two-phase media was implemented in [1] with Abaqus/Explicit. The problem was solved with a single Abaqus job using two identical meshes: one for the fluid phase and the other one for the skeleton. This corresponds to $\mathbf{v}^{(1)} = \mathbf{v}_s$, $\mathbf{v}^{(2)} = \mathbf{v}_f$. The mass-force terms $b_{ij}\xi\mathbf{v}^{(j)}$ were introduced with the help of connector elements acting between the collocated nodes. In relation to the algorithm described in [1], two advantages of the present approach can be noticed. First, the use of connector elements requires sophisticated programming prior to the solution of the problem. Second, for a given time integration method (explicit or implicit), the calculation of one time increment with a double mesh as in [1] requires more time than the calculation of one increment with two Abaqus jobs operating in parallel with single meshes.

5 Numerical Verification

The present approach implemented with Abaqus/Standard has been tested on the solution of a two-dimensional plane-strain dynamic problem for an elastic medium for the domain shown in Fig. 2. The medium is at rest at $t = 0$ with zero stresses and pore pressure. Boundary conditions for $t > 0$ (BC1, BC2 and BC3) are prescribed as follows.

- BC1 represents an impermeable boundary with a given velocity of the solid phase

$$v_{s1}(t) = 0, \tag{51}$$

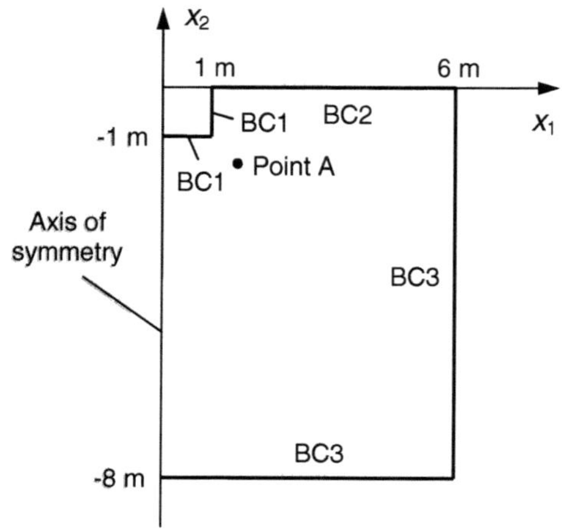

Fig. 2 Computational domain of the boundary value problem

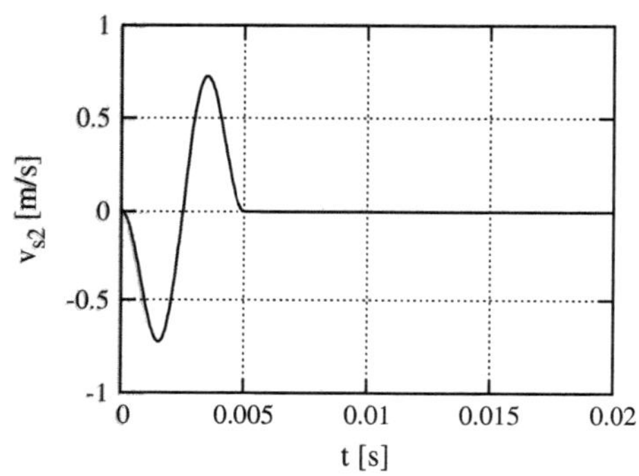

Fig. 3 Boundary condition (52) for the vertical velocity

$$v_{s2}(t) = \begin{cases} v_{amp} \sin(2\pi t/t_0) \sin(\pi t/t_0) & \text{if } t \le t_0, \\ 0 & \text{if } t > t_0, \end{cases} \tag{52}$$

where $v_{amp} = -0.3\pi$ m/s and $t_0 = 5 \times 10^{-3}$ s. The function $v_{s2}(t)$ is shown in Fig. 3. This boundary condition imitates the prescribed motion of a rigid foundation on a half space $x_2 < 0$.

- BC2 corresponds to a free surface with zero pore pressure and zero traction.
- BC3 describes a non-reflecting boundary as discussed in Sect. 3.

The solid skeleton is assumed to be linearly elastic and isotropic with the Lamé constants λ and μ. The parameters of the medium are shown in Table 1. The compression modulus of the pore fluid is taken to be equal to the modulus of pure water and is much higher than the shear modulus of the skeleton (a situation typical of soils). The size of the finite elements in the mesh varies between 3 and 9 cm. The time step is equal to 10^{-5} s.

Figures 4 and 5 show the solution to this boundary value problem as a function of time at Point A with the coordinates $x_1 = 1.5$ m, $x_2 = -1.5$ m, see Fig. 2. For

Table 1 Parameters of the medium

λ (MPa)	μ (MPa)	K_f (MPa)	k (m/s)	ϱ_s (kg/m^3)	ϱ_f (kg/m^3)	n
120	80	2200	10^{-2}	2650	1000	0.4

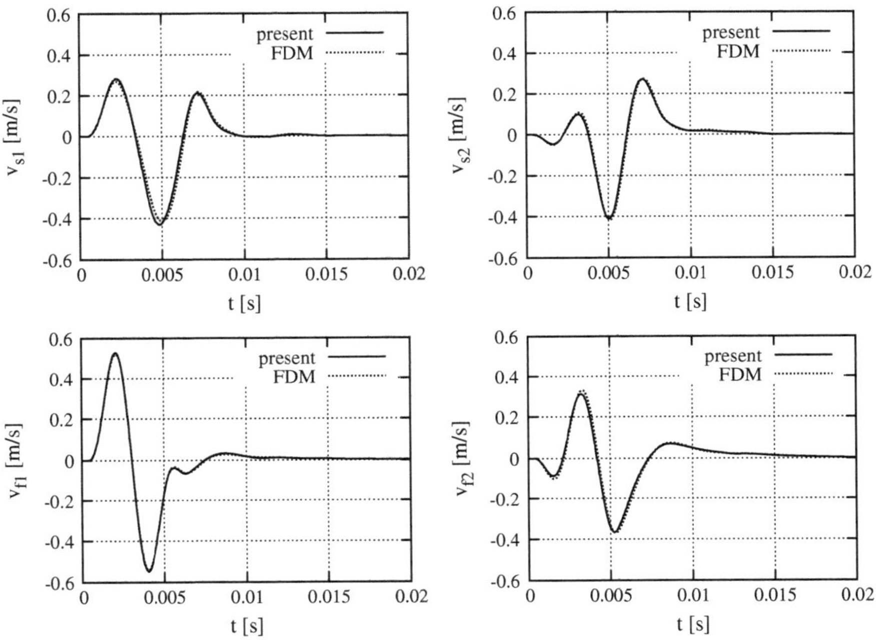

Fig. 4 Solutions obtained with the present approach and with the finite-difference method (FDM). Velocity components at Point A (see Fig. 2) are shown as functions of time

comparison, the figures also show the solution obtained independently by the finite-difference method for the original system (1)–(3), (6) without the decomposition into two coupled problems.

Calculations with the present method have shown that numerical solutions may exhibit instabilities in the form of spurious oscillations increasing with time. Stability has been found to depend on the time integration method and on the choice of the coefficients a_{ij}, see (8)–(11). Abaqus/Standard offers two implicit time integration schemes: the Hilber-Hughes-Taylor and the backward Euler schemes. Numerical experiments with the boundary value problem presented in this section revealed instability of the Hilber-Hughes-Taylor scheme for all combinations of a_{ij} that were tested. The solution shown in Figs. 4 and 5 could be obtained only with the backward Euler method. The dependence of stability of the backward Euler method on the coefficients a_{ij} is shown in Fig. 6 for some combinations of the coefficients. In the stable cases, there is no dependence of the solutions on the choice of a_{ij}. Note that stability of the proposed method in its general form presented in Sect. 2 may depend

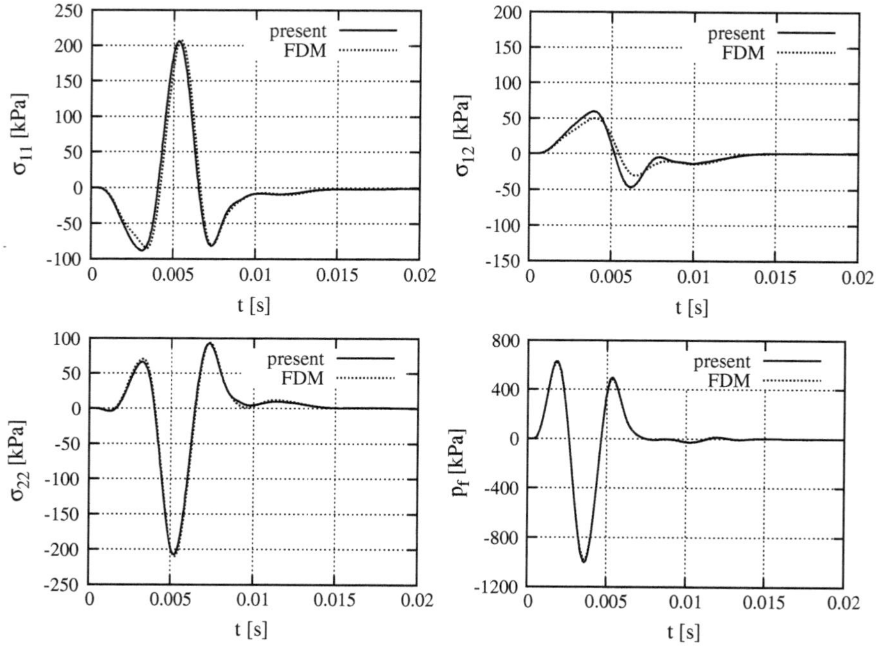

Fig. 5 The same as in Fig. 4 for the stresses

Fig. 6 Stable and unstable combinations of the coefficients a_{ij}

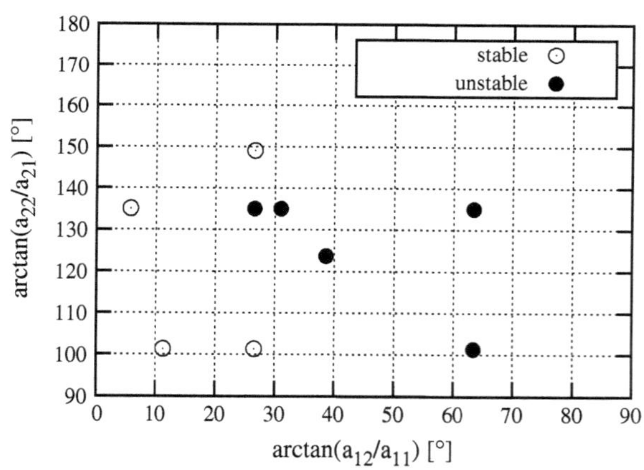

on the numerical implementation of the coupling and the solution method for the one-phase problems. The present findings apply to the algorithm described in Sect. 4.

6 Concluding Remarks

The system of dynamic equations for a porous fluid-saturated solid can be written as two coupled systems, each of them describing the deformation of a one-phase medium. This decomposition offers the possibility of solving boundary value problems for two-phase fluid-saturated solids with a computer program which can perform a dynamic analysis for one-phase solids only (as, for instance, the commercial program Abaqus). The necessary condition for the numerical implementation of this approach is the ability to establish the required coupling between the two problems within the available software and to solve them concurrently. The approach has been realized in the present study with the finite-element program Abaqus/Standard and verified by the comparison of the solution with that obtained independently by a different method.

Acknowledgments The study was financed by the Deutsche Forschungsgemeinschaft as part of the Research Unit FOR 1136 'Simulation of geotechnical construction processes with holistic consideration of the stress strain soil behaviour', Subproject 6 'Soil deformations close to retaining walls due to vibration excitations'.

Appendix

As mentioned in Sect. 3, the use of the same boundary condition in the two coupled boundary value problems would lead to a wrong numerical solution. Here we will show this fact by a simple example.

Write a boundary condition in the form

$$g^{(1)}(t) + g^{(2)}(t) = f(t), \tag{53}$$

where the unknown terms $g^{(1)}, g^{(2)}$ contain quantities of the first and the second boundary value problems, and $f(t)$ is a given function. Obviously, two unknown functions $g^{(1)}(t), g^{(2)}(t)$ cannot be found from one Eq. (53). They are determined not only by this boundary condition but also from the solution of the two boundary value problems. Suppose that boundary condition (53) is used in both boundary value problems. Let the numerical scheme be such that the values of $g^{(1)}, g^{(2)}$ at time $t+\Delta t$ are calculated as

$$g^{(1)}(t + \Delta t) = f(t + \Delta t) - g^{(2)}(t), \tag{54}$$

$$g^{(2)}(t + \Delta t) = f(t + \Delta t) - g^{(1)}(t), \tag{55}$$

where $g^{(1)}(t), g^{(2)}(t)$ at time t are known, and Eqs. (54), (55) are used as boundary conditions in the first and the second boundary value problems, respectively. As a result of the numerical solution, we will find two functions $g^{(1)}, g^{(2)}$ from one Eq. (53). Considering the limit $\Delta t \to 0$, it is easy to see that the functions $g^{(1)}, g^{(2)}$

obtained in such a way approximate the solution

$$g^{(1)}(t) = \frac{1}{2}f(t) - \frac{1}{2}f(0) + g^{(1)}(0), \tag{56}$$

$$g^{(2)}(t) = \frac{1}{2}f(t) - \frac{1}{2}f(0) + g^{(2)}(0) \tag{57}$$

to Eq. (53), where $g^{(1)}(0)$, $g^{(2)}(0)$, $f(0)$ are given initial values satisfying (53). The numerical scheme with

$$g^{(2)}(t + \Delta t) = f(t + \Delta t) - g^{(1)}(t + \Delta t) \tag{58}$$

instead of (55) gives a different solution to (53):

$$g^{(1)}(t) = f(t) - f(0) + g^{(1)}(0), \tag{59}$$

$$g^{(2)}(t) = g^{(2)}(0). \tag{60}$$

We see that the use of one boundary condition twice results in the numerical determination of two unknown functions involved in the boundary condition independently of the solutions to the two boundary value problems.

References

1. Ye, F., Goh, S.H., Lee, F.H.: Dual-phase coupled u-U analysis of wave propagation in saturated porous media using a commercial code. Comput. Geotech. **55**, 316–329 (2014)
2. Biot, M.A.: Theory of propagation of elastic waves in a fluid-saturated porous solid. I. Low-frequency range. J. Acoust. Soc. Am. **28**(2), 168–178 (1956)
3. Zienkiewicz, O.C., Chan, A.H.C., Pastor, M., Schrefler, B.A., Shiomi, T.: Computational geomechanics with special reference to earthquake engineering. Wiley, Chichester (1999)
4. Osinov, V.A.: Non-reflecting boundary conditions for plane waves in anisotropic elasticity and poroelasticity. Acta Mech. **223**, 593–607 (2012)

Neohypoplasticity—Estimation of Small Strain Stiffness

I. Loges and A. Niemunis

Abstract Behaviour of soils under small cycles is examined in the triaxial apparatus and the results are used for the calibration of several constitutive relations. The small strain relation is not exactly linear and stiffness E_{ijkl} in $\dot{\sigma}_{ij} = E_{ijkl}\dot{\varepsilon}_{kl}$ is not constant. The popular hypoplastic (HP) model describes the small strain behaviour using the intergranular strain (Niemunis, Herle, Mech Cohesive-Frictional Mater 2(4):279–299 1997). However, this idea with an additional strain has several shortcomings. A better approach is the paraelastic (PE) model (Niemunis et al, Acta Geotech 6(2):67–80 2011; Prada Sarmiento, Paraelastic description of small-strain soil behaivour 2012). In this study the paraelasticity has been used already while evaluating of the raw data from triaxial test results. Similarly a simplified high cycle accumulation (HCA) formula (Niemunis et al, Comput Geotech 32(4):245–263 2005) and a simple assumption of stress dependence of E_{ijkl} have been used to purify the measured test data. A general curve-fitting strategy for testing of different constitutive models is developed. Some shortcomings of PE and HCA could be observed.

Keywords Small strain stiffness · Paraelasticity · Hyperelasticity · Curve fitting

1 Introduction

1.1 Notation

For evaluation of triaxial tests the Cartesian components σ_{ij}, ϵ_{ij} or Roscoe invariants $p, q, \varepsilon_{\text{vol}}, \varepsilon_q$ of stress $\boldsymbol{\sigma}$ and strain $\boldsymbol{\epsilon}$ (compression is positive) are less convenient than the isomorphic components:

I. Loges (✉) · A. Niemunis
Institute of Soil Mechanics and Rock Mechanics,
Karlsruhe Institute of Technology, Karlsruhe, Germany
e-mail: iryna.loges@gmail.com

Th. Triantafyllidis (ed.), *Holistic Simulation of Geotechnical Installation Processes*,
Lecture Notes in Applied and Computational Mechanics 77,
DOI 10.1007/978-3-319-18170-7_9

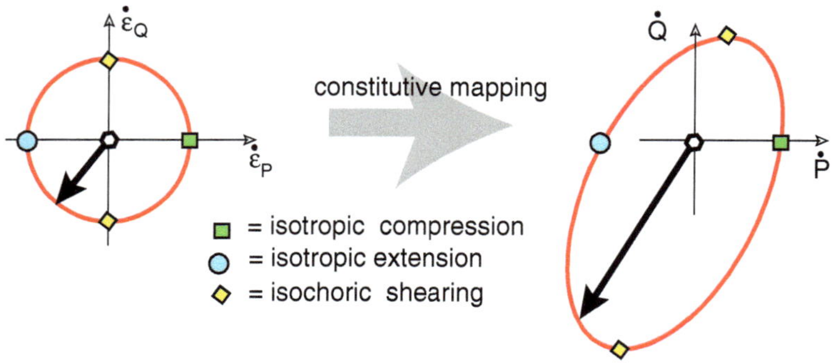

Fig. 1 Concept of response envelopes. *Left* applied strain increments (equal in all directions). *Right* stress increments obtained as the material response

$P = (\sigma_{11} + \sigma_{22} + \sigma_{33})/\sqrt{3} = p\sqrt{3}$, $Q = (2\sigma_{11} - \sigma_{22} - \sigma_{33})/\sqrt{3} = \sqrt{\frac{3}{2}}q$, $\varepsilon_P = (\varepsilon_{11} + \varepsilon_{22} + \varepsilon_{33})/\sqrt{3}$, $\varepsilon_Q = (2\varepsilon_{11} - \varepsilon_{22} - \varepsilon_{33})/\sqrt{3}$. Isomorphic variables preserve the orthogonality and distance contrarily to Roscoe invariants, e.g. $\|\boldsymbol{\sigma}\| = \sqrt{P^2 + Q^2} \neq \sqrt{p^2 + q^2}$, wherein $\| \quad \|$ is the Euclidean norm. For brevity we denote the isomorphic components of stress, strain and stiffness as $\sigma_A, \varepsilon_A, E_{AB}$ with the usual summation convention, e.g. $\sigma_A = E_{AB}\varepsilon_B$. Accordingly, δ_{AB} is a 2×2 version of the Kronecker symbol.

In the paraelastic (PE) model [7, 9] the reversals of stress and strain paths are of essential importance. The abbreviation $\overset{\triangle}{\sqcup} = \sqcup - \sqcup^R$ denotes the span of $\sqcup$ measured from the most recent reversal value $\sqcup^R$ and $h_\sqcup = \| \overset{\triangle}{\sqcup} \|$ is the size of the span.

1.2 On Evaluation of Stiffness

Stiffness E_{ijkl} used to describe stress rates $\dot{\sigma_{ij}} = E_{ijkl}\dot{\varepsilon_{kl}}$ at small strains is an important element of the hypoplastic model [5, 6]. Triaxial data dealing with the small deformation response allows for evaluation of the components of the small strain stiffness. At first, the triaxial testing programme was intended to reproduce experimentally the so-called response envelopes, to compare them (graphically) with the ones obtained from the HP and to conclude some improvements of the model from this comparison. Eventually a different approach based on the error of curve fitting has been proposed.

The polar representation of stiffness with the *response envelopes* [2, 4] is a well established graphic tool for modelling of soil behaviour. At a given state, a constitutive model obtains small perturbations, usually strain increments $\dot{\epsilon}_A$ of the same length $\|\dot{\epsilon}\| = \text{const}$ but in different directions. The resulting stress increments $\dot{\sigma}_A$ are plotted, Fig. 1. The obtained response envelope[1] is a valuable graphic representation of stiffness. For example, the response envelopes should be continuous,[2] otherwise the jumps would render the constitutive model unstable.

Tests with very small amplitudes ($\varepsilon^{\text{ampl}} < 10^{-5}$) are technically not feasible in a standard triaxial apparatus. A direct measurement of stiffness is too strongly blurred by inaccuracies from the testing device. At too large amplitudes various non-linear effects as accumulation, hysteretic phenomena, barotropy, anisotropy etc. overly distort the response envelope and unreliable small strain models are concluded. Therefore, we use cycles with intermediate amplitudes ($\varepsilon^{\text{ampl}} \approx 5 \times 10^{-4}$) and apply various purification procedures to remove the nonlinearities and the accumulation.

Primarily, the constitutive curve-fitting procedure is used to purify the results. After different purifying operation the net small-strain response remains. Beside the components of the elastic stiffness various material constants of PE and HCA etc. are calibrated in the course of the purification. Judging by the fitting error (after optimal calibration), we may also easily discover, which versions of these model are better suited for a given material and a given test. The error of fitting seems to be a convenient quantitative criterion for testing of a constitutive model. A graphical comparison of experimental and theoretical response envelopes is very subjective. Nevertheless, the response envelopes may be used for general visualization of the final results.

A reliable curve fitting requires an extensive experimental database. We use stress and strain paths from several triaxial stress controlled low-cycle tests (asterisk tests) performed at different initial stress levels with different polarizations. For example, Fig. 2 shows a typical asterisk (around the average stress $P^{\text{av}} = 521.05\,\text{kPa}$ and $Q^{\text{av}} = 124.05\,\text{kPa}$). The prescribed stress path, Fig. 2 left, consists of six polarizations α_{π}, each composed of regular stress cycles with 180° reversals. Such reversals are chosen to simplify the pertaining PE equation. If the average stress $\boldsymbol{\sigma}^{\text{av}}$ was overlaid by a chaotic stress path (a random walk) we could still use the curve fitting procedures but the complicated full versions of PE and HCA models would be indispensable.

The high quality laboratory tests were performed on the cubical triaxial samples with local strain measurement by Espino [1]. A single asterisk is intended for evaluation of a single small strain stiffness $E_{AB}^{\max}$ at $\boldsymbol{\sigma}^{\text{av}}$. The calibration of the material constants of PE and HCA and evaluation of these models is obtained as a by-product just for this stress level.

The curve-fitting algorithm is based on the minimization of error which is the sum of discrepancies between the experimental points and proposed constitutive relation. This is described in Sect. 2. In Sect. 3 we discuss some remedies of some problems appearing due to inaccuracies in raw data. The results of curve fitting can be found in Sect. 4. Having determined the constitutive relation between stress and strain, the response envelopes are drawn at different stress points (for illustration).

---

[1] It is common to consider that the form of response-envelope is an ellipse, but in general it is not.

[2] Some authors require also the convexity of response-envelopes, but the necessity of this condition is not proven.

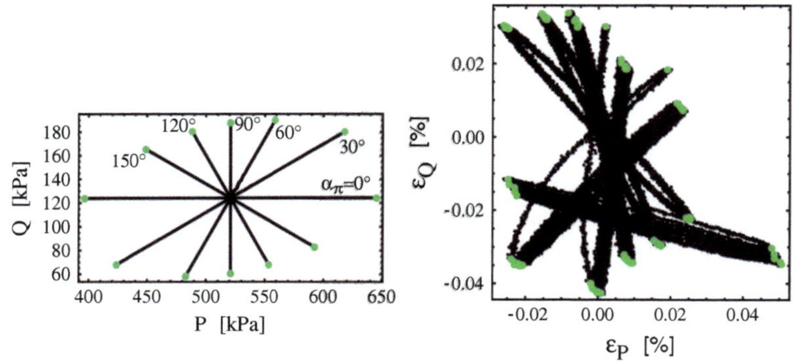

Fig. 2 Stress controlled low-cyclic asterisk test and corresponding strain response (provided in master thesis by Espino [1]). Reversal points are marked in *green*. *Left* Stress cycles. *Right* Strain response (problem with accumulation)

2 Linear Approximation of the Small Strain Stiffness

In order to estimate the small strain stiffness from the triaxial test we consider stress and strain values measured from the most recent reversal point R. These quantities are termed spans and are denoted as $\overset{\Delta}{\sigma}_{ij} = \sigma_{ij} - \sigma_{ij}^R$ and $\overset{\Delta}{\varepsilon}_{ij} = \varepsilon_{ij} - \varepsilon_{ij}^R$. A reversal is established at the current state whenever the distance h of this state from the most recent reversal begins to decrease, see PE [7, 9]. Only uniaxial cycles with 180° reversals are applied in tests, so that the general definition of distance given in [7, 9] simplifies to the Euclidean distance. Only two stress and two strain components can be measured in the triaxial tests and therefore it suffices to use isomorphic (with sign) components ε_P, ε_Q, P and Q. For uniaxial cycles we use the Euclidean measures of distance:

$$h_\epsilon = \sqrt{(\overset{\Delta}{\epsilon}_P)^2 + (\overset{\Delta}{\epsilon}_Q)^2} \quad \text{or} \quad h_\sigma = \sqrt{(\overset{\Delta}{P})^2 + (\overset{\Delta}{Q})^2}. \tag{1}$$

Given the first reversal $R = (P^R, Q^R, \epsilon_P^R, \epsilon_Q^R)$ and the test results in the form of a list of numbered $m = 1, 2, \ldots M$ records (quadruples) $(P, Q, \epsilon_P, \epsilon_Q)^m$ one can determine all subsequent reversal points (updating sequentially the record R) using following MATHEMATICA script:

```
findRever[data_,AnzahlP_:100]:=Module[{m,n,k,i,j,l,P,Q,EP,EQ,hS,spanS,begin,
                                       iReverAll,P0,Q0,jdel},
   m = Length[data]; iReverAll = {{1, data[[1, 1 ;; 4]]}}; {P0, Q0} = data[[1, 1 ;; 2]];
    hSlast = 0;
   Do[{P, Q, EP, EQ} = data[[i, 1 ;; 4]];
   Label[begin]; spanS = {P, Q} - iReverAll[[-1, 2, 1 ;; 2]]; hS = Norm[spanS];
      If[hS < hSlast, AppendTo[iReverAll, {i - 1, data[[i - 1, 1 ;; 4]]}];
       hSlast = 0; Goto[begin], hSlast = hS];
   , {i, 1, m}]; iReverAll = Drop[iReverAll, 1]; n = Length[iReverAll]; jdel = {};
   (*Delete not true reversal points*)
```

```
  Do[If[iReverAll[[j+1,1]]-iReverAll[[j, 1]]<AnzahlP,
     If[Flatten[Position[Ordering[{Norm[iReverAll[[j, 2, 1 ;; 2]]-{P0,Q0}],
                         Norm[iReverAll[[j+1, 2, 1;;2]]-{P0,Q0}]}],2]][[1]]==1,
      jdel = AppendTo[jdel, j + 1], jdel = AppendTo[jdel, j]]
       (*jdel = index of reversal point to be deleted*)];
  ,{j, 1, n - 1}]; k = Length[jdel];
  Do[iReverAll = Drop[iReverAll, {jdel[[-1 - 1]]}], {l, 0, k - 1}];
iReverAll];
```

Note that the reversals of the stress and strain path should coincide. Some discrepancies appear presumably due to measurement errors in the lab. Next we may append to each quadruple a list of spans $\overset{\triangle}{P}$, $\overset{\triangle}{Q}$, $\overset{\triangle}{\epsilon}_P$, $\overset{\triangle}{\epsilon}_Q$ and their sizes h_σ, h_ϵ followed by the void ratio e:

$$(P, Q, \epsilon_P, \epsilon_Q, \overset{\triangle}{P}, \overset{\triangle}{Q}, \overset{\triangle}{\epsilon}_P, \overset{\triangle}{\epsilon}_Q, h_\sigma, h_\epsilon, e)^m. \tag{2}$$

This is done with the following MATHEMATICA procedure:

```
createFullList[data_, Rever_] := Module[{n,P,Q,EP,EQ,hS,hE,spanS,spanE,i,outline,
                                          dataf,ReversAll,iR},
   n = Length[data]; ReversAll = Rever; dataf = {}; iR = 1;
   ReversAll = PrependTo[ReversAll, {1, data[[1, 1 ;; 4]]}];
   ReversAll = AppendTo[ReversAll, ReversAll[[-1]]];
   Do[{P, Q, EP, EQ} = data[[i, 1 ;; 4]]; If[i == ReversAll[[iR+1,1]],iR += 1;];
      spanS = {P, Q} - ReversAll[[iR, 2, 1 ;; 2]]; hS = Norm[spanS];
      spanE = {EP, EQ} - ReversAll[[iR, 2, 3 ;; 4]]; hE = Norm[spanE];
      outline = Join[data[[i, 1 ;; 4]], Flatten[{spanS, spanE, hS, hE}]];
      AppendTo[dataf, outline];, {i, 1, n}];
{dataf, Rever}];
```

In this paper we are considering a nearly linear (paraelastic) relation between the stress and strain spans. For triaxial case it is described with isomorphic components as

$$\left\{\begin{matrix} \overset{\triangle}{P} \\ \overset{\triangle}{Q} \end{matrix}\right\} = \begin{bmatrix} E_{PP} & E_{PQ} \\ E_{QP} & E_{QQ} \end{bmatrix} \cdot \left\{\begin{matrix} \overset{\triangle}{\epsilon}_P \\ \overset{\triangle}{\epsilon}_Q \end{matrix}\right\} \quad \text{or briefly} \quad \overset{\triangle}{\sigma}_A = E^s_{AB} \overset{\triangle}{\varepsilon}_B \tag{3}$$

with $E^s_{AB} \neq$ const. The stiffness tensor is reduced to three independent (and unknown) components E_{PP}, E_{QQ} and $E_{PQ} = E_{QP}$. If these components were constant we could determine them defining the squared error

$$F_0 = \sum_m (\overset{\triangle}{\sigma}_A - E^s_{AB} \overset{\triangle}{\varepsilon}_B)(\overset{\triangle}{\sigma}_A - E^s_{AC} \overset{\triangle}{\varepsilon}_C), \tag{4}$$

wherein the sum is taken over the records $m = 1, 2, \ldots M$ of interest. The components E_{AB} are found using the stationarity condition

$$(\partial F_0 / \partial E^s_{RS}) = -2 \sum_m \overset{\triangle}{\varepsilon}_S (\overset{\triangle}{\sigma}_R - E^s_{RC} \overset{\triangle}{\varepsilon}_C) = 0_{RS} \quad \text{and} \qquad \text{hence} \tag{5}$$

$$E^s_{RC} = \left[\sum_m \overset{\triangle}{\varepsilon}_C \overset{\triangle}{\varepsilon}_S\right]^{-1} \left[\sum_m \overset{\triangle}{\varepsilon}_S \overset{\triangle}{\sigma}_R\right]. \tag{6}$$

Equation (6) has been implemented in the MATHEMATICA routine `getConstApprox`

```
getConstApprox[espan_, sspan_] := Module[{ee, es, m, Estiffav, F},
  m = Length[espan];
  ee = Sum[Outer[Times, espan[[i]], espan[[i]] ], {i, 1, m}]; (*dyad eps x eps*)
  es = Sum[Outer[Times, espan[[i]], sspan[[i]] ], {i, 1, m}]; (*dyad eps x sig*)
  Estiffav = (Inverse[ee].es ) // Transpose;
  F = Sum[(sspan[[i]] - Estiffav.espan[[i]]).(sspan[[i]] - Estiffav.espan[[i]]), {i, 1, m}];
{F, Estiffav}];
```

Note that expressions in the square brackets are sums of dyads (having only one non-zero eigenvalue) so it is essential to have spans $\overset{\triangle}{\varepsilon}_C$ in different *directions* to avoid singularity. Of course, the first expression is symmetric. If the material response is elastic then the second expression should be also symmetric.

Usually, the secant stiffness E^s is not a constant within the experimental data. The stress components $\sigma_K = (P, Q)$, especially in the first cycle, may considerably deviate from the average value $\sigma_K^{\text{av}} = (P^{av}, Q^{av})$ during the test. Therefore the (linear, in Eq. (7)) influence of stress σ_K on the stiffness should be also considered:

$$E^s_{ijkl}(\boldsymbol{\sigma}) \approx E^0_{ijkl} + \frac{\partial E^s_{ijkl}}{\partial \sigma_{rs}} \left(\sigma_{rs} - \sigma^{av}_{rs}\right). \tag{7}$$

With isomorphic components and $\mathsf{E}^{s(av)} = \mathsf{E}^s(P^{av}, Q^{av})$ we obtain

$$E^s_{AB}(P, Q) \approx E^{s(av)}_{AB} + \frac{\partial E^s_{AB}}{\partial P}\left(P - P^{av}\right) + \frac{\partial E^s_{AB}}{\partial Q}\left(Q - Q^{av}\right). \tag{8}$$

In Eq. (8) we allow the resulting secant stiffness E^s_{RC} to vary depending on the amplitude, on stress and on the void ratio only

$$E^s_{RC} = E^s_{RC}(h_\epsilon, P, Q, e). \tag{9}$$

E^s_{RC} does not depend on the direction of deformation (a so-called polarization). Suppose, we have an average stress level $(P^{\text{av}}, Q^{\text{av}})$ around which the test has been done. We may approximate stiffness in the vicinity of $(P^{\text{av}}, Q^{\text{av}})$ using the following Taylor series (here only the barotropy):

$$\begin{bmatrix} E_{PP} & E_{PQ} \\ E_{QP} & E_{QQ} \end{bmatrix}^s = \begin{bmatrix} E_{PP} & E_{PQ} \\ E_{QP} & E_{QQ} \end{bmatrix}^{\text{av}} + (P - P^{\text{av}}) \begin{bmatrix} E_{PP,P} & E_{PQ,P} \\ E_{QP,P} & E_{QQ,P} \end{bmatrix} \quad \text{or} \quad E^s_{AB} = E^{\text{av}}_{AB} + \tilde{P}\, E'_{AB} \tag{10}$$

the unknown derivatives $E_{PP,P}$, $E_{PQ,P}\ldots$ of stiffness components with respect to pressure P are abbreviated with E'_{AB}. The average stiffness E^{av}_{AB} (components E_{PP},

$E_{PQ}, \ldots$) is taken from (6) and it is assumed to hold exactly at the average stress (P^{av}, Q^{av}). Additional parameters $E_{PP,P}, E_{PQ,P} \ldots$ should improve the minimization of error F. The error function (4) takes now the form

$$F_1 = \sum_m (s_A - \tilde{P}\, E'_{AB} \overset{\triangle}{\varepsilon}_B)(s_A - \tilde{P}\, E'_{AC} \overset{\triangle}{\varepsilon}_C), \tag{11}$$

wherein $s_A = \overset{\triangle}{\sigma}_A - E^{\text{av}}_{AB} \overset{\triangle}{\varepsilon}_B$ and $\tilde{P} = P - P^{\text{av}}$ denotes the deviation of pressure P from P^{av} (with sign) for each record m. The components E'_{AB} are found using the stationarity condition

$$(\partial F_1 / \partial E'_{RS}) = 0_{RS} = -2 \sum_m \tilde{P}\, \overset{\triangle}{\varepsilon}_S (\overset{\triangle}{s}_R - \tilde{P}\, E'_{RC} \overset{\triangle}{\varepsilon}_C) \quad \text{hence} \tag{12}$$

$$E'_{RC} = \left[\sum_m \tilde{P}^2 \overset{\triangle}{\varepsilon}_C \overset{\triangle}{\varepsilon}_S \right]^{-1} \left[\sum_m \tilde{P} \overset{\triangle}{\varepsilon}_S \overset{\triangle}{s}_R \right]. \tag{13}$$

Equation (13) is implemented in `getPApprox` MATHEMATICA routine:

```
getPApprox[espan_,sspan_,Estiffav_,Ppoint_,Pav_]:=Module[{ee,es,m,F,EstiffP,sPspan},
  m = Length[espan];
  sPspan=Table[sspan[[i]]-Estiffav.espan[[i]],{i,1,m}];
  ee=Sum[(Ppoint[[i]]-Pav)^2*Outer[Times,espan[[i]],espan[[i]]],{i,1,m}]; (*dyad*)
  es=Sum[(Ppoint[[i]]-Pav)*Outer[Times,espan[[i]],sPspan[[i]]],{i,1,m}]; (*dyad*)
  EstiffP = (Inverse[ee].es ) // Transpose;
  F=Sum[(sPspan[[i]]-(Ppoint[[i]]-Pav)*EstiffP.espan[[i]]).(sPspan[[i]]-(Ppoint[[i]]-Pav)*
                      EstiffP.espan[[i]]), {i, 1, m}];
{F, EstiffP}];
```

Having found the influence of P we may apply the analogous strategy to investigate the influence of Q. These improvements can be made one by one.

```
getQApprox[espan_,sspan_,Estiffav_,EstiffP_,Ppoint_,Pav_,Qpoint_,Qav_]:=
                                        Module[{ee, es, m, F, EstiffQ, sPspan, sQspan},
  m = Length[espan];
  sPspan = Table[sspan[[i]] - Estiffav.espan[[i]], {i, 1, m}];
  sQspan = Table[sPspan[[i]] - (Ppoint[[i]] - Pav)*EstiffP.espan[[i]], {i, 1, m}];
  ee=Sum[(Qpoint[[i]]-Qav)^2*Outer[Times,espan[[i]],espan[[i]]],{i, 1, m}];  (*dyad*)
  es=Sum[(Qpoint[[i]]-Qav)*Outer[Times,espan[[i]],sQspan[[i]]],{i,1,m}];  (*dyad*)
  EstiffQ=(Inverse[ee].es)//Transpose;
  F=Sum[(sQspan[[i]]-(Qpoint[[i]]-Qav)*EstiffQ.espan[[i]]).(sQspan[[i]]-(Qpoint[[i]]-Qav)*
                              EstiffQ.espan[[i]]), {i, 1, m}];
{F, EstiffQ}];
```

Finally we obtain the constitutive stiffness function

$$E^s_{RC}(\sigma_K) = E^{\text{av}}_{RC} + \tilde{P} E'_{RC} + \tilde{Q} E'_{RC}, \tag{14}$$

which is performed in MATHEMATICA as:

```
findStiffnessPQ[dataff_,{Pav_,Qav_},{P_:P,Q_:Q}]:=Module[{F1,F2,F3,Estiffav,EstiffP,EstiffQ,
                        Estiff, EstiffP4, EstiffQ4, F4, Pexp, Qexp, spanE, spanS },
   spanE=dataff[[All,7;;8]]; spanS=dataff[[All,5;;6]];
   {Pexp,Qexp}=dataff[[All,1;;2]]//Transpose;
   {F1, Estiffav} = getConstApprox[spanE, spanS];
   {F2, EstiffP}  = getPApprox[spanE, spanS, Estiffav, Pexp, Pav] // FullSimplify;
   {F3, EstiffQ}  = getQApprox[spanE, spanS, Estiffav, EstiffP, Pexp, Pav, Qexp, Qav];
   Estiff = Estiffav + (P - Pav)*EstiffP + (Q - Qav)*EstiffQ // FullSimplify;
{F3, Estiff}];
```

3 Problems with Small-Strain Measurements and their Remedies

Apart from the stress dependence the raw data may include a number of other undesirable effects, which can affect the evaluation of small strain stiffness. All fitting procedures described below have been carried out using MATHEMATICA internal functions `FindFit`, `NMinimize` or `FindMinimum`.

3.1 Strain Accumulation

It is evident, e.g. Fig. 2, that during first cycles of each polarization material experiences strong accumulation of strain. It becomes less pronounced in subsequent cycles. The decline of accumulation rate in the first polarization is shown in Fig. 3. Individual cycles have been plotted with different colour. The smaller the strain amplitude ε^{ampl} the less accumulation is observed, see Fig. 3 for comparison.

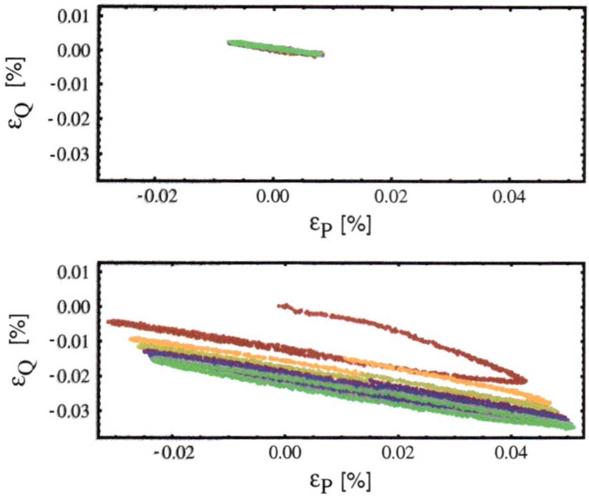

Fig. 3 Strain accumulation within the first polarization depending on the loading amplitude. *Top* No strain accumulation within 5 cycles with lower stress amplitude. *Bottom* Strain accumulation within 7 cycles with higher stress amplitude

In order to purify the experimental data from the cumulative phenomena all irregular cycles (from the so-called conditioning phase) were manually removed so that the subsequent ones could have been approximated with the simplified HCA-model [8, 11]. This operation is needed in each polarization ($\alpha_\pi = 0°, 30°, 60°, 90°, 120°, 150°$) defined in Fig. 2. The simplified expression for the accumulate strain rate is:

$$\dot{\epsilon}_{AB}^{\pi(\text{acc})}(t_i) = \mathrm{m}_{AB}^{\pi}(C_1^{\pi} + C_2^{\pi} t_i). \tag{15}$$

In order to find unknown flow direction m_{AB}^{π} and HCA parameters C_1^{π} and C_2^{π} we approximate experimental strain points at t_i with following expression:

$$\varepsilon_P^{\pi}(t_i) = A_P^{\pi}\cos\left(\omega^{\pi} t_i + \theta_P^{\pi}\right) + \left(A_P^{\pi(\text{acc})} + B_P^{\pi(\text{acc})}\, t_i/2\right) t_i, \tag{16}$$

$$\varepsilon_Q^{\pi}(t_i) = A_Q^{\pi}\cos\left(\omega^{\pi} t_i + \theta_Q^{\pi}\right) + \left(A_Q^{\pi(\text{acc})} + B_Q^{\pi(\text{acc})}\, t_i/2\right) t_i, \tag{17}$$

where the second summand in each line corresponds to the cumulative part, i.e. $A_{(P,Q)}^{\pi(\text{acc})} = m_{(P,Q)}^{\pi} C_1^{\pi}$ and $B_{(P,Q)}^{\pi(\text{acc})} = m_{(P,Q)}^{\pi} C_2^{\pi}$. In order to remove the accumulation we minimize the squared error between the experimental data and the fitting expressions (16, 17). This is implemented in the following MATHEMATICA procedure:

```
detrendAcct[{eP_, eQ_}, time_] := Module[{m,err=0,aP,aQ,w,APacc,BPacc,AQacc,BQacc,thetaP,
                                       thetaQ, eExp, eApp, i, out, solu, time0},
  m = Length[eP ]; time0 = time[[1]];
  Do[eExp={eP[[i]],eQ[[i]]};
    eApp={aP*Cos[w*(time[[i]]-time0)+thetaP],aQ*Cos[w*(time[[i]]-time0)+thetaQ]}
    +{APacc+BPacc*(time[[i]]-time0)/2,AQacc+BQacc*(time[[i]]-time0)/2}*(time[[i]]-time0);
    err += (eExp - eApp).(eExp - eApp),
  {i, 1, m}];
  solu = FindMinimum[err, {aP, aQ, {w, 0.0063}, thetaP,thetaQ, APacc, BPacc, AQacc, BQacc}];
  out = {aP, aQ, w, thetaP, thetaQ, APacc, BPacc, AQacc,BQacc} /. solu[[2]];
{solu[[1]]/m, out}];
```

The calculated components $\{m_P, m_Q\}$ and the parameters C_1 and C_2 are given in Table 1 for each polarization. It can be seen that the flow direction changes from polarization to polarization. This behaviour contradicts the assumption of HCA model that

Table 1 Approximated HCA parameters for each polarization

Polarization	Flow direction		HCA parameters	
Number	m_P	m_Q	$C_1 * 10^7$	$C_2 * 10^{11}$
Polarization $\alpha_\pi = 0°$	0.5761	−0.8174	2.3174	−7.3276
Polarization $\alpha_\pi = 30°$	0.1931	−0.9812	1.1320	−3.4493
Polarization $\alpha_\pi = 60°$	0.3190	−0.9477	0.9378	−2.8483
Polarization $\alpha_\pi = 90°$	0.0069	−0.9999	3.0960	−0.1260
Polarization $\alpha_\pi = 120°$	0.3473	−0.9377	−0.1507	0.6563
Polarization $\alpha_\pi = 150°$	0.5286	−0.8489	−0.3552	1.2726

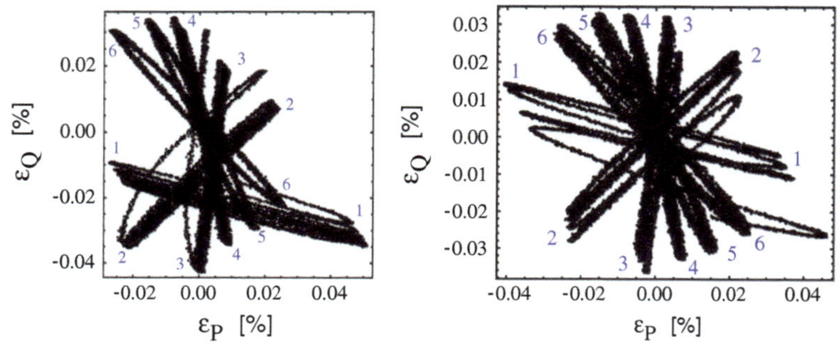

Fig. 4 Strain path before and after subtraction of accumulated strain. *Left* Original strain path. *Right* Strain path with subtracted strain accumulation according to Eq. (15)

the flow rule is a function of stress only. Moreover, it has been shown that C_1 does not decrease from polarization to polarization. According to the HCA model such decrease is a consequence of the cyclic pre-loading history of the sample.

The raw experimental data and experimental data with subtracted strain accumulation are compared in Fig. 4.

3.2 Measurement Noise

Apart from accumulation described in Sect. 3.1 we need to remove the noise in the raw data due to the measurement technology. The smaller is the strain amplitude the higher is the data impurity with respect to the signal. The removal of the noise is provided by a convolution of the raw data with a discrete Gauss distribution as a smoothing function. This is implemented in the following MATHEMATICA procedure:

```
smoothData[data_, nGlPoints_: 2] := Module[{ker, t, n, i, EPtime, EQtime,
                                     smoothD, dataExtendB, dataExtendE, smoothDstep},
   ker = Table[Exp[-n^2/10.0]/Sqrt[10.0 \[Pi]], {n, -nGlPoints, nGlPoints}];
   ker /= (Plus @@ ker);
```

Fig. 5 Measurement noise: original (in *blue*) and smoothed (in *red*) strain $\varepsilon_P(t)$ diagram

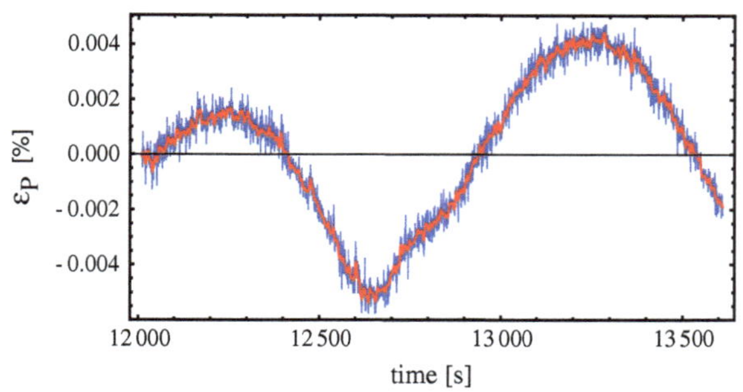

```
  smoothD = ( ListCorrelate[ker , #]  & /@ Transpose[data] ) // Transpose ;
  dataExtendB = Array[{0, 0} &, nGlPoints]; dataExtendE = Array[{0, 0} &, nGlPoints];
  Do[dataExtendB={data[[nGlPoints-i+1,1]],Interpolation[smoothD,data[[nGlPoints-i+1,1]]]};
     dataExtendE={data[[Length[data]-nGlPoints+i,1]],
                  Interpolation[smoothD,data[[Length[data]-nGlPoints+i,1]]]};
  PrependTo[smoothD, dataExtendB]; AppendTo[smoothD, dataExtendE];
 ,{i, 1, nGlPoints}];
smoothD];
```

We avoid smoothing across the reversal points. Therefore all reversals have to be determined a priori. Figure 5 shows the experimental data with subtracted strain accumulation befor and after the smoothing procedure.

In order to speed up the calculation one may decrease the number of experimental points. In the following example each *n*th point is processed:

```
pickPoints[data_, time_, step_: 1] := Module[{dataRed = {}, timeRed = {}},
  Do[AppendTo[dataRed,data[[i]]];AppendTo[timeRed,time[[i]]],{i,1,Length[time],step}];
{dataRed, timeRed}];
```

3.3 Hysteretic σ–ε Behaviour

Triaxial tests with intermediate strain amplitudes show hysteretic σ–ε behaviour (see e.g. Fig. 6). It means that for the same stress state we obtain different strain states depending on loading direction (or vice versa). In order to take this effect into account we use the paraelasticity. The PE formulation for 1D cycles can be simplified as a secant [7, 9] stiffness decreasing with the amplitude (or span $h_\epsilon = \left\| \varepsilon - \varepsilon^R \right\|$):

$$E^s_{RC} \approx (1 - f h_\epsilon^\chi) E^{\max}_{RC} \tag{18}$$

The unknown material constants f and χ can be determined in the course of the fitting of experimental data. For $h_\epsilon < 0.02\,\%$ this paraelastic dependence should work well. The desired small stain stiffness $\mathsf{E}^{\max}$ is obtained from the extrapolation at $h_\epsilon \to 0$.

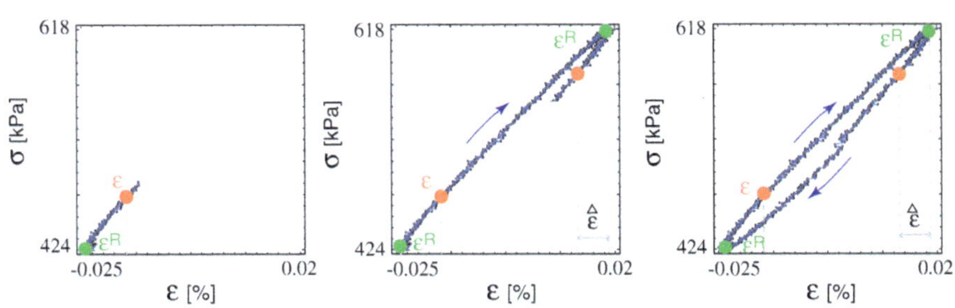

Fig. 6 Hysteretic behaviour simulated for low cyclic tests

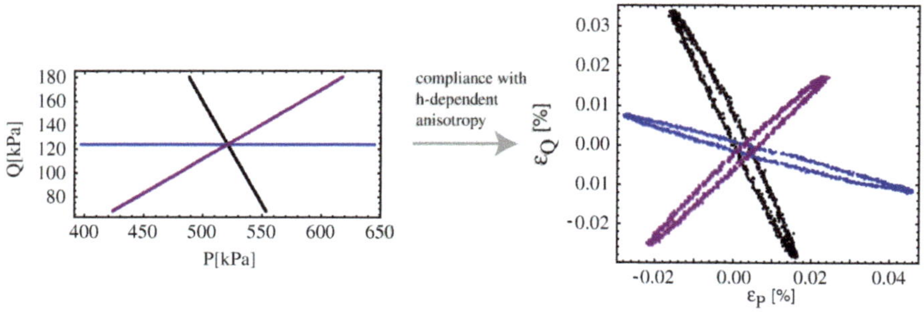

Fig. 7 Anisotropy of stiffness varies with h_ϵ

3.4 Anisotropy of Stiffness Depending on h_ϵ

In some tests, despite uniaxial stress path along the prescribed polarization, an occurrence of oval strain paths could be observed, Fig. 7. This effect should not be mixed up with the commonly observed σ–ε hysteresis. This ovality described as a variable anisotropy of stiffness. This phenomenon is not accounted for in any extant small strain model.

In order to capture the h_ϵ-dependent anisotropy of stiffness a minor modification of PE is proposed by introducing of a small h_ϵ-dependent tensorial correction $\mathsf{C}(h_\epsilon)$ to secant stiffness:

$$\mathsf{E}^s \approx (1 - fh_\epsilon^\chi)\mathsf{E}^{\max} + \mathsf{C}(h_\epsilon). \tag{19}$$

Unlike the scalar modification of E^s proposed in PE model (19) allows for a h_ϵ-dependent anisotropy. Equation (19) can be rewritten in the form

$$\mathsf{E}^s \approx \mathsf{E}^{\max} - fh_\epsilon^\chi\, \mathsf{E}^{corr} \tag{20}$$

with non-proportional tensors $\mathsf{E}^{\max} \not\sim \mathsf{E}^{corr}$. Here, apart from f and χ, we have 8 unknown components of matrices $\mathsf{E}^{\max}$ and E^{corr} which should be found by fitting. The effect of variable stiffness anisotropy can be alternatively captured using a "rotation" tensor R:

$$\mathsf{E}^s \approx \mathsf{E}^{\max} - fh_\epsilon^\chi\, R^T : \mathsf{E}^{\max} : R \tag{21}$$

In the case of isomorphic components we obtain:

$$E_{RC}^s \approx (\delta_{RA}\delta_{CB} - fh_\epsilon^\chi\, R_{RA}R_{CB})\, E_{AB}^{\max} \tag{22}$$

which generates strain loops from linear stress path. The improvement can be concluded judging by the significant reduction of fit errors. The rotation tensor $R = R(h_\epsilon)$ is defined as:

$$R(h_\epsilon) = \begin{bmatrix} \cos(kh_\epsilon) & \sin(kh_\epsilon) \\ -\sin(kh_\epsilon) & \cos(kh_\epsilon) \end{bmatrix}. \tag{23}$$

Beside four (or three due to symmetry) unknowns components of matrix $E_{AB}^{\max}$ we need just one additional parameter k to be found while fitting procedure.

4 Numerical Results and Discussions

The experimental data provided for this analysis by Espino [1] consist of 7 asterisk-shaped stress paths performed at different stress levels as shown in Fig. 8. Material parameters are fitted for each stress level separately. Numerical results and error ranges are therefore independent.

First we read the experimental raw data including information about the time for each measured point from the input file and reorganize it to the form of quadruples $(P, Q, \varepsilon_P, \varepsilon_Q)$. After subtracting strain accumulation (if necessary), the reversal points are found using `findRever`. Between the reversals the smoothing procedure `smoothData` is carried out. The essential steps for determination of unknown material parameters, among them the parameters of small strain stiffness itself, are provided below:

a. **Filtering of experimental points.** In order to determine the dependence of E^s on the span h_ϵ the measured records are grouped depending on the size of the span h_ϵ. A filter function `filterExperPoints` picks out the records containing spans (=increments) of the required size h_ϵ (`hDesired` $\pm$ `Acc`):

```
filterExperPoints[dataf_, hEdesired_, Accuracy_: 10.^(-8)] := Module[{dataff,i,Pav,Qav},
   dataff = Select[dataf, Abs[#[[10]] - hEdesired] < Accuracy &];
   If[Length[dataff] == 0 || Length[dataff] == 1,
```

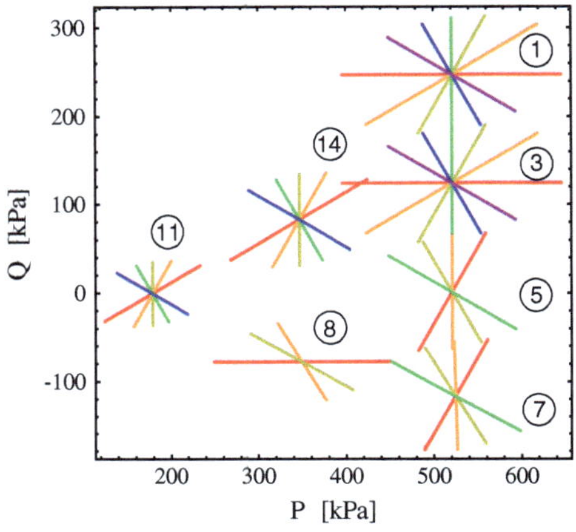

Fig. 8 Stress asterisks at different stress levels chosen for small strain analyse

Fig. 9 Filtered strain points

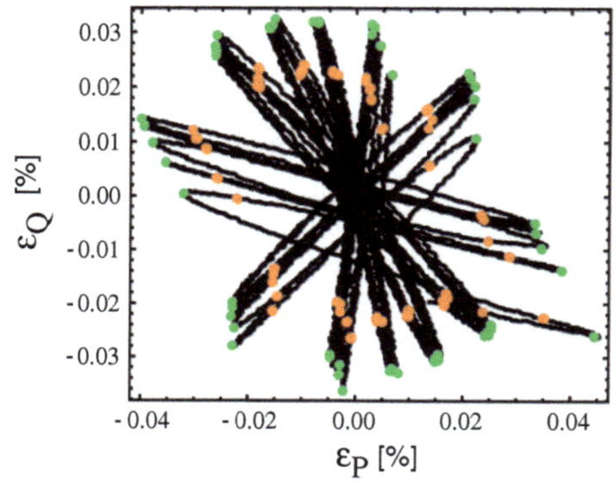


```
    Print[Style["Error: Zero records passed the filter. Analysis will be stopped.
    Check the defined value of hEdesired or the accuracy", 14, Red, Bold]]; Abort[]];
  If[Length[dataff] < 10,
    Print["Warning: Only", Length[dataff] - 1, "records passed the filter \n"]];
  {Pav, Qav} = Mean[dataff[[2 ;; -1, 1 ;; 2]]]; Print["{Pav,Qav} =", {Pav, Qav}];
{dataff[[2 ;; -1]], {Pav, Qav}}];
```

For example, in Test 3 (see Fig. 8) this procedure could filter out 159 strain points at the distance $h_\epsilon = 0.0001 \pm 2 \times 10^{-6}$ from the most recent reversals. They are shown in orange in ε_P–ε_Q diagram in Fig. 9. The number of the experimental points (m_f) after filtering procedure available for further analysis for the chosen test and h_ϵ is given in Table 2. The accuracy increases with number of points, of course.

b. **Calculating stiffness.** Linear approximation of the small strain secant stiffness $E^s_{AB}(P, Q)$ for a given h_ϵ is carried out according to Eq. (14). For example, for $h_\epsilon = 0.9 \times 10^{-4} \pm 2 \times 10^{-6}$ the values of stiffness components from different tests are given in Table 3 for the average stress level (P^{av}, Q^{av}). Squared errors, e.g. F_1 form Eq. (11), can be expressed as a ratio of standard error deviation per pure measurement to h_σ:

Table 2 Number of filtered points available for small stiffness analysis for chosen h_ϵ range

	Number of filtered points (m_f) for						
Chosen $h_\epsilon \pm 2 \times 10^{-6}$	Test 1	Test 3	Test 5	Test 7	Test 8	Test 11	Test 14
$h_\epsilon = 0.9 \times 10^{-4}$	161	175	125	96	51	113	231
$h_\epsilon = 2.1 \times 10^{-4}$	122	132	84	75	32	88	176
$h_\epsilon = 3.3 \times 10^{-4}$	110	110	73	58	28	78	163
$h_\epsilon = 4.5 \times 10^{-4}$	119	123	80	91	38	76	174
$h_\epsilon = 5.1 \times 10^{-4}$	166	138	94	32	26	58	203

Table 3 The components of small strain stiffness tensor $E^s_{AB} = E^s_{AB}(P, Q)$ for $h_\epsilon = 0.9 \times 10^{-4} \pm 2 \times 10^{-6}$

Test number	Stress level (kPa)		Small strain stiffness (kPa)			Fitting error (%)
	P	Q	E^s_{PP}	$E^s_{PQ} = E^s_{QP}$	E^s_{QQ}	F
Test 1	521.06	246.96	351 547	110 778	282 050	1.26
Test 3	521.05	124.05	430 211	66 077.6	246 056	0.65
Test 5	521.12	1.09	477 436	−6 762.28	218 573	0.95
Test 7	525.01	−116.51	471 363	−122 052	161 716	1.83
Test 8	349.35	−77.43	343 273	−86 897.9	113 112	1.52
Test 11	179.42	−0.38	268 915	−34 837.7	109 468	2.20
Test 14	347.97	83.06	361 681	156.29	179 567	0.45

$$F = \frac{\sqrt{F_1}}{m_f\, h_\sigma}. \tag{24}$$

Commonly used in statistic coefficient of determination R^2 (e.g. [10]) is obtained as $R^2 \approx 1 - m_f F^2$. Calculated here average relative fitting error (F) varies from 0.5 to 2.2 % per measurement for all tests. Some discrepancies mean non-linear dependence of the small strain stiffness on stress and they are to be removed taking into account hysteretic σ–ε behaviour (18) and the variable anisotropy of stiffness approximated by (23).

c. **Determination of paraelastic material parameters f and χ.** Paraelastic parameters f and χ are determined separately for all stress levels from Fig. 8. For this purpose stiffness components for different h_ϵ must be evaluated. Given the following MATHEMATICA skript we may find the paraelastic parameters from results similar to Table 3 but for different h_ϵ:

```
FindFit[{hList,EnormsList},(1-f*h^\[Kappa])*Emax,{{f,-3},{\[Kappa],0.2},{Emax,1.5}},h]
```

The results of fitting analysis are presented in Table 4. Except for the Test 5 the values of paraelastic parameters are considerably lower than expected from earlier calibration of the PE model (e.g. $f \approx 300$ and $\chi \approx 0.9$ in [9]). It means that for chosen a h_ϵ range (from $h_\epsilon = 0.9 \times 10^{-4}$ to $h_\epsilon = 5.1 \times 10^{-4}$) the material is

Table 4 Paraelastic parameters f, χ and $\|E^{\max}\|$

Parameter	Test 1	Test 3	Test 5	Test 7	Test 8	Test 14
f	0.48	3.15	154.74	0.84	1.40	0.68
χ	0.05	0.40	0.93	0.02	0.01	0.03
$\|E^{\max}\|$	1.42	1.08	1.03	3.57	1.86	2.10

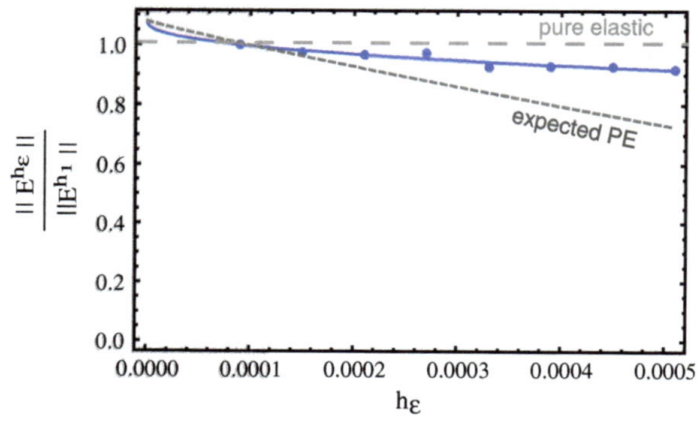

Fig. 10 The secant stiffness $\|E^{h_\epsilon}\| / \|E^{h_1}\|$ decreases with the distance h_ϵ from the most recent reversal here shown for Test 3 only. The expected PE curve is shown for $f = 300$ and $\chi = 0.9$

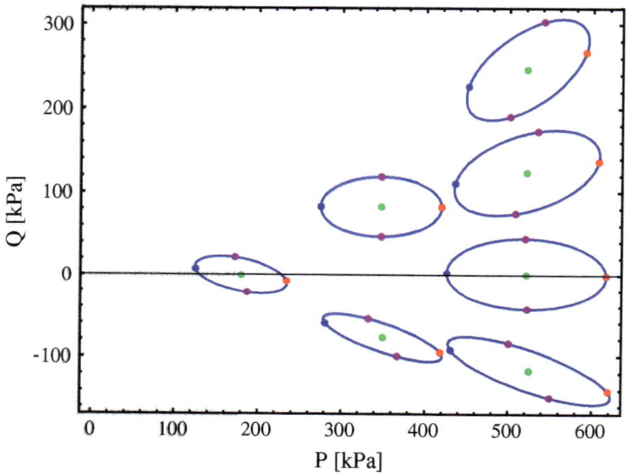

Fig. 11 The stress response envelopes in isomorphic *P*-*Q* plane for different tests. The *red* points correspond to the isotropic compression, *blue* ones to the isotropic extension and the *purple* ones to the deviatoric shearing. *Green* points represent the initial stress state

nearly elastic with a very weak hysteretic effect. A comparison is shown in Fig. 10 for Test 3 only.
We presume that the weak hysteretic behaviour is a consequence of the experimental errors (relatively high cyclic frequency, low number of cycles, temperature effects during loading, etc.)

d. **Response envelopes**. The elastic stiffness has been evaluated from experimental data for each stress level. The constitutive response at small strain can be visualized in the form of response envelopes, Fig. 11. As expected, the ellipses rotate and change their size depending on P and Q. The ellipses also become narrow for triaxial extension.

5 Summary

A general algorithm for determination of stress-dependent small strain stiffness has been presented. This algorithm is insensitive to the method how the tests are performed and controlled (stress-, strain- or mixed-controlled tests can be used). The assumption of a linear dependence of the secant stiffness on stress can be easily extended to non-linear forms. It has been shown that problems of data analysis, such as an accompanying measurement noise or a residual strain accumulation, can be removed using the proposed algorithms. Implementing paraelastic model with some minor modification for variable anisotropy allows to reduce number of laboratory tests, instead concentrating only on one stress (strain) amplitude. The small strain stiffness evaluated from the raw data by fitting the constitutive model can be graphically represented in the isomorphic stress diagram in the form of response envelopes. However, it seems that the direct implementation of constitutive equations into fitting algorithm is of advantage compared with the graphical method, in particular when advanced and well established models are available. We do not recommend the response envelopes as intermediate results between laboratory tests and the constitutive modelling. In future we intend to apply the proposed algorithm to the more experimental data of better quality (e.g. from SP 8, DFG FOR1136 [3]) and to compare our stiffness calculation method with a graphical description of small strain stiffness.

Acknowledgments The authors are grateful to the M. Sc. E. Espino Cedro for his careful experimental work and provision of interesting test data used in this paper. The authors acknowledge financial support of SP1, DFG FOR1136.

References

1. Espino, E.: Quasi-statische Untersuchungen zur Elastizität von Sand als Grundlage eines neuen hypoplastischen Stoffmodells. Master's thesis, Institut für Boden- und Felsmechanik, Karlsruher Institut für Technologie, Mai (2014)
2. Gudehus, G.: A comparison of some constitutive laws for soils under radially symmetric loading and unloading. In: Proceedings of the 3rd International Conference on Numerical Methods in Geomechanics, Aachen, Aachen, 1979. Balkema
3. Hettler, A., Danne, St.: Strain response envelopes for low cycle loading pro-cesses. In: International Conference on Soil Mechanics and Geotechnical Engineering, Paris, vol. 2, pp. 1491–1494 (2013)
4. Lewin, P.I., Burland, J.B.: Stress-probe experiments on saturated normally consolidated clay. Géotechnique **20**, 38–56 (1970)
5. Niemunis, A.: Extended hypoplastic models for soils. Politechnika Gdańska, 2003. Habilitation, Monografia 34
6. Niemunis, A., Herle, I.: Hypoplastic model for cohesionless soils with elastic strain range. Mech. Cohesive-Frictional Mater. **2**(4), 279–299 (1997)
7. Niemunis, A., Prada-Sarmiento, L.F., Grandas-Tavera, C.E.: Paraelasticity. Acta Geotech. **6**(2), 67–80 (2011)

8. Niemunis, A., Wichtmann, T., Triantafyllidis, T.: A high-cycle accumulation model for sand. Comput. Geotech. **32**(4), 245–263 (2005)
9. Prada Sarmiento, L.F.: Paraelastic description of small-strain soil behaivour. PhD thesis, IBF, Karlsruher Institut für Technologie, Nr. 173 (2012)
10. Sachs, L.: Angewandte Statistik. Anwendung statistischer Methoden. Springer, New York (1996)
11. Wichtmann, T.: Explicit accumulation model for non-cohesive soils under cyclic loading. PhD thesis, Ruhr-University Bochum, Heft 38 (2005)

Improved Integration of High-Cycle Accumulated Strain Using Hierarchical and EAS Finite Elements

A. Niemunis and I. Melikayeva

Abstract The spatial variability of high-cycle accumulated strain is larger than the variability obtained from monotonic loading. Gradients of strain cause that the fit of a given strain field $\epsilon_{ij}^{\rm acc}$ is difficult for the conventional elements (if no refinement is used). Usage of identical mesh for monotonic and cumulative deformations leads to numerical self-stresses in elements. More flexible elements are therefore examined. They can reduce the self-stresses considerably.

Keywords High-cycle accumulation · Self-stress · EAS-element · Hierarchical element · Locking

1 Introduction

1.1 Notation

A fixed orthogonal Cartesian coordinate system x_1, x_2, x_3 is used. A repeated (dummy) index implies summation. We use the Kronecker's symbol δ_{ij} and the permutation symbol e_{ijk}. The fourth order identity tensor is $(\mathsf{J})_{ijkl} = \delta_{ik}\delta_{jl}$ and its symmetrizing part is $I_{ijkl} = \frac{1}{2}(\delta_{ik}\delta_{jl} + \delta_{il}\delta_{jk})$. Proportionality of tensors is denoted by tilde, $\sim$. The components of diagonal matrices (with zero off-diagonal components) are written as diag[, ,], for example 2nd order identity tensor can be expressed as diag[1, 1, 1]. The operator $(\sqcup)^{\rightarrow}$ or $\vec{\sqcup}$ denotes normalization of $\sqcup$, for example $\vec{\epsilon}_{ij} = \epsilon_{ij}/\sqrt{\epsilon_{kl}\epsilon_{kl}}$. The superposed dot $\dot{\sqcup}$ denotes the material time derivative of $\sqcup$. The superposed breve $\breve{\sqcup}$ denotes the prescribed value of $\sqcup$. The superscript $\sqcup^e$ refers to $\sqcup$ inside the element number e. Virtual displacements, strains or stresses are denoted

A. Niemunis (✉) · I. Melikayeva
Institute of Soil Mechanics and Rock Mechanics,
Karlsruhe Institute of Technology, Karlsruhe, Germany
e-mail: andrzej.niemunis@kit.edu

Th. Triantafyllidis (ed.), *Holistic Simulation of Geotechnical Installation Processes*,
Lecture Notes in Applied and Computational Mechanics 77,
DOI 10.1007/978-3-319-18170-7_10

as $\delta\sqcup$. Finite increments of $\sqcup$ are denoted by $\overset{\Delta}{\sqcup}$. The superscripts $\sqcup^{\text{acc}}$ and $\sqcup^{\text{ampl}}$ denote the cumulative part of $\sqcup$ or the amplitude of $\sqcup$, respectively. Small strains $\epsilon_{ij} = u_{(i,j)} = \frac{1}{2}\left(u_{i,j} + u_{j,i}\right)$ and mechanical sign convention (tension positive) are used. The lower Greek indices refer to the number of node or a number of a nodeless variable. The lower Latin indices refer to the number of the Cartesian component, e.g. $u_{\alpha i}$ denotes a component i of displacement of the node α. For both Greek and Latin letters apply the summation convention over repeated dummy indices. In the case of nodes the summation runs over all nodes of a given element.

The following symbols are of importance:

$A_{\omega i}$	$= \int_{V^e} N_{\omega,j} E_{ijkl} \epsilon_{kl}^{\text{acc}}$, force-RHS due to initial strain $\epsilon_{ij}^{\text{acc}}$
$\bar{A}_\alpha$	$= \int_{V^e} M_{\alpha ij} E_{ijkl} \epsilon_{kl}^{\text{acc}} \text{d}V$, stress-RHS due to initial strain $\epsilon_{ij}^{\text{acc}}$
$B_{ij\alpha k}$	Strain-displacement matrix for $\epsilon_{ij} = B_{ij\alpha k} u_{\alpha k}$
$c_{\alpha i}$	Nodeless displacement parameter
$c_{\alpha i}$	Nodeless displacement parameter
$D_{\alpha\beta}$	$= \int_{V^e} M_{\alpha ij} E_{ijkl} M_{\beta kl} \text{d}V$, stress-EAS stiffness
E_{ijkl}	$= \partial\sigma_{ij}/\partial\epsilon_{kl}$ stiffness (links relaxation with creep in HCA)
f_i	Volume force
J	Jacobian for dimensionless coordinates $J = \det(@\mathbf{x}/@\boldsymbol{\xi})$
n_i	Unit outer normal vector to the boundary
$K_{\omega i\gamma k}$	$= \int_{V^e} N_{\omega,j} E_{ijkl} N_{\gamma,l} \text{d}V$, force-displacement stiffness
$L_{\omega i\alpha}^T$	$= \int_{V^e} N_{\omega,j} E_{ijkl} M_{\alpha kl} \text{d}V$, force-EAS stiffness
$L_{\alpha\gamma k}$	$= \int_{V^e} M_{\alpha ij} E_{ijkl} N_{\gamma,l} \text{d}V$, stress-displacement stiffness
$N_\alpha, N_{\alpha,i}$	Shape function of node α or its Cartesian gradient wrt x_i
$M_{\alpha ij}$	Interpolator for assumed strain field $\tilde{\epsilon}_{ij} = M_{\alpha ij}\tilde{\epsilon}_\alpha$
$M_{\alpha ij}^a$	$M_{\alpha ij}$ scaled by J_0/J for orthogonality
P_α	Interpolator for stress, $P_1(\boldsymbol{\xi}) \equiv 1$ in EAS-Element
$P^n(\xi), P^n(\eta)$	Legendre polynomial of degree n wrt ξ or η
R_{ω_i}	$= \int_{V^e} N_\omega f_i \text{d}V + \int_{S^{et}} N_\omega \check{t}_i \text{d}S$ from tractions and body forces
s_{ij}	Difference between stress as primary function and stress as a slave function of u_i and $\tilde{\epsilon}_{ij}$
t_i	$= \sigma_{ij} n_j$, stress vector
$u_i, u_{\alpha i}$	Field of ith components of displacement or its nodal value
γ_{12}	$= \epsilon_{12} + \epsilon_{21}$, shear strain (Voigt notation)
ϵ_{ij}	Strain (small, tension positive)
$\tilde{\epsilon}_{ij}, \tilde{\epsilon}_\alpha$	Enhanced strain field and its "nodal" value
$\boldsymbol{\xi} = (\xi, \eta)$	Dimensionless coordinates in an element , $\xi, \eta \in [-1, 1]$
$\xi_\alpha = \xi_{\alpha 1}$	$= \pm 1$, a dimensionless coordinate of node α
$\sigma_{ij}, \sigma_{\alpha ij}$	Cauchy stress (tension positive) field and its "nodal" value

1.2 FE Implementation of the High Cycle Accumulation Model

According to the high cycle constitutive model [4] (HCA) for soils

$$\dot{\sigma}_{ij} = E_{ijkl}(\dot{\epsilon}_{kl} - \dot{\epsilon}^{\rm acc}_{kl} - \dot{\epsilon}^{\rm pl}_{kl}) \tag{1}$$

the rate of accumulation $\dot{\epsilon}^{\rm acc}$ is proportional to the square of the amplitude.

For a typical punch problem of a shallow foundation in soil mechanics the stress amplitude due to a vertical point load $P^{\rm ampl}$ on the surface of the elastic half-space can be roughly estimated using the well known formulas by Boussinesq (3D case) or by Flamant line load $\bar{P}^{\rm ampl}$ (2D)

$$\sigma^{\rm ampl}_{rr} = -\frac{3P^{\rm ampl}}{2\pi r^2}\cos\theta \quad \text{or} \quad \sigma^{\rm ampl}_{rr} = -\frac{3\bar{P}^{\rm ampl}}{2\pi r}\cos\theta, \tag{2}$$

wherein $\cos\theta = \vec{\mathbf{x}} \cdot \mathrm{diag}\{0, -1\}$. In plane strain problems $\epsilon^{\rm ampl}$ decreases linearly in space and hence $\dot{\epsilon}^{\rm acc}_{ij}$ decreases with the square of the distance r from the foundation. According to some recent triaxial tests [7] the exponent n in $\epsilon^{\rm acc}_{ij} \sim \left(\epsilon^{\rm ampl}\right)^n$ can be smaller than two, typically $1.4 < n < 2.0$. On the other hand, however, some additional spatial variability of $\epsilon^{\rm acc}_{ij}$ results from the barotropic stiffness (deformation amplitude is larger at the ground surface), from the stress obliquity (in the vicinity of the foundation the accumulation is larger due to larger mobilized friction angle) and from the barotropy of the accumulation ($\epsilon^{\rm acc}_{ij}$ is larger at low effective mean stress, i.e. at the ground surface). Taking into account all these effects we may conclude that the plane stain finite elements used with HCA should be able to reproduce at least all strain fields which vary proportionally to r^2. It turns out that this requirement is of crucial importance for the results of the FE-analysis with HCA and it is not just a matter of inaccurate predictions of settlements. Another important problem is a feed back phenomenon caused by self-stresses which appear in overly constrained elements. The numerical errors seem to amplify themselves.

Choosing a FE formulation we should eliminate numerical self-stresses. However we should not eliminate *any* self-stress appearing in the FE solution. For example, under conditions $\dot{\epsilon}_{ij} = 0$ and at a small stress obliquity that guarantees $\epsilon^{\rm pl}_{kl}$ any high cycle loading leads, according to Eq. (1), to the relaxation $\dot{\sigma}_{ij} = -E_{ijkl}\epsilon^{\rm acc}_{kl}$ which should be preserved in the FE simulations. Similarly, some self-stresses may appear due to kinematic incompatibility

$$\eta_{ij} = e_{imk}e_{jnl}\epsilon^{\rm acc}_{kl,mn} \neq 0.$$

These self-stresses are of physical nature and they should not be artificially eliminated. In all examples presented in this paper the prescribed 2D strain fields $\epsilon^{\rm acc}_{ij}$ are

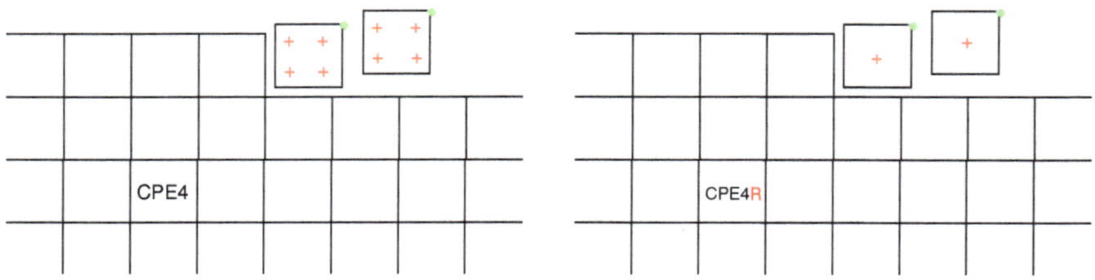

Fig. 1 Number of degrees of freedom (DOF) and number of constraints per element. *Left* Full integration with 2 DOFS (u_1, u_2) and 12 constraints $\epsilon_{11} = \epsilon_{11}^{\mathrm{acc}}$, $\epsilon_{22} = \epsilon_{22}^{\mathrm{acc}}$ and $\gamma_{12} = \gamma_{12}^{\mathrm{acc}}$ in four integration points. *Right* Reduced integration with 2 DOFS (u_1, u_2) and 3 constraints $\epsilon_{11} = \epsilon_{11}^{\mathrm{acc}}$, $\epsilon_{22} = \epsilon_{22}^{\mathrm{acc}}$, and $\gamma_{12} = \gamma_{12}^{\mathrm{acc}}$ in one integration point

assumed to be kinematically compatible, that is they satisfy[1]

$$\epsilon_{11,22}^{\mathrm{acc}} + \epsilon_{22,11}^{\mathrm{acc}} - \gamma_{12,12}^{\mathrm{acc}} = 0 \tag{3}$$

Moreover, only statically determinate supports are chosen so that the only self-stresses appearing in our examples are of *numerical* origin.

1.3 Spatial Integration of Prescribed a Strain Field

For prediction of settlements due to high-cycle loading we calculate the fields ϵ^{ampl}, $\epsilon_{ij}^{\mathrm{acc}}$ and then we need to integrate $\epsilon_{ij}^{\mathrm{acc}}(\mathbf{x})$ to a displacement field $u_i^{\mathrm{acc}}(\mathbf{x})$. The problem of artificial self-stresses in conventional elements can be expected by counting the number of DOFs and the number of constraints (Fig. 1).

Discrepancies $u_{(i,j)} = \epsilon_{ij} \neq \epsilon_{ij}^{\mathrm{acc}}$ result in stresses

$$\overset{\triangle}{\sigma}_{ij} = E_{ijkl}(\overset{\triangle}{\epsilon}_{kl} - \overset{\triangle}{\epsilon}{}_{kl}^{\mathrm{acc}}).$$

Further in the text we omit $\overset{\triangle}{\sqcup}$ for simplicity. We deal with increments only. The average quadratic discrepancy can be declared to be a target function T

$$T = \sum_e \int_{V^e} \frac{1}{2}(\overset{\triangle}{u}_{(i,j)} - \overset{\triangle}{\epsilon}{}_{ij}^{\mathrm{acc}})(\overset{\triangle}{u}_{(i,j)} - \overset{\triangle}{\epsilon}{}_{ij}^{\mathrm{acc}})\,\mathrm{d}V \tag{4}$$

[1]Obtained from $u_{1,122}^{\mathrm{acc}} + u_{2,211}^{\mathrm{acc}} - (u_{1,212}^{\mathrm{acc}} + u_{2,212}^{\mathrm{acc}}) = 0$.

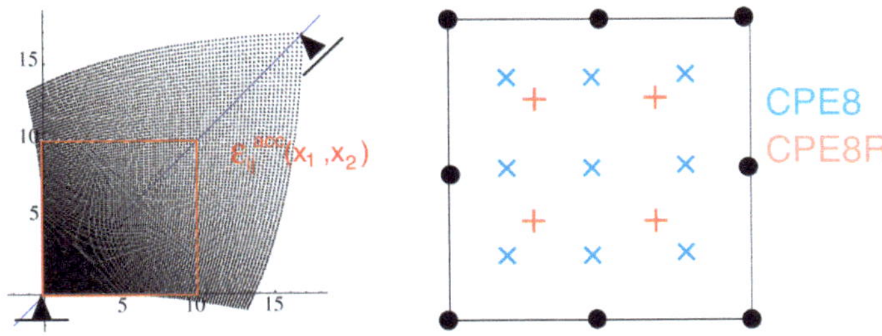

Fig. 2 Example: integration of the accumulation field $\epsilon_{ij}^{\text{acc}}(x_1, x_2)$ given in (5)

can be discretized with nodal displacements $u_{\alpha i}$:

$$\overset{\triangle}{u}_i(\mathbf{x}) = N_\alpha(\mathbf{x})\overset{\triangle}{u}_{\alpha i}$$
$$\overset{\triangle}{u}_{(i,j)} = B_{ij\alpha k}\overset{\triangle}{u}_{\alpha k} = \frac{1}{2}N_{\alpha,n}(\delta_{ni}\delta_{jk} + \delta_{nj}\delta_{ik})\overset{\triangle}{u}_{\alpha k}$$

and minimized under Dirichlet BC. Minimization of error $\partial T/\partial u_{\alpha k} = 0$ leads to a simple FE problem

$$\sum_{Elem.}\int_{V^{(e)}} \left[B_{ij\alpha k}B_{ij\beta l}\right] \mathrm{d}V \;\; u_{\beta l} = \sum_{Elem.}\int_{V^{(e)}} B_{ij\alpha k}\epsilon_{ij}^{\text{acc}}\, \mathrm{d}V = 0$$

corresponding to a static equilibrium problem with elastic parameters $E = 1$ and $\nu = 0$.

As an example we consider a kinematically compatible, freely supported strain field, Fig. 2, with

$$\begin{cases} \epsilon_{11}^{\text{acc}} = 0.001\; x_1(x_1 + x_2), \\ \epsilon_{22}^{\text{acc}} = 0.001\; x_2(x_1 + x_2), \\ \gamma_{12}^{\text{acc}} = 0.001\; (x_1 + x_2) \end{cases} \tag{5}$$

It will be spatially integrated using plane strain 8-node elements CPE8 and CPE8R with full and reduced Gauss quadrature. Analytical integration leads (without rotation) to (6).

$$\begin{cases} u_1 = 0.001\left(\frac{x_1^3}{3} + \frac{x_1^2 x_2}{2} - \frac{x_2^3}{6} + \frac{x_2^2}{2}\right) \\ u_2 = 0.001\left(-\frac{x_1^3}{6} + \frac{x_1^2}{2} + \frac{x_1 x_2^2}{2} + \frac{x_2^3}{3}\right) \end{cases} \tag{6}$$

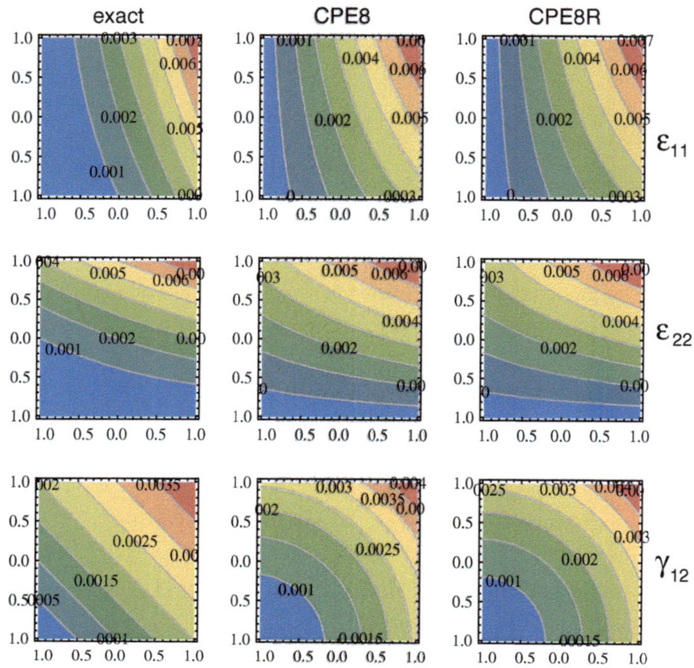

Fig. 3 Components of strain field (5), in columns: analytically, with CPE8 and with CPE8R. The displacement of the upper right corner was exact = 0.010371, CPE8 = 0.010285, CPE8R = 0.010370. The fields in CPE8 and CPE8R are quantitative almost identical, but the integration error is significantly different

Minimization of the target function T with a single CPE8 element with full or reduced Gauss integration result in

$$\frac{\int_V (u_{(i,j)} - \epsilon_{ij}^{\text{acc}})^2 \mathrm{d}V}{\int_V (\epsilon_{ij}^{\text{acc}})^2 \mathrm{d}V} \approx \begin{cases} 1\,\% & \text{if } 9\,\text{GPs} \\ 0\,\% & \text{if } 4\,\text{GPs} \end{cases}$$

a relatively small error because our element has no neighbours in this test. The strain fields from CPE8 and CPE8R in Fig. 3 are similarly poor. In the reduced integration in CPE8R elements the errors are overlooked, which is of advantage because they are not converted to stresses (Figs. 4, 5, 6, 7).

The HCA model implemented as a user's material routine works with the conventional finite element. Such elements integrate the accumulated strain (a fields which often vary proportional to r^2, given in integration points) into the displacement field (given as nodal variables). The essential problem is not the inaccuracy of the resulting predictions of settlement but the fact that a portion of accumulated strain cannot be approximated by nodal displacements and this portion is converted to stress. This

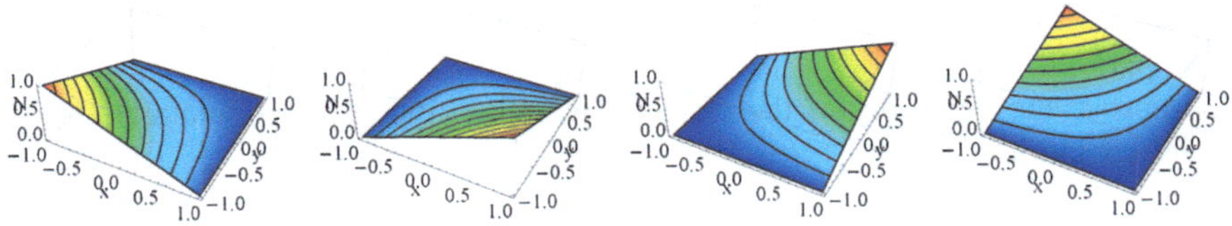

Fig. 4 Vertex shape functions N_1, N_2, N_3 and N_4

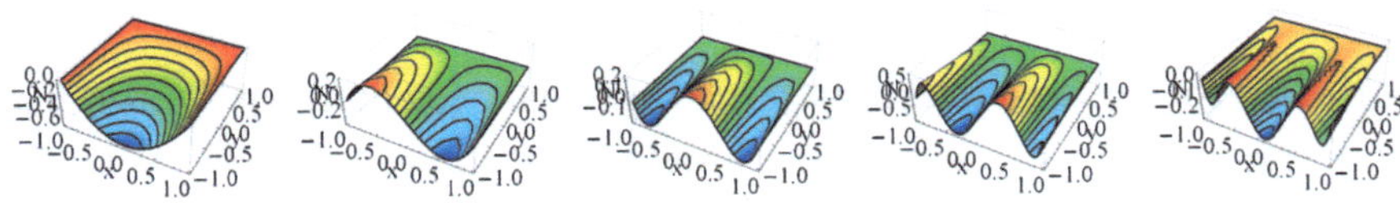

Fig. 5 Interpolation functions N_5, N_9, N_{13}, N_{17} and N_{21} with non-vanishing values on edge 1-2. Further 15 functions for edges 2-3, 3-4 and 4-1 are also used. They are not shown here

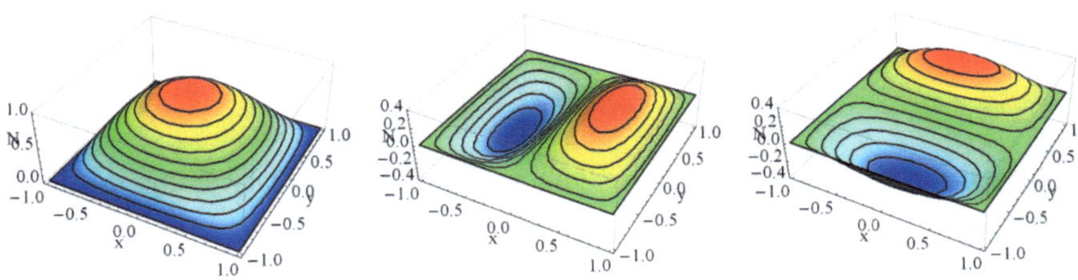

Fig. 6 Bubble functions N_{25}, N_{26} and N_{27} vanishing on all edges

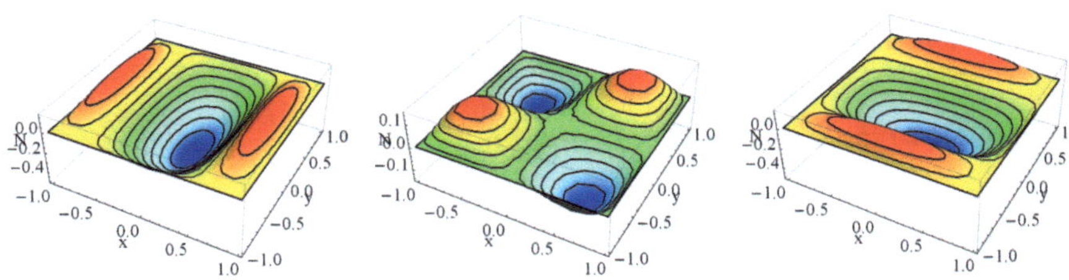

Fig. 7 And some more bubble functions (N_{28}, N_{29} and N_{30}) built of polynomials of higher degree

self-stress is of numerical origin and may systematically grow. In subsequent increments the errors in stress may become large so that they can decisively affect the rate of accumulation and completely distort the results. Even positive feedback phenomenon may be observed. Therefore the first step towards a better implementation of HCA should be the elimination of the numerical self stress rather than the improvement of the accuracy of strain integration. Two numerical strategies are considered

1. weaken the link between strain and displacement (allowing for errors in $\epsilon_{ij} \approx u_{(i,j)}$)
2. improve the flexibility of the element by using high order polynomials in the interpolation functions

2 Three-Field Potential Energy

Given a field of accumulated strains $\epsilon_{ij}^{\text{acc}}$ we solve a static equilibrium problem to find the stresses σ_{ij}, strains ϵ_{ij} and displacements u_i (or their increments $\overset{\triangle}{\sqcup}$). Assuming elasticity $\sigma_{ij} = E_{ijkl}(u_{k,l} - \epsilon_{kl}^{\text{acc}})$ in which stiffness tensor satisfies minor symmetries ($E_{ijkl} = E_{ijlk}$ and $E_{ijkl} = E_{jikl}$) the static equilibrium equation $\sigma_{ij,j} + f_i = 0_i$ with Neumann BC $\check{t}_i = E_{ijkl}(u_{k,l} - \epsilon_{kl}^{\text{acc}})n_j$ on S^t and Dirichlet BC $u_i = \check{u}_i$ on S^u has the weak form

$$\int_V \varphi_i\,(\sigma_{ij,j} + f_i)\mathrm{d}V = 0 \quad \text{with} \quad u_i = \check{u}_i \quad \text{and} \quad \varphi_i = 0_i \quad \text{on} \quad S_u \tag{7}$$

i.e. the test functions $\varphi_i(\mathbf{x})$ must vanish on the Dirichlet boundary. Using the divergence theorem this weak form is conveniently expressed by

$$\int_V \varphi_{i,j} E_{ijkl}(u_{k,l} - \epsilon_{kl}^{\text{acc}})\mathrm{d}V = \int_{S^t} \varphi_i \check{t}_i \mathrm{d}S + \int_V \varphi_i f_i \mathrm{d}V \tag{8}$$

An equivalent variational form can be obtained from the potential energy

$$\Pi(u_i) = \int_V \frac{1}{2}(u_{i,j} - \epsilon_{ij}^{\text{acc}})E_{ijkl}(u_{k,l} - \epsilon_{kl}^{\text{acc}}) - u_i f_i \mathrm{d}V - \int_{S^t} u_i \check{t}_i \mathrm{d}S \tag{9}$$

with a single unknown field $u_i(x_j)$. The stationarity condition $\delta\Pi = 0$ under assumption $E_{ijkl} = E_{klij}$ leads to

$$\delta\Pi = \int_V \delta u_{i,j} E_{ijkl}(u_{k,l} - \epsilon_{kl}^{\text{acc}}) - \delta u_i f_i \mathrm{d}V - \int_{S^t} \delta u_i \check{t}_i \mathrm{d}S = 0 \tag{10}$$

with the virtual displacements $\delta u_i = \varphi_i$ and with $\delta u_{i,j} = \delta(u_{i,j} - \epsilon_{ij}^{\text{acc}})$.

Following the general concept by Hu [3] and Washizu [6] we introduce strain $\epsilon_{ij}(\mathbf{x})$ as an *independent* unknown field (master field, to be discretized with FE with additional nodal variables). It should be possibly similar (but not necessarily identical) to the strain $\epsilon_{ij}^u = u_{(i,j)}$ obtained[2] from the displacement field. We say that $u_i(\mathbf{x})$ is the master field for the slave field ϵ_{ij}^u and that they are strongly connected $\epsilon_{ij}^u = u_{(i,j)}$. The desired compatibility conditions $\epsilon_{ij} - \epsilon_{ij}^u \approx 0$ will be treated as a weak link to be added using Lagrange multipliers λ_{ij}^ϵ to the potential energy Π. This

[2]The brackets in the index part denote symmetrization $u_{(i,j)} = \frac{1}{2}\left(u_{i,j} + u_{j,i}\right)$.

means that the compatibility condition is required only via stationarity condition $\delta\Pi^* = 0$ of the extended potential $\Pi^* = \Pi - \lambda^{\epsilon}_{ij}(\epsilon_{ij} - \epsilon^{u}_{ij})$ of the form

$$\begin{aligned}\Pi^*(u_i, \epsilon_{ij}, \lambda^{\epsilon}_{ij}) = &\int_V \left[\frac{1}{2}\epsilon^{el}_{ij} E_{ijkl}\epsilon^{el}_{kl} - u_i f_i\right] \mathrm{d}V - \int_{S^t} u_i \breve{t}_i \mathrm{d}S \\ &- \int_V \lambda^{\epsilon}_{ij}(\epsilon_{ij} - u_{(i,j)})\mathrm{d}V \end{aligned} \tag{11}$$

and with unknown Lagrange multiplier $-\lambda^{\epsilon}_{ij}$. For brevity $\epsilon^{el}_{ij} = \epsilon_{ij} - \epsilon^{\mathrm{acc}}_{ij}$ is used. In (11) we have obtained a three-field $(u_i, \epsilon_{ij}, \lambda^{\epsilon}_{ij})$ potential.[3] From the stationarity condition $\delta\Pi^* = 0$ follows

$$\begin{aligned}&\int_V \left[\delta\epsilon_{ij} E_{ijkl}\epsilon^{el}_{kl} - \delta u_i f_i\right] \mathrm{d}V \\ &- \int_{S^t} \delta u_i \breve{t}_i \mathrm{d}S - \int_V \left[\lambda^{\epsilon}_{ij}(\delta\epsilon_{ij} - \delta u_{(i,j)}) + \delta\lambda^{\epsilon}_{ij}(\epsilon_{ij} - u_{(i,j)})\right] \mathrm{d}V = 0 \end{aligned} \tag{12}$$

in which $\delta\epsilon_{ij} = \delta\epsilon^{el}_{ij}$ is used. We may rewrite (12) collecting coefficients at $\delta\epsilon_{ij}, \delta u_i, \delta\lambda^{\epsilon}_{ij}$. Each of these coefficients should vanish independently because the variations $\delta\epsilon_{ij}, \delta u_i, \delta\lambda^{\epsilon}_{ij}$ are independent. Choosing all variations equal zero except $\delta\epsilon_{ij} \neq 0$ simplifies (12) to

$$\int_V E_{ijkl}\epsilon^{el}_{kl} - \lambda^{\epsilon}_{ij}\mathrm{d}V = 0 \tag{13}$$

and provides an interpretation[4] $\lambda^{\epsilon}_{ij} = E_{ijkl}\epsilon^{el}_{kl} = \sigma_{ij}$. Further we can either introduce the stress field $\lambda^{\epsilon}_{ij} = \sigma_{ij}$ as a new master field $\sigma_{ij}(\mathbf{x})$ or continue using ϵ^{el}_{kl} and replace $\lambda^{\epsilon}_{ij} = E_{ijkl}\epsilon^{el}_{kl}$. We choose $\sigma_{ij}(\mathbf{x})$ as an independent field.

---

[3]Similarly one could also weaken the constitutive relationship and instead of the usual strong connection $\sigma_{ij} = E_{ijkl}(u_{k,l} - \epsilon^{\mathrm{acc}}_{kl})$ we could introduce an independent unknown field of stress σ_{ij} which should be compatible with the constitutive stress $\sigma^{\epsilon}_{ij} = E_{ijkl}(\epsilon_{kl} - \epsilon^{\mathrm{acc}}_{kl})$. This σ^{ϵ}_{ij} would be treated as a slave of the master field ϵ_{kl}. The weak for of the connection $\sigma_{ij} - \sigma^{\epsilon}_{ij}$ is $\lambda^{\sigma}_{ij}(\sigma_{ij} - \sigma^{\epsilon}_{ij})$ with the Lagrangian multipliers λ^{σ}_{ij}.

Similarly, we may also proceed with the essential BCs in the form $u_i - \breve{u}_i = 0$ defined on S_u. They will be added to the functional with the help of Lagrange multipliers λ^{u}_{i} in the form $-\int_{S_u} \lambda^{u}_{i}(u_i - \breve{u}_i)\mathrm{d}S$

This term should be added to the extended potential. The result is called the Hu-Washizu functional.

[4]Note that $\lambda^{\epsilon}_{ij} = \sigma_{ij}$ must be symmetric and hence $\lambda^{\epsilon}_{ij}\delta u_{(i,j)} = \lambda^{\epsilon}_{ij}\delta u_{i,j}$ i.e. symmetrization may be omitted in (11), (12).

Equation (12) with $\lambda^{\epsilon}_{ij} = \sigma_{ij}$ generates a system of three equations obtained by setting different two of three independent variations δu_i, $\delta\epsilon_{ij}$, $\delta\sigma_{ij}$ to zero:

$$-\int_V \delta u_i f_i \mathrm{d}V - \int_{S^t} \delta u_i \breve{t}_i \mathrm{d}S + \int_V \delta u_{i,j}\sigma_{ij}\mathrm{d}V = 0 \tag{14}$$

$$\int_V \delta\sigma_{ij}(\epsilon_{ij} - u_{i,j})\mathrm{d}V = 0 \tag{15}$$

$$\int_V \delta\epsilon_{ij}\left[E_{ijkl}\epsilon^{el}_{kl} - \sigma_{ij}\right]\mathrm{d}V = 0 \tag{16}$$

3 Enhanced Assumed Strain

Following Simo and Rifai [5] we decompose the strain field within an element into

$$\epsilon_{ij} = u_{(i,j)} + \tilde{\epsilon}_{ij} = \epsilon^u_{ij} + \tilde{\epsilon}_{ij} \tag{17}$$

The superscript in ϵ^u_{ij} the portion of strain obtained from the strong link $\epsilon^u_{ij} = u_{(i,j)}$. The independent strain field ϵ_{ij} in (11) may be replaced by (17) with the independent portion $\tilde{\epsilon}_{ij}$. The stationarity condition (12) written out with $\delta\epsilon^{el}_{ij} = \delta\epsilon_{ij} = \delta u_{(i,j)} + \delta\tilde{\epsilon}_{ij}$, with $\lambda^{\epsilon}_{ij} = \sigma_{ij}$ and with $\epsilon^{el}_{ij} = u_{(i,j)} + \tilde{\epsilon}_{ij} - \epsilon^{\mathrm{acc}}_{ij}$ takes the form

$$\int_V (\delta u_{(i,j)} + \delta\,\tilde{\epsilon}_{ij})E_{ijkl}\epsilon^{el}_{kl}\mathrm{d}V - \int_V \delta u_i f_i \mathrm{d}V - \int_{S^t} \delta u_i \breve{t}_i \mathrm{d}S - \int_V \delta\,\tilde{\epsilon}_{ij}\sigma_{ij}\mathrm{d}V - \int_V \tilde{\epsilon}_{ij}\delta\sigma_{ij}\mathrm{d}V = 0 \tag{18}$$

Choosing different combinations of the independent variations $\delta u_{(i,j)}$, $\delta\,\tilde{\epsilon}_{ij}$ or $\delta\sigma_{ij}$ Eq. (18) may generate the following system

$$\int_V \delta u_{(i,j)} E_{ijkl}\epsilon^{el}_{kl}\mathrm{d}V = \int_V \delta u_i f_i \mathrm{d}V + \int_{S^t} \delta u_i \breve{t}_i \mathrm{d}S \tag{19}$$

$$\int_V \delta\sigma_{ij}\tilde{\epsilon}_{ij}\mathrm{d}V = 0 \tag{20}$$

$$\int_V \delta\,\tilde{\epsilon}_{ij}E_{ijkl}\epsilon^{el}_{kl}\mathrm{d}V - \int_V \delta\,\tilde{\epsilon}_{ij}\sigma_{ij}\mathrm{d}V = 0 \tag{21}$$

Following Simo and Rifai [5], we make the following restrictions:

1. Bubnov-Galerkin interpolation method is used, i.e. the solution function of a variable $\sqcup$ and the test function of the corresponding variation $\delta\sqcup$ have identical basis functions, e.g. $\sigma_{ij} = P_\alpha\sigma_{\alpha ij}$ and $\delta\sigma_{ij} = P_\alpha\delta\sigma_{\alpha ij}$ with the same interpolation functions P_α.

2. The enhancement $\tilde{\epsilon}_{ij}$ of strain and (virtual) stress $\delta\sigma_{ij}$ are expressed with their own basis functions, namely

$$\tilde{\epsilon}_{ij} = M_{\alpha ij}\ \tilde{\epsilon}_{\alpha} \quad \text{and} \quad \delta\sigma_{ij} = P_{\beta ij}\ \delta\sigma_{\beta} \tag{22}$$

The orthogonality (20) is satisfied

$$\int_V \tilde{\epsilon}_{ij}\delta\sigma_{ij}\mathrm{d}V = \tilde{\epsilon}_{\alpha}\left[\int_V M_{\alpha ij}P_{\beta ij}\mathrm{d}V\right]\sigma_{\beta} = 0 \tag{23}$$

by choice of orthogonal basis functions alone, $\int_V P_{\beta ij}M_{\alpha ij}\mathrm{d}V = 0$ for any combination of α and β (also for $\alpha = \beta$) and *irrespectively of the coefficients* $\tilde{\epsilon}_{\beta}$ and σ_{α}.

3. At least the constant distribution stress across the element should exist, i.e. $P_1(\mathbf{x}) = 1$. Let it be the only basis function for the independent stress field. For the basis functions of $\tilde{\epsilon}_{ij}$ this implies[5]

$$\int_{\square} M_{\alpha ij}(\boldsymbol{\xi})J\mathrm{d}\boldsymbol{\xi} = 0, \tag{24}$$

wherein $\boldsymbol{\xi} = \{\xi, \eta\}$ is the pair of natural coordinates and $J(\boldsymbol{\xi}) = \det(\partial\mathbf{x}/\partial\boldsymbol{\xi})$ is the Jacobian. Finding $M_{\alpha ij}(\boldsymbol{\xi})$ can be eased introducing $M^a_{\alpha ij}(\boldsymbol{\xi})$ such that

$$M_{\alpha ij} = M^a_{\alpha ij}\frac{J_0}{J} \tag{25}$$

The Jacobian $J(\boldsymbol{\xi}) = \det(\partial\mathbf{x}/\partial\boldsymbol{\xi})$ varies inside the element but $J_0 = J|_{\xi=\eta=0}$ is its (fixed) value evaluated in the middle of the element. If we find $M^a_{\alpha ij}$ that satisfies

$$\int_{\square} M^a_{\alpha ij}\mathrm{d}\boldsymbol{\xi} = 0 \tag{26}$$

then (24) will be also satisfied. Note that only $M_{\alpha ij}$ are the basis functions and $M^a_{\alpha ij}$ are not. By the definition of $M^a_{\alpha ij}$ we have simply cancelled the usual J-multiplication of Gauss quadrature[6] shown in (24).

4. No basis functions $M_{\alpha ij}$ may reappear as one of the derivatives of N_{β}, in particular, no $M_{\alpha ij} = \text{const}$ is allowed for.

With these restrictions (20) is satisfied a priori and need not be further considered. Moreover we may drop the second integral in (21). We write out $\epsilon^{el}_{kl} = u_{(k,l)} + \tilde{\epsilon}_{kl} - \epsilon^{\text{acc}}_{kl}$ in (19) and (21) obtaining

---

[5]The enforcement of full orthogonality of $\tilde{\epsilon}_{ij}$ with $\sigma_{ij} = E_{ijkl}\epsilon^{el}_{kl}$ would imply $\tilde{\epsilon}_{ij} = 0$.

[6]$\int_{\square} \ldots \mathrm{d}\boldsymbol{\xi}$ represents $\int_{-1}^{1}\int_{-1}^{1} \ldots \mathrm{d}\xi\mathrm{d}\eta$.

$$
\begin{cases}
\int_V \delta u_{(i,j)} E_{ijkl} \left(u_{k,l} + \tilde{\epsilon}_{kl}\right) \mathrm{d}V = \int_V \delta u_i f_i \mathrm{d}V + \int_{S^t} \delta u_i \breve{t}_i \mathrm{d}S \\
\qquad\qquad\qquad\qquad\qquad\qquad + \int_V \delta u_{(i,j)} E_{ijkl} \epsilon_{kl}^{\mathrm{acc}} \\
\int_V \delta \tilde{\epsilon}_{ij} E_{ijkl} \left(u_{k,l} + \tilde{\epsilon}_{kl}\right) \mathrm{d}V \quad = \int_V \delta \tilde{\epsilon}_{ij} E_{ijkl} \epsilon_{kl}^{\mathrm{acc}} \mathrm{d}V
\end{cases}
\tag{27}
$$

The following discretization is used

$$
u_i = N_\omega u_{\omega i}, \quad u_{i,j} = N_{\omega,j} u_{\omega i} \quad \text{and} \quad \tilde{\epsilon}_{ij} = M_{\alpha ij} \; \tilde{\epsilon}_\alpha \tag{28}
$$

wherein $N_{\omega,j}$ denotes the cartesian derivative $N_{\omega,j} = \frac{\partial N_\omega}{\partial \xi} \frac{\partial \xi}{\partial x_j} + \frac{\partial N_\omega}{\partial \eta} \frac{\partial \eta}{\partial x_j}$ in the 2D case. Substituting the discretized fields into (27) we obtain

$$
\begin{cases}
\delta u_{\omega i} \int_V \left(N_{\omega,j} E_{ijkl} N_{\omega,l} u_{\omega k} + N_{\omega,j} E_{ijkl} M_{\alpha kl} \tilde{\epsilon}_\alpha\right) \mathrm{d}V = \delta u_{\omega i} \left[\int_V N_\omega f_i \mathrm{d}V \right. \\
\qquad\qquad\qquad\qquad\qquad\qquad \left. + \int_{S^t} N_\omega \breve{t}_i \mathrm{d}S + \int_V N_{\omega,j} E_{ijkl} \epsilon_{kl}^{\mathrm{acc}}\right] \\
\delta \tilde{\epsilon}_\alpha \int_V \left(M_{\alpha ij} E_{ijkl} N_{\omega,l} u_{\omega k} + M_{\alpha ij} E_{ijkl} M_{\beta kl} \tilde{\epsilon}_\beta\right) \mathrm{d}V = \delta \tilde{\epsilon}_\alpha \int_V M_{\alpha ij} E_{ijkl} \epsilon_{kl}^{\mathrm{acc}} \mathrm{d}V
\end{cases}
\tag{29}
$$

provided E_{ijkl} have both minor symmetries.[7] Due to the arbitrariness of the variations $\delta u_{\omega i}$ and $\delta \tilde{\epsilon}_{\alpha ij}$ of the nodal values we may bracket them out so that (29) generates the following system of equations (here written in a block form for brevity)

$$
\begin{cases}
K_{\omega i \gamma k} u_{\gamma k} + L^T_{\omega i \alpha} \tilde{\epsilon}_\alpha \quad = R_{\omega i} + A_{\omega i} \\
L_{\alpha \gamma k} u_{\gamma k} + D_{\alpha\beta} \tilde{\epsilon}_\beta \quad = \bar{A}_{\alpha ij}
\end{cases}
\tag{30}
$$

with abbreviations

$$
K_{\omega i \gamma k} = \int_V N_{\omega,j} E_{ijkl} N_{\gamma,l} \mathrm{d}V \tag{31}
$$

$$
L^T_{\omega i \alpha} = \int_V N_{\omega,j} E_{ijkl} M_{\alpha kl} \mathrm{d}V \tag{32}
$$

$$
L_{\alpha \gamma k} = \int_V M_{\alpha ij} E_{ijkl} N_{\gamma,l} \mathrm{d}V \tag{33}
$$

$$
D_{\alpha\beta} = \int_V M_{\alpha ij} E_{ijkl} M_{\beta kl} \mathrm{d}V \tag{34}
$$

$$
R_{\omega_i} = \int_V N_\omega f_i \mathrm{d}V + \int_{S^t} N_\omega \breve{t}_i \mathrm{d}S \tag{35}
$$

$$
A_{\omega i} = \int_V N_{\omega,j} E_{ijkl} \epsilon_{kl}^{\mathrm{acc}} \tag{36}
$$

$$
\bar{A}_\alpha = \int_V M_{\alpha ij} E_{ijkl} \epsilon_{kl}^{\mathrm{acc}} \mathrm{d}V \tag{37}
$$

[7] hence we may omit symmetrization brackets in $u_{(i,j)}$.

with $L_{\alpha\gamma k} = L^T_{\gamma k\alpha}$ from comparison of (33) and (32) using minor and major symmetries of E_{ijkl}. Evidently the transposition pertains to the index $_\alpha$ and the pair $_{\gamma k}$.

Sometimes we may want to eliminate the discrete enhancement of strain $\tilde{\epsilon}_{\beta kl}$ from the set of unknowns. For this purpose we use static condensation. First we find

$$\tilde{\epsilon}_\beta = D^{-1}_{\alpha\beta}\bar{A}_\alpha - D^{-1}_{\alpha\beta}L_{\alpha\gamma k}u_{\gamma k} \tag{38}$$

from the second equation in (30) and then we substitute $\tilde{\epsilon}_\beta$ into the first equation in (30)

$$K_{\omega i\gamma k}u_{\gamma k} + L^T_{\omega i\beta}\left(D^{-1}_{\eta\beta}\bar{A}_\eta - D^{-1}_{\eta\beta}L_{\eta\gamma k}u_{\gamma k}\right) = R_{\omega i} + A_{\omega i} \quad \text{or} \tag{39}$$

$$\left[K_{\omega i\gamma k} - L^T_{\omega i\beta}D^{-1}_{\eta\beta}L_{\eta\gamma k}\right]u_{\gamma k} = R_{\omega i} + A_{\omega i} - L^T_{\omega i\beta}D^{-1}_{\eta\beta}\bar{A}_\eta \tag{40}$$

The EAS-element stiffness is given in the square brackets. Static condensation is not obligatory but it is useful in elastic problems because judging by the degrees of freedom the element looks like a conventional (single-field) displacement element.

3.1 Smart Enhancements $\tilde{\epsilon}_{ij}(x)$ to the Strain Field

In order to satisfy the restriction (24) we propose $M^a_{\alpha ij}$ that satisfies (26) for example $M^a_{\alpha ij} = \xi$ or $= \eta$ or $= \xi\eta$ etc. Indeed, they guarantee the orthogonality condition with a constant stress thanks to

$$\int_\square M^a_{\alpha ij}\mathrm{d}\xi = 0 \tag{41}$$

Further restriction on $M_{\alpha ij}$ was that it should differ all from derivatives $N_{\alpha,J}$ of interpolation functions N_α of displacements. This pertains to $M_{\alpha ij}$ as well as to $M^a_{\alpha ij}$ because of $J =$ const in rectangular elements. The well-known basis functions for displacement at node α of a quadrilateral element are

$$N_\alpha(\xi, \eta) = (1 + \xi\xi_\alpha)(1 + \eta\eta_\alpha)/4 \quad \text{with} \quad \xi_\alpha, \eta_\alpha = \pm 1 \quad \text{for} \quad \alpha = 1, 2, 3, 4 \tag{42}$$

They contain components linear in ξ and η so $M^a_{\alpha ij} =$ const are forbidden (except for $M^a_{\alpha ij} = 0$, of course). The basis functions (42) contain also the product $\xi\eta$ and hence the derivatives $N_{\alpha,1}$ appearing in ϵ_{11} contain η and the derivatives $N_{\alpha,2}$ appearing in ϵ_{22} contains ξ. In order to preclude identical basis functions in $\tilde{\epsilon}_{ij}$ as obtained for $u_{(i,j)}$ one cannot use $M^a_{\alpha 11} = \eta$ or $M^a_{\alpha 22} = \xi$.

The following base functions can be found in the literature for four nodeless parameters $\tilde{\epsilon}_\alpha$, $\alpha = 1, 2, 3, 4$

$$\begin{bmatrix} \leftarrow M^a_{\alpha 11} \rightarrow \\ \leftarrow M^a_{\alpha 22} \rightarrow \\ \leftarrow M^a_{\alpha 12} \rightarrow \end{bmatrix} \begin{Bmatrix} \tilde{\epsilon}_{\alpha=1} \\ \tilde{\epsilon}_2 \\ \tilde{\epsilon}_3 \\ \tilde{\epsilon}_4 \end{Bmatrix} = \begin{bmatrix} \xi\, 0\, 0\, 0 \\ 0\, \eta\, 0\, 0 \\ 0\, 0\, \xi\, \eta \end{bmatrix} \begin{Bmatrix} \tilde{\epsilon}_1 \\ \tilde{\epsilon}_2 \\ \tilde{\epsilon}_3 \\ \tilde{\epsilon}_4 \end{Bmatrix} (= M^a_{\alpha ij}\tilde{\epsilon}_\alpha) \tag{43}$$

Some authors use seven nodeless parameters $\tilde{\epsilon}_\alpha$, $\alpha = 1, 2, \dots 7$

$$\begin{bmatrix} \leftarrow M^a_{\alpha 11} \rightarrow \\ \leftarrow M^a_{\alpha 22} \rightarrow \\ \leftarrow M^a_{\alpha 12} \rightarrow \end{bmatrix} = \begin{bmatrix} \xi\, 0\, 0\, 0\, \xi\eta & 0 & 0 \\ 0\, \eta\, 0\, 0 & 0 & \xi\eta & 0 \\ 0\, 0\, \xi\, \eta & 0 & 0 & \xi\eta \end{bmatrix} \tag{44}$$

3.2 Stress Recovery

After an FE calculation we are given the nodal displacements $u_{\alpha i}$ and hence the approximation of the displacement function u_i. Using these values we will find the optimal stress field σ_{ij}. The discrepancy

$$s_{ij} = \sigma_{ij} - E_{ijkl}(u_{(k,l)} + \tilde{\epsilon}_{kl} - \epsilon^{\text{acc}}_{kl}) \tag{45}$$

between the stress σ_{ij} understood as an independent (element-wise constant) field and the slave stress field $E_{ijkl}(u_{(k,l)} + \tilde{\epsilon}_{kl} - \epsilon^{\text{acc}}_{kl})$ obtained from enhanced strains should give possibly small complementary energy

$$\int_{V^e} \frac{1}{2} s_{ij} E^{-1}_{ijkl} s_{ij} \mathrm{d}V \rightarrow \min \text{ or } \int_{V^e} \delta s_{ij} E^{-1}_{ijkl} s_{kl} \mathrm{d}V = 0 \tag{46}$$

Substituting (45) into (46) and choosing $\delta\sigma_{ij} \neq 0$ and $\delta\tilde{\epsilon}_{ij} = 0$ or vice versa we obtain

$$\int_{V^e} \delta\sigma_{ij} E^{-1}_{ijkl} s_{kl} \mathrm{d}V = 0 \quad \text{or} \quad \int_{V^e} \delta\sigma_{ij} E^{-1}_{ijkl} \left[\sigma_{kl} - E_{klrs}(u_{(r,s)} + \tilde{\epsilon}_{rs} - \epsilon^{\text{acc}}_{rs})\right] \mathrm{d}V = 0 \tag{47}$$

and

$$\int_{V^e} \delta\tilde{\epsilon}_{mn} E_{ijmn} E^{-1}_{ijkl} s_{kl} \mathrm{d}V = 0 \text{ or} \tag{48}$$

$$\int_{V^e} \delta\tilde{\epsilon}_{mn} E_{mnij} E^{-1}_{ijkl} \left[\sigma_{kl} - E_{klrs}(u_{(r,s)} + \tilde{\epsilon}_{rs} - \epsilon^{\text{acc}}_{rs})\right] \mathrm{d}V = 0 \tag{49}$$

respectively. These results may be simplified to

$$\int_{V^e} \delta\sigma_{ij} \left[E^{-1}_{ijkl}\sigma_{kl} - (u_{(i,j)} + \tilde{\epsilon}_{ij} - \epsilon^{\text{acc}}_{ij}) \right] \mathrm{d}V = 0 \tag{50}$$

$$\int_{V^e} \delta\tilde{\epsilon}_{kl} \left[\sigma_{kl} - E_{klrs}(u_{(r,s)} + \tilde{\epsilon}_{rs} - \epsilon^{\text{acc}}_{rs}) \right] \mathrm{d}V = 0 \tag{51}$$

and making use of the orthogonality condition (20) further to

$$\int_{V^e} \delta\sigma_{ij} \left[E^{-1}_{ijkl}\sigma_{kl} - (u_{(i,j)} - \epsilon^{\text{acc}}_{ij}) \right] \mathrm{d}V = 0 \tag{52}$$

$$\int_{V^e} \delta\tilde{\epsilon}_{kl} E_{klrs}(u_{(r,s)} + \tilde{\epsilon}_{rs} - \epsilon^{\text{acc}}_{rs}) \mathrm{d}V = 0 \tag{53}$$

We use the discretized forms $\delta\sigma_{ij} = P_\alpha \delta\sigma_{\alpha ij}$ and $\sigma_{ij} = P_\beta \sigma_{\beta ij}$ with the single basis function $P_1 = 1$. Moreover, we approximate $u_{(i,j)} = N_{\alpha,j} u_{\alpha i}$ obtaining the nodeless stress parameter σ_{1ij} from (52)

$$\delta\sigma_{1ij} \int_{V^e} \left[E^{-1}_{ijkl}\sigma_{1kl} - (N_{\alpha,j} u_{\alpha i} - \epsilon^{\text{acc}}_{ij}) \right] \mathrm{d}V = 0 \quad \text{or} \tag{54}$$

$$\sigma_{1kl} = \frac{1}{V^e} E_{ijkl} \left[\int_{V^e} N_{\alpha,j} u_{\alpha i} \mathrm{d}V - \int_V \epsilon^{\text{acc}}_{ij} \mathrm{d}V \right] \tag{55}$$

The remaining additional strain field $\tilde{\epsilon}_{rs} = M_{\alpha ij}\tilde{\epsilon}_\alpha$ can be found from using the nodeless parameters $\tilde{\epsilon}_\alpha$ from (38). In the Mma script calculating an EAS-element the names of variables follow the notation used in this paper.

```
 getStiffEAS[e_]:=Module[{xxe,xxG,Emod,nu,Estiff,BB,   KK,DD,LL, Na,Ma,x,y, w,
nnode = Length[elements[[e]] ];   nGauss =4;
 xxe = xx[[#]] & /@ elements[[e]];
maxG=1; {Na,dNadx,dNady,dV0} = getCartD[xxe,1, maxG] ;(* jacobian of shape in the middle *)
maxG=4; maxE=4;
DD = Array[0&, {maxE,maxE}];
LL = Array[0&, {maxE,nnode*2}] ;
KK = Array[0&,{ nnode*2,   nnode*2}] ;
A =  Array[0&, nnode *2 ] ;
Ab=  Array[0&, maxE  ] ;
{Emod,nu}=getMatConstants[e,  (Plus @@xxe)/nnode  ];
Estiff=getElasticPlaneStrainStiffness[Emod,nu] ;
For[iG=1,iG<=nGauss,iG++, (*  Print["iG=",iG];*)
 {Na,dNadx,dNady,dV} = getCartD[xxe,iG, maxG] ;
xxG =  Na. xxe;
 BB =   {{dNadx[[1]],0,   dNadx[[2]],0,dNadx[[3]],0, dNadx[[4]],0},
             {0,dNady[[1]], 0,dNady[[2]],0,dNady[[3]],0,dNady[[4]]  },
            { dNady[[1]],dNadx[[1]] ,  dNady[[2]],dNadx[[2]]  ,   dNady[[3]],
                                  dNadx[[3]] , dNady[[4]],dNadx[[4 ]]  }};
{{x,y}, w} = getGP[iG, nGauss] ;
If[maxE==4, Ma =  {{x,0,0,0},{0,y,0,0},{0,0,x,y} } ] ; (*4 enhancements  rows: e11,e22,g12 *)
If[maxE==7,Ma  =  {{x,0,0,0 ,x* y, 0,0},{0,y,0,0,0, x *y, 0},{0,0,x,y,0 ,0, x *y} } ];
If[maxE==13,Ma  =  {{x,0,0,0 ,x* y, 0,0,x^2 Sign[x],0,0,y^2 Sign[y],0,0},
                                {0,y,0,0,0, x *y, 0,0,x^2Sign[x],0,0,y^2Sign[y],0},
                                {0,0,x,y,0 ,0, x *y,0,0,x^2Sign[x],0,0,y^2Sign[y]} } ];
Ma *= (dV0/dV);
```

```
 KK +=  Transpose[BB].Estiff.BB *dV;
  DD += Transpose[Ma]. Estiff. Ma * dV;
 LL += Transpose[Ma].Estiff.BB *dV;
  A += Transpose[BB]. Estiff. getEpsilon0[xxG]*dV ;
Ab +=  Transpose[Ma]. Estiff.getEpsilon0[xxG]*dV;
] ; (* iG *)

 KK -= Transpose[LL].Inverse[DD].LL  ;          (* static condensation *)
A -=  Transpose[LL].Inverse[DD].Ab;               (* static condensation *)
{splitToSquareBlocks[ KK , 2 ]  ,Partition[A,2] }
 ]
```

4 Hierarchical Element

Flexibility of a finite element can be improved using the concept of hierarchical element (HE) also known as *p*-refinement [1, 2, 9]. A single displacement field is used. Its approximation within an element is improved using additional interpolation functions rather than splitting the element into smaller ones (so called *h*-refinement). The major advantage of the *p*-refinement results from the fact that we continue using of the initial coarse mesh (in our case mesh of CPE4 elements). Contrarily to the alternative remeshing techniques, no additional nodes will be added in the HE and the geometry of the root element remains unchanged. The refinement of an existing element is achieved by adding nodeless interpolation polynomials to the basic set of interpolation functions. An indispensable procedure in the HE is the static condensation in which all nodeless parameters are eliminated before aggregation. Although the hierarchical elements are computationally rather expensive they are expected to reduce the self-stress obtained from the HCA model because of the high order of polynomial approximation.

Therefore we try out a 4-nodal plane strain element with hierarchical expansion of the displacement field by the polynomials up the $p = 6$th order. In place of the conventional nodal approximation

$$u_i(\xi) = \sum_{\alpha=1}^{4} N_\alpha(\xi)\, u_{\alpha i} \quad \text{with} \quad \xi = (\xi, \eta) \quad \text{and} \tag{56}$$

$$N_\alpha(\xi, \eta) = \frac{1}{4}(1 + \xi_\alpha \xi)(1 + \eta_\alpha \eta) \quad \text{with} \quad \alpha = 1, \ldots 4 \tag{57}$$

of the CPE4 element we use the extended superparametric form

$$u_i(\xi) = \sum_{\alpha=1}^{4} N_\alpha(\xi) u_{\alpha i} + \sum_{\alpha=5}^{30} N_\alpha(\xi) c_{\alpha i} \tag{58}$$

$$\mathbf{x}(\xi) = \sum_{\alpha=1}^{4} N_\alpha(\xi) \mathbf{x}_\alpha \tag{59}$$

where

- $u_{\alpha i}$ = conventional nodal displacements (at node α in direction i) for shape functions $N_\alpha(\xi)$, $\alpha = 1, 2, 3, 4$
- $c_{\alpha i}$ = internal multipliers of polynomials $N_\alpha(\xi)$, $\alpha = 5, \ldots 4p - 4$ allowing for displacements on edges
- $c_{\alpha i}$ = internal multipliers of polynomials $N_\alpha(\xi)$, $\alpha = 4p-3, \ldots (p-2)(p-3)/2$ vanishing on edges

Summing up, we use in our CPE4HE element $(4 + 20 + 6) \cdot 2 = 60$ parameters and 30 interpolation functions.

Note that the displacements need not be continuous across the element borders anymore. The jumps in **u** may be caused by different c_β coefficients in neighbouring elements. Therefore it is necessary to perform a patch test to be sure that the elements can reproduce arbitrary constant strain fields.

1. Nodal (vertex) shape functions N_α are associated with vertrices η by $N_\alpha(\boldsymbol{\xi}_\eta) = \delta_{\alpha\eta}$. From this condition one obtains

$$N_\alpha(\xi, \eta) = \frac{1}{4}(1 + \xi_\alpha \xi)(1 + \eta_\alpha \eta) \quad \text{for} \quad \alpha = 1, \ldots 4$$

```
nb1[xi_]:= (1 - xi)/2;      nb2[xi_] := (1 + xi)/2;
nb[xi_] := {nb1[xi], nb2[xi]};
Nv1[xi_, eta_] := Outer[Times, nb[xi] , nb[eta]];
{eN[[1]], eN[[2]], eN[[3]], eN[[4]]} = {Nv1[xi,eta][[1,1]], Nv1[xi,eta][[2,1]],
                                          Nv1[xi,eta][[2,2]], Nv1[xi,eta][[1,2]] };
```

2. Additional hierarchical functions N_β with non-zero edge values (so called edge functions) are defined by:

$$N_\beta(\xi, \eta) = \frac{1}{2}(1 + \xi_\beta \xi)(1 + \eta_\beta \eta) F_\beta(\xi, \eta) \quad \text{for} \quad \beta = 5, \ldots 24$$

wherein $F(\xi, \eta)$ are integrated from some Legendre polynomials $P^n(\xi)$ and $P^m(\eta)$ wherein n, m denote the degree of the polynomial

```
P[xi_]:= Array[LegendreP[#, xi] &, 5];
F[xi_]:= Table[ If[ k==1, 0, Sqrt[(2k-1)/2] Integrate[P[t][[k-1]],{t,-1,xi}]],{k,1,6}];
Nek31 = nb1[eta]F[xi];  Nek32 = nb2[eta]F[xi];
Nek13 = nb1[xi] F[eta]; Nek23 = nb2[xi] F[eta];
For[k=1, k<=p-1, k++,
   eN[[5+4(k-1)]]   = Nek31[[k+1]];
   eN[[5+4(k-1)+1]]= Nek23[[k+1]];
   eN[[5+4(k-1)+2]]= Nek32[[k+1]];
   eN[[5+4(k-1)+3]]= Nek13[[k+1]];];
```

Exemplarily one get N_9 (3rd order of approximation, not vanishing on the edge 1-2) and N_{15} (4th order of approximation, not vanishing on the edge 3-4) as

following:
$N_9(\xi,\eta) = \frac{1}{4}\sqrt{\frac{5}{2}}\xi(1-\eta)(\xi^2-1)$ and $N_{15}(\xi,\eta) = \frac{1}{16}\sqrt{\frac{7}{2}}\xi(1+\eta)(5\xi^4-6\xi^2+1)$

3. Hierarchical bubble functions N_γ vanish on all element edges and have the form

$$N_\gamma(\xi,\eta) = (1-\eta^2)(1-\xi^2)P^n(\xi)P^m(\eta) \quad \text{for} \quad \gamma = 25,\dots 30$$

with some additional Legendre polynomials $P^n(\xi)$ and $P^m(\eta)$ of degree n and m.

```
eN[[25]] = (1-xi^2)(1-eta^2);                    eN[[26]] = eN[[25]] P[xi][[1]];
eN[[27]] = eN[[25]] P[eta][[1]];                 eN[[28]] = eN[[25]] P[xi][[2]];
eN[[29]] = eN[[25]] P[xi][[1]]P[eta][[1]];       eN[[30]] = eN[[25]] P[eta][[2]];
```

Exemplarily one can get for internal approximation functions N_{27} and N_{30} following: $N_{27}(\xi,\eta) = \eta(1-\eta^2)(1-\xi^2)$ and $N_{30}(\xi,\eta) = \frac{1}{2}(3\eta^2-1)(1-\eta^2)(1-\xi^2)$

The discretization of the balance of momentum with the constitutive relation

$$\overset{\triangle}{\sigma}_{ij} = \overset{\triangle}{\sigma}{}^{0}_{ij} + E_{ijkl}(\overset{\triangle}{\epsilon}_{kl} - \overset{\triangle}{\epsilon}{}^{\text{acc}}_{kl}) \tag{60}$$

gives us the usual system of linear equations:

$$K_{\eta i\beta j}a_{\beta j} = R_{\eta i}, \text{ with } a_{\beta j} = \begin{cases} u_{\beta j}, & \text{for } \beta = 1,\dots 4 \\ c_{\beta j}, & \text{for } \beta = 5,\dots 30 \end{cases} \tag{61}$$

that is the generalized list of coefficients $a_{\alpha i}$ consists of nodal parameters $u_{\alpha i}$ and nodeless parameters $c_{\alpha i}$ (both for $i = 1, 2$). Moreover we have

$$K_{\eta a\beta b} = \int_V B_{ij\eta a}E_{ijkl}B_{kl\beta b}\mathrm{d}V \quad \text{with} \quad \eta,\beta = 1,\dots 30 \text{ and } a,b = 1,2 \tag{62}$$

$$R_{\eta a} = \int_V B_{ij\eta a}E_{ijkl}\epsilon^{acc}_{kl}\mathrm{d}V + \int_V B_{ij\eta a}\sigma^0_{ij}\mathrm{d}V + \int_V N_\eta \breve{b}_a\mathrm{d}V + \int_S N_\eta \breve{t}_a\mathrm{d}S. \tag{63}$$

The matrix $B_{ij\eta k}$ computes strain from $\epsilon_{ij}(\boldsymbol{\xi}) = B_{ij\eta k}(\boldsymbol{\xi})\, a_{\eta k}$, in particular under plane strain condition with deformation $(\epsilon_{11}, \epsilon_{22}, \gamma_{12})$ we could use

$$B_{11\eta k} = N_{\eta,1}\delta_{1k} \tag{64}$$

$$B_{22\eta k} = N_{\eta,2}\delta_{2k} \tag{65}$$

$$B_{12\eta k} + B_{21\eta k} = N_{\eta,2}\delta_{1k} + N_{\eta,1}\delta_{2k} \tag{66}$$

Matrix $B_{ij\eta k}$ is built at each Gauss integration point $\boldsymbol{\xi}_G$. At this point it should be noted that the order of Gauss quadrature integration has to be chosen to provide a sufficient precision. It is especially important for integrals of higher order polynomials. In our case of $p = 6$ the Gauss quadrature of at least 7th order is required. In other words, we need at least 49 integration points per element. Their position and their weighting coefficients can be found with the following script

```
getAllGPs[maxG_,  maxDim_ , edge_: 1] :=
   Module[ {Pn, p,  x, w, coordsGP, weightsGP, CoordEdge1m, CoordEdge1p},
   Pn = 1/(2^maxG*maxG!)*D[(p^2 - 1)^maxG, {p, maxG}];  (* Legendre polynomial *)
   x = p /.( Solve[Pn == 0, p]// N // Chop // Sort );   (* 1-D coordinate *)
   w = 2./((1 - x^2)*(D[Pn, p] /.{p -> x})^2);          (* weight coefficient *)
     If[ maxDim == 1,
       CoordEdgeM = Array[-1 &, maxG]; CoordEdgeP = Array[1 &, maxG];
       If[edge == 1, coordsGP = Transpose[{x, CoordEdge1m} ]];
       If[edge == 2, coordsGP = Transpose[{CoordEdge1p, x} ]];
       If[edge == 4, coordsGP = Transpose[{CoordEdge1m, x} ]];
       weightsGP = w; ];
     If[maxDim == 2 ,
        coordsGP  = Flatten[ Outer[List, x, x], 1];
        weightsGP = Flatten[ Outer[Times, w, w]];];
    Transpose[{coordsGP , weightsGP }]];
```

Equation (61) can be rewritten in the block-matrix form:

$$\left[\begin{array}{c|c} K_{\alpha i\beta j} & K_{\alpha i\gamma j} \\ \hline K_{\eta i\beta j} & K_{\eta i\gamma j} \end{array}\right] \cdot \begin{Bmatrix} u_{\beta j} \\ c_{\gamma j} \end{Bmatrix} = \begin{Bmatrix} R_{\alpha i} \\ R_{\eta i} \end{Bmatrix} \quad \text{with} \quad \alpha, \beta = 1, ..4; \gamma, \eta = 5, ..30 \tag{67}$$

and with summation over all dummy indices. Using single-indexed variables u_A and c_B we obtain

$$\left[\begin{array}{c|c} K^{uu}_{AB} & K^{uc}_{AC} \\ \hline K^{cu}_{DB} & K^{cc}_{DC} \end{array}\right] \cdot \begin{Bmatrix} u_B \\ c_C \end{Bmatrix} = \begin{Bmatrix} R^u_A \\ R^c_D \end{Bmatrix} \quad \text{with} \quad A, B = 1, ..8; C, D = 1, ..52 \tag{68}$$

in which we distinguish the respective parts of stiffness matrix K and loading R with superscripts.

```
Kuu = K[[1;;8, 1;;8]];        Kuc = K[[1;;8, 9;;60]];
Kcu = K[[9;;60,1;;8]];        Kcc = K[[9;;60 9;;60]];
Ru  = R[[1;;8]];              Rc  = R[[9;;60]];
```

In order to decrease the number of unknowns, the static condensation procedure is performed on the element level. System (67) of 60 equations can be reduced to a system

$$K^{\text{red}}_{AB}\, u_A = R^{\text{red}}_A \tag{69}$$

of 8 equations in which only the nodal displacements appear and the 52 variables c_C have been eliminated from

$$K^{\text{red}}_{AB} = K^{uu}_{AB} - K^{uc}_{AM} \left(K^{cc}\right)^{-1}_{MN} K^{cu}_{NB} \tag{70}$$

$$R^{\text{red}}_A = R^u_A - K^{uc}_{AM} \left(K^{cc}\right)^{-1}_{MD} R^c_D \tag{71}$$

Reduced stiffness K^{red}_{AB} and reduced loading R^{red}_A obtained from (70) and (71) can be aggregated to the respective global quantities from which, under consideration of Dirichlet BC the nodal displacements u_A or $u_{\alpha i}$ and hence

$$c_C = (K^{cc})^{-1}_{CD} \left(R^c_D - K^{cu}_{DA} u_A\right) \tag{72}$$

can be found. Typical Dirichlet BCs, for example by ones which fix the whole edge of an element are now briefly discussed. According to (58) some $u_{\alpha i}$ and some $c_{\beta i}$ influence the displacement of the edge and hence all of them must be restricted.

For example, let the displacement $u_i(\xi, -1) = \breve{u}_i(\xi)$ be prescribed on the element edge $\eta = -1$, and hence

$$\sum_{\alpha=1}^{4} N_\alpha(\xi, -1)\, u_{\alpha i} + \sum_{\alpha=5}^{30} N_\alpha(\xi, -1)\, c_{\alpha i} = \breve{u}_i(\xi) \tag{73}$$

One may note that displacement of the first edge depends only on N_α for $\alpha = 1, 2, 5, 9, 13, 17, 21$, which simplifies (73). For practically important case $\breve{u}_i(\xi) = 0$ on the element edge (fixed displacement in i-direction) we can find solution for $c_{\beta i}$ with the script:

```
uBC = 0; edgeBC = 1; iBC = 2;
 If[edgeBC == 1, {xiBC, etaBC} = {xi, -1};  v = xi;];
 If[edgeBC == 2, {xiBC, etaBC} = {1, eta};  v = eta;];
 If[edgeBC == 3, {xiBC, etaBC} = {xi, 1};   v = xi;];
 If[edgeBC == 4, {xiBC, etaBC} = {-1, eta}; v = eta;];
Cbetai = Array[c, {26, 2}, {5, 1}]; Cedge = {};
For[k = 1, k < 6, k++, Cedge = Append[Cedge, Cbetai[[1 + 4*(k - 1) + (edgeBC - 1)]]];];
cb = {Transpose[Cbetai][[1]], Transpose[Cbetai][[2]]};
{NN, Nx, Ny} = getNshape4ddHE[xiBC, etaBC];
Eq = CoefficientList[NN[[5 ;; 30]].cb[[iBC]], v];
unknowns = Transpose[Cedge][[iBC]];
Solve[{Eq == uBC}, unknowns]
```

Obtaining following restrictions on the edge 1-2: $c_{5i} = c_{9i} = c_{13i} = c_{17i} = c_{21i} = 0$. The remaining parameters $c_{\beta i}$ are free. Analogously, conditions for fixed displacement in i-direction can be obtained on the remaining edges:

- edge 2-3: $c_{6i} = c_{10i} = c_{14i} = c_{18i} = c_{22i} = 0$
- edge 3-4: $c_{7i} = c_{11i} = c_{15i} = c_{19i} = c_{23i} = 0$
- edge 4-1: $c_{8i} = c_{12i} = c_{16i} = c_{20i} = c_{24i} = 0$

Note that in the case of uniform displacement of the element edge ($\breve{u}_i(\zeta)=$const) Eq. (73) reduces due to the "partition of unity" property $\left(\sum_{\alpha=1}^{4} N_\alpha(\xi) = 1\right)$ to

$$\sum_{\alpha=5}^{30} N_\alpha(\xi, -1)\, c_{\alpha i} = 0, \tag{74}$$

which leads to the results mentioned above for the case of fixed displacements.

4.1 Stress Recovery

After an FE calculation the nodal displacements $u_{\alpha i}$ and internal parameters $c_{\alpha i}$ are known and so the approximation of the displacement function u_i. The strain ϵ and stress fields σ can be calculated as

$$\epsilon_{ij}(\xi) = B_{ij\eta k}(\xi)\, a_{\eta k} - \epsilon_{ij}^{acc}(\boldsymbol{\xi}) \tag{75}$$

$$\sigma_{kl}(\xi) = E_{klij}\epsilon_{ij}(\xi) + \sigma_{kl}^{0}(\xi) \tag{76}$$

where matrix $B_{ij\eta k}$ is defined by (64)–(66) and vector $a_{\eta k}$ by (61).

5 Test Results

The comparison of the EAS and HE elements are carried out using a prescribed strain field ϵ^{acc}

$$\begin{cases} \epsilon_{11}^{acc} = 0.01\, x_2^2/4, \\ \epsilon_{22}^{acc} = 0.01\, x_1^2/4, \\ \gamma_{12}^{acc} = 0.01\, x_1 x_2 \end{cases}$$

defined upon a freely supported quadratic area presented in Fig. 8. In this example all strain components vary quadratically in space as expected in HCA applications.

The strain field is compatible and hence no self stress appears in analytical solution. The exact displacement field (without rotation) can be easily integrated assuming $u_{1,2} = u_{2,1}$ and $u_i(0) = 0$

$$u_1^{acc} = 0.01(x_1 x_2^2)/4 \quad \text{and} \quad u_2^{acc} = 0.01(x_1^2 x_2)/4 \tag{77}$$

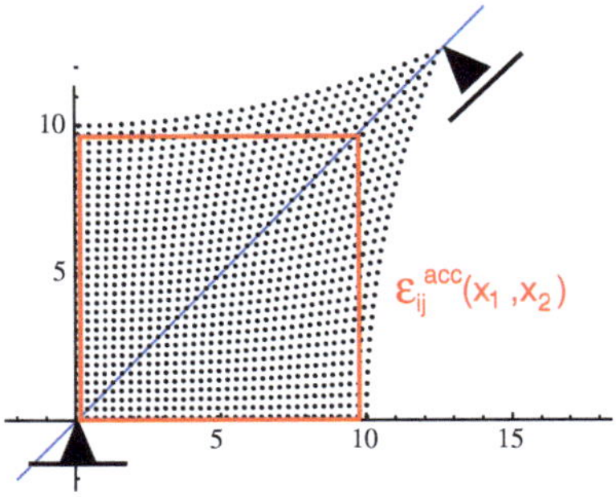

Fig. 8 Analytically integrable strain field

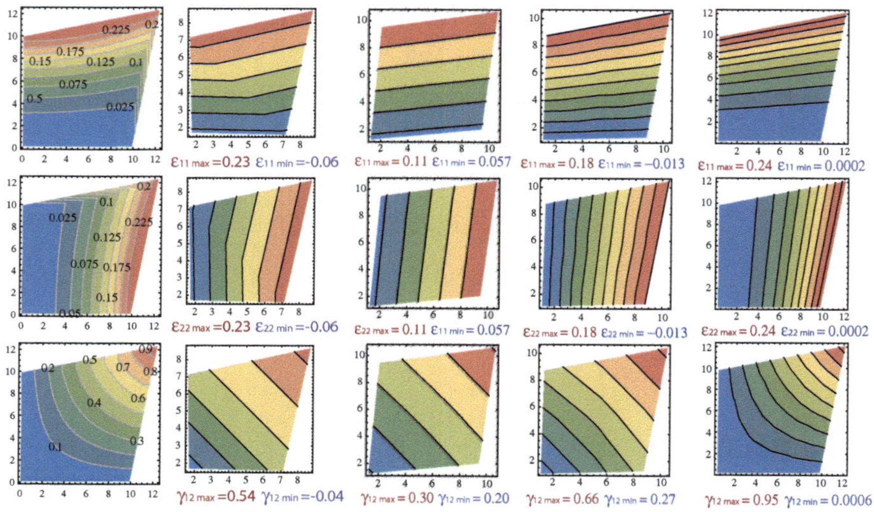

Fig. 9 Strain components from 1 element. In columns: exact, CPE4, CPE4R, EAS. HE

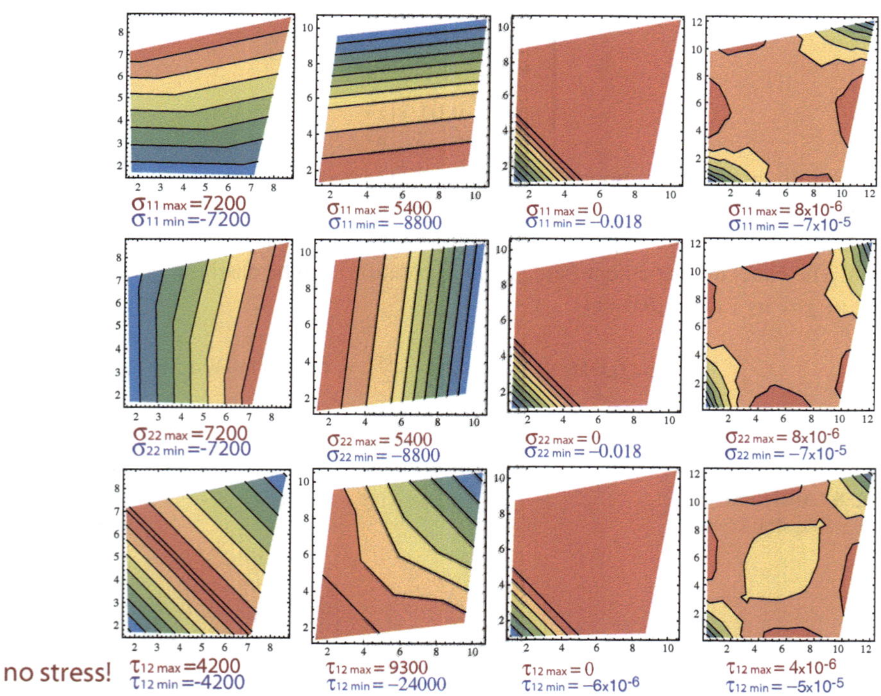

Fig. 10 Stress components from 1 element. In columns: exact, CPE4, CPE4R, EAS. HE

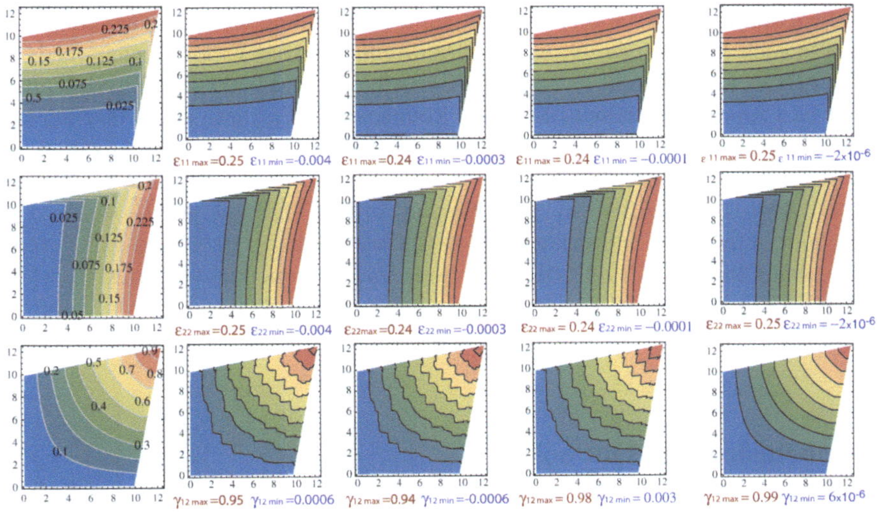

Fig. 11 Strain components from 100 elements. In columns: exact, CPE4, CPE4R, EAS. HE

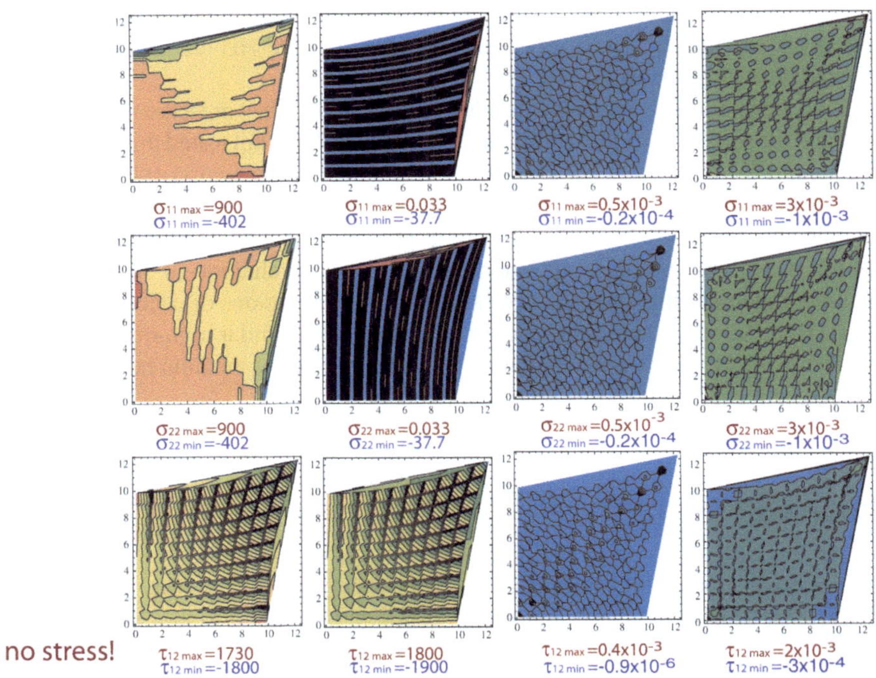

Fig. 12 Stress components from 100 elements. In columns: exact, CPE4, CPE4R, EAS. HE

The displacement of the upper right support according to analytical solution is 2.5 m in each direction ($u_1 = u_2$). Discretization with one CPE4 element leads to $u_1 = 2.0833$, CPE4R to $u_1 = 1.875$, CPE4EAS to $u_1 = 2.0833$ and CPE4HE gives us the best approximation with $u_1 = 2.5$. Refinement of the mesh up to 100 elements provides better displacement approximations: in the case of CPE4 discretization the displacement of the upper right node is $u_1 = 2.504$, from CPE4R discretization $u_1 = 2.494$, for CPE4EAS $u_1 = 2.496$ and for CPE4HE one get $u_1 = 2.5$. The behaviour of the stress and strain fields are shown in Figs. 9, 10, 11 and 12. One can notice that CPE4 and CPE4R discretizations lead to numerical self-stresses, while CPE4(EAS) and CPE4(HE) provide good approximation of displacements and allow to avoid non-physical high stress gradients.

6 Conclusion and Outlook

We have demonstrated that both HE and EAS elements generate much less artificial stress than the conventional CPE4 and CPE4R elements. Unfortunately, HE elements fail the patch test so they cannot be recommended. Moreover, the definition of Dirichlet BC may appear troublesome and the HE elements seen to be computationally expensive.

EAS is a promising method to develop FEs for HCA. They preserve simple geometry and the self-stresses are negligible. However, we have not tested this elements in the ABAQUS® environment yet. However, one can find an application of EAS-element to geotechnical plasticity problems (including consolidation) in [8]. MATHEMATICA is a pleasant tool for preliminary testing of new FEs. Compared with ABAQUS® we have a possibility to control all intermediate results, evaluation of stiffness terms including their spectral analysis and fast analytical derivations of necessary expressions. Both ABAQUS® and MATHEMATICA can produce similar high-quality graphic results. In future we intend to implement soil as a two-phase medium but our present $u - p - w$ formulation is not stable yet.

References

1. Babuska, I., Griebel, M., Pitkaranta, J.: The problem of selecting the shape functions for a p-type finite element. Technical report final report, Institute for Physical Science and Technology, University of Maryland (1988)
2. Flaherty, J.E., Eaton, A.: Finite element analysis. Lecture notes. Technical report, CSCI, Math 6860, Rensselaer Polytechnic Institute (2000)
3. Hu, H.C.: On some variational principles in the theory of elasticity and plasticity. Sci. Sin **4**, 33–54 (1955)
4. Niemunis, A., Wichtmann, T., Triantafyllidis, T.: A high-cycle accumulation model for sand. Comput. Geotech. **32**(4), 245–263 (2005)
5. Simo, J.C., Rifai, M.S.: A class of mixed assumed strain methods and the method of incompatible modes. Int. J. Numer. Meth. Eng. **29**, 1595–1638 (1990)

6. Washizu, K.: On the variational principles of elasticity and plasticity. Technical report, 25–18, Aeroelastic and Structures Research Laboratory, Massachusetts Institute of Technology, Cambridge (1955)
7. Wichtmann, T., Niemunis, A., Triantafyllidis, Th.: Towards the FE prediction of permanent deformations of offshore wind power plant foundations using a high-cycle accumulation model. In: International Symposium: Frontiers in Offshore Geotechnics, pp. 635–640, Perth, Australia (2010)
8. Zhou, X.X., Chow, Y.K., Leung, C.F.: Application of enhanced assumed strain finite element method to predict collapse loads of undrained geotechnical problems. Int. J. Numer. Anal. Meth. Geomech. **31**, 1033–1043 (2007)
9. Zienkiewicz, O.C., Taylor, R.L., Zhu, J.Z.: The Finite Element Method, 6th edn. Elsevier, Oxford (2005)

Simulation of Soils Under Rapid Cyclic Loading Conditions

W. Ehlers, M. Schenke and B. Markert

Abstract The stability of structures strongly relies upon the strength and stiffness of the foundation soil underneath. If fluid-saturated or nearly saturated soils are subjected to rapid cyclic loading conditions, for instance, during earthquakes, the intergranular frictional forces might be dramatically reduced. Subsequently, the load-bearing capacity decreases or even vanishes, if the soil grains loose contact to each other. This phenomena is often referred to as soil liquefaction. Drawing our attention to fluid-saturated granular materials with heterogeneous microstructures, the modelling is carried out within a continuum-mechanical framework by exploiting the macroscopic Theory of Porous Media (TPM) together with thermodynamically consistent constitutive equations. In this regard, the present contribution proceeds from a fully saturated soil, composed of an elasto-plastic solid skeleton and a materially incompressible pore fluid. The governing material parameters of the solid skeleton have been identified for the research-unit sand. The underlying equations are used to simulate soils under rapid cyclic loading conditions. In this regard, the semi-infinite domain is split into a near field, which usually the domain of interest, and a far field, which extents the simulated domain towards infinity. In order to avoid wave reflections at the near-field boundaries an energy-absorbing layer is introduced. Finally, several simulations are carried out. Firstly, a parametric study of the particular far-field treatment is performed and, secondly, soil liquefaction is simulated, where the underlying initial-boundary-value problem is inspired by practically relevant scenarios.

W. Ehlers (✉) · M. Schenke
Institute of Applied Mechanics, University of Stuttgart, Stuttgart, Germany
e-mail: ehlers@mechbau.uni-stuttgart.de

B. Markert
Institute of General Mechanics, RWTH Aachen University, Aachen, Germany

Th. Triantafyllidis (ed.), *Holistic Simulation of Geotechnical Installation Processes*,
Lecture Notes in Applied and Computational Mechanics 77,
DOI 10.1007/978-3-319-18170-7_11

1 Introduction

From a continuum-mechanical point of view, granular materials, such as soils, can neither be classified as solids nor fluids. Their macroscopic observed state (solid- or fluid-like) is a direct consequence of the microstructural intergranular frictional forces and, thus, strongly depends on the loading conditions. If fluid-saturated soils are subjected to rapid cyclic loading conditions, depending on the amplitude and the frequency of the excitation, its load-bearing capacity may decrease dramatically causing the soil to exhibit a fluid-like behaviour, i. e. it liquefies. For instance, buildings on the surface may tilt, which is referred to as structural overturning, or even entirely collapse. In the related literature, the general term "liquefaction" comprises more specific liquefaction-related phenomena, in particular, "flow liquefaction" and "cyclic mobility" [1]. The term "flow liquefaction" addresses an instability phenomenon, which is associated with loose soils with a low shear strength. Therein, the intergranular frictional forces are reduced dramatically by an increasing pore pressure until the residual shear strength cannot sustain static equilibrium anymore. In contrast, the term "cyclic mobility" is associated with medium dense to dense soils and refers to a limited plastic deformation under cyclic loading conditions, where the overall stability of the granular assembly is maintained.

When describing liquefaction phenomena, on the one hand, a comprehensive understanding of the mutual interactions of the various components, in particular, the solid skeleton, composed of the grains, and the pore fluid, which itself can be a mixture of various interacting components, is decisive. On the other hand, special attention also needs to be paid to the description of the contractant (densification) and dilatant (loosening) behaviour of the solid skeleton under pure shear deformation, which is a consequence of the sliding and the rolling of the grains. In particular, depending on the initial density, the soil exhibits a macroscopically contractant (loose soil) or dilatant behaviour (medium-dense to dense soil) under shear loading. Note that the dilatant behaviour of medium-dense and dense soils is preceded by a slightly contractant behaviour [2]. As a consequence, medium-dense and dense soils exhibit a contractant behaviour if they are subjected to small shear deformations. In order to explain, soil liquefaction, attention is drawn to a fluid-saturated soil with a low *Darcy* permeability subjected to rapid cyclic shear deformations. Therein, in contrast to a dry soil, the materially incompressible pore fluid (here water) has no time to evacuate from the reducing pore space. As a consequence, an excess of pore pressure accumulates, thereby reducing the intergranular normal forces, and thus, the intergranular frictional forces. Therefore, the load-bearing capacity of the whole fluid-saturated soil is weakened or might be lost entirely.

Aiming at the simulation of liquefaction phenomena, there are several models available (e.g. [3–7]) most of which are based on *Biot*'s theory [8]. However, these models proceed from different approaches in order to describe the behaviour of the solid skeleton. In this regard, some are associated with the Cam-Clay-based descriptions (e.g. [9, 10]) and others with the hypoplasticity framework (e.g. [11]). Furthermore, it is also worth to mention the more phenomenological approaches, such as [12, 13], which employ a direct stress-strain relation that distinguishes between loading and unloading stages.

The present contribution proceeds from a thermodynamically consistent approach based on the Theory of Porous Media (TPM) (e.g. [14–16]), where the solid skeleton is described as an elasto-plastic material including isotropic hardening and a stress-dependent failure surface. The governing equations comprise the balance equations of [17] and the elasto-plastic solid-skeleton description of [18]. Following this, the governing balance equations are discretised with respect to space and time, thereby accounting for the transient loading conditions. In this regard, the semi-infinite half-space is spatially discretised, by splitting the analysed domain into a near field, which is, in general, the domain of interest, and a far field, which extents towards infinity. However, truncating the semi-infinite half-space at the near field, which is often sufficient in quasi-static simulations, introduces artificial boundaries at which, in a dynamic analysis, the incident waves are reflected back into the domain of interest. In order to overcome this problem, several methods have been proposed in the literature. In general, they can be classified into so-called coupling methods, such as, for instance, the combined finite-element-infinite-element-method (FEM-IEM), and so-called absorbing-boundary-condition (ABC) methods. The present contributions proceeds from the approach proposed in [19]. Therein, the near and the far field are spatially discretised by mixed *Taylor-Hood* finite elements (FE) and infinite elements (IE), respectively, and, additionally, an energy-absorbing layer at the FE-IE interface has been introduced. In the next step, the temporal discretisation is carried out, thereby, accounting for the special requirements of the global (spatial discretisation) and the local system (plastic evolution). In particular, the *Hilber-Hughes-Taylor* method is used for the global system, whereas, the implicit (backward) *Euler* is used for the local system. Following this, the discretised governing equations are implemented into the coupled finite-element solver PANDAS, which is linked to the commercial FE package Abaqus via a general interface. This coupling allows the definition of complex initial-boundary-value problems through Abaqus, thereby using the material models of PANDAS. The material parameters of the solid-skeleton model have been identified for the sand used in the research-unit FOR 1136, by the commonly used *Least-Squares* approximation. The numerical model will be used, firstly, to perform a parametric study of the far-field treatment and, secondly, to simulate flow liquefaction in a loose soil. Finally, future aspects are addressed in the conclusions.

2 Description of the Fluid-Saturated Soil

A suitable framework for the description of fully-saturated soils is provided by the TPM. Following this, the individual components are described separately through their respective mass and momentum balances, but joined together to a holistic formulation by incorporating suitable production terms.

Within the macroscopic TPM approach, one assumes a homogeneous distribution of overlaid individual components φ^α, which, in the present case, are the materially incompressible solid skeleton ($\alpha = S$) and the materially incompressible pore fluid ($\alpha = F$), both within a representative elementary volume (REV) $\mathrm{d}v$. The composition of the bulk volume element is defined through respective volume fractions $n^\alpha = \mathrm{d}v^\alpha/\mathrm{d}v$, where $\mathrm{d}v^\alpha$ is the partial volume of the component φ^α within the REV. Note that the saturation condition $\sum_\alpha n^\alpha = n^S + n^F = 1$ must hold. Following this, two density functions are defined. The material (realistic or effective) density $\rho^{\alpha R} = \mathrm{d}m^\alpha/\mathrm{d}v^\alpha$ relates the components local mass $\mathrm{d}m^\alpha$ to its volume $\mathrm{d}v^\alpha$, while the partial (global or bulk) density $\rho^\alpha = \mathrm{d}m^\alpha/\mathrm{d}v$ is associated with the bulk volume. Moreover, both density definitions are related to each other through $\rho^\alpha = n^\alpha \rho^{\alpha R}$. As we assume materially incompressible and uncrushable grains, the realistic density of the solid remains constant under the prescribed isothermal conditions, but the bulk density can still change through a changing volume fraction n^α.

In the framework of the TPM, the solid φ^S and the pore fluid φ^F are treated as superimposed continua where each spatial point is simultaneously occupied by particles of both components and each components particle is moving according to its own motion function and, thus, have their own velocity and acceleration field. In this regard, it is convenient to express the solid motion in the *Lagrange*an or material setting through the solid displacement $\mathbf{u}_S$ and the fluid motion in the *Euler*ian or spatial setting through the seepage velocity $\mathbf{w}_F$ relative to the solid motion. Following this, the displacement, velocity and acceleration functions read [17]:

$$\begin{aligned}
&\bullet \text{ solid motion: } \mathbf{u}_S = \mathbf{x} - \mathbf{X}_S,\ \mathbf{v}_S = (\mathbf{u}_S)'_S = \overset{\prime}{\mathbf{x}}_S,\ (\mathbf{v}_S)'_S = \overset{\prime\prime}{\mathbf{x}}_S,\\
&\bullet \text{ fluid motion: } \qquad \mathbf{w}_F = \overset{\prime}{\mathbf{x}}_F - \overset{\prime}{\mathbf{x}}_S, \quad (\mathbf{v}_F)'_F \approx (\mathbf{v}_F)'_S.
\end{aligned} \tag{1}$$

Therein, $\mathbf{X}_S$ denotes the position of a solid material point in the reference configuration ($t = t_0$), while $\mathbf{x}$ is the position of a point in the current configuration ($t > t_0$). Moreover, $(\cdot)'_S$ and $(\cdot)'_F$ denote material time derivatives following the motion of the solid skeleton and the pore fluid, respectively. Note that according to [8], for the lower frequency range ($f \leq 30\,\mathrm{Hz}$), which is the case within the scope of the present contribution, the convective terms can be neglected. Thus, $(\mathbf{v}_F)'_F = (\mathbf{v}_F)'_S + \operatorname{grad}\mathbf{v}_F\,\mathbf{w}_F \approx (\mathbf{v}_F)'_S$.

According to [17], the governing balance equations are the convective-less total momentum balance of the overall porous material, the convective-less momentum balance of the pore fluid and the total volume balance of the overall porous material. They read:

$$\rho^S(\mathbf{v}_S)'_S + \rho^F(\mathbf{v}_F)'_S = \operatorname{div}(\mathbf{T}^S_E - p\,\mathbf{I}) + (\rho^S + \rho^F)\mathbf{b}, \tag{2}$$

$$\rho^F(\mathbf{v}_F)'_S = \operatorname{div}(-n^F p\,\mathbf{I}) + \rho^F\,\mathbf{b}\frac{(n^F)^2\gamma^{FR}}{k^F}\mathbf{w}_F + p\operatorname{grad} n^F, \tag{3}$$

$$0 = \operatorname{div}(\mathbf{v}_S + \frac{k^F}{\gamma^{FR}}\{\rho^{FR}[\mathbf{b} - (\mathbf{v}_F)'_S] - \operatorname{grad} p\}). \tag{4}$$

Therein, $\mathbf{b}$ is the unique mass-specific body force, k^F is the hydraulic conductivity (*Darcy* permeability) and $\gamma^{FR} = g\rho^{FR}$ is the effective fluid weight with $g = |\mathbf{b}| =$ const. as the scalar gravitational acceleration. Moreover, $\mathbf{T}^S_E$ is the effective solid stress, which is associated with the intergranular forces, p is the pore-fluid pressure and $\mathbf{I}$ is the second-order identity tensor. The corresponding primary variables of the resulting three-field formulation are the solid displacement $\mathbf{u}_S$, the pore-fluid velocity $\mathbf{v}_F$ and the pore-fluid pressure p.

In order to complete the model, a constitutive description of the effective solid stress $\mathbf{T}^S_E$ is necessary. In extension of [17], which proceeds from a purely elastic description, we continue with an elasto-plastic model, in particular, with an elasto-(visco)plastic solid skeleton including isotropic hardening and a stress-dependent failure surface (cf. [18] for details). Restricting the presentation to the small-strain regime, the linear solid strain tensor is given by

$$\varepsilon_S = \frac{1}{2}(\operatorname{grad}\mathbf{u}_S + \operatorname{grad}^T\mathbf{u}_S) \qquad \longrightarrow \qquad \varepsilon_S = \varepsilon_{Se} + \varepsilon_{Sp}, \tag{5}$$

which in the framework of elasto-plasticity is additively split into an elastic ε_{Se} and a plastic part ε_{Sp}. Following this, the solid volume fraction can be written as (cf. [20]),

$$n^S = n^S_{0S}(1 - \varepsilon^V_S) = n^S_{0S}(1 - \varepsilon^V_{Sp})(1 - \varepsilon^V_{Se}) = n^S_p(1 - \varepsilon^V_{Se}). \tag{6}$$

Therein, n^S_{0S} denotes the initial solid volume fraction and $\varepsilon^V_S = \operatorname{div}\mathbf{u}_S = \varepsilon_S \cdot \mathbf{I}$ is the volumetric solid strain, which is split into its corresponding elastic part $\varepsilon^V_{Se} = \varepsilon_{Se} \cdot \mathbf{I}$ and plastic part $\varepsilon^V_{Sp} = \varepsilon_{Sp} \cdot \mathbf{I}$. Note that, as we proceed from a continuum-mechanical framework, in contrast to geomechanics, volumetric compression corresponds to negative volumetric quantities, i.e. $\mathbf{T}^S_E \cdot \mathbf{I} < 0$ and $\varepsilon^V_S < 0$, whereas volumetric expansion corresponds to positive volumetric quantities, i.e. $\mathbf{T}^S_E \cdot \mathbf{I} > 0$ and $\varepsilon^V_S > 0$.

2.1 Elastic Domain

In order to capture the non-linear behaviour of sand, even in the geometrically linear regime, the following stress-strain relation, based on a non-linear elastic potential, has been introduced [18]:

$$\mathbf{T}^S_E = 2\,\mu^S\,\varepsilon^D_{Se} + \left[k^S_0 + k^S_1\left(\frac{\varepsilon^V_{Se\,\mathrm{crit}}}{\varepsilon^V_{Se\,\mathrm{crit}} - \varepsilon^V_{Se}} - 1\right)\right]\varepsilon^V_{Se}\,\mathbf{I}. \tag{7}$$

Therein, $\varepsilon^D_{Se} = \varepsilon_{Se} - 1/3\;\varepsilon^V_{Se}\,\mathbf{I}$ denotes the deviator of the elastic strain tensor. Moreover, μ^S is the constant elastic shear modulus, k^S_0 and k^S_1 are volumetric bulk moduli, and $\varepsilon^V_{Se\,\mathrm{crit}}$ is the critical volumetric strain, which is given by

$$\varepsilon^V_{Se\,\mathrm{crit}} = 1 - \frac{n^S_{\max}}{n^S_P}, \tag{8}$$

where $n^S_{\max}$ is a material parameter defining the densest packing.

2.2 Plastic Domain

Within the framework of elasto-plasticity, the elastic domain is bounded by an appropriate yield surface. For soils, or granular matter in general, a suitable criterion is provided in [21]. It reads:

$$F = \sqrt{\Gamma\,\mathrm{II}^D + \frac{1}{2}\alpha\mathrm{I}^2 + \delta^2\mathrm{I}^4} + \beta\mathrm{I} + \epsilon\mathrm{I}^2 - \kappa = 0,$$
$$\text{where} \quad \Gamma = (1 + \gamma\frac{\mathrm{III}^D}{(\mathrm{II}^D)^{3/2}})^m. \tag{9}$$

Therein, I, II^D and III^D are the first principal invariant of $\mathbf{T}^S_E$, and the (negative) second and third principal invariants of the effective stress deviator $(\mathbf{T}^S_E)^D$. The material parameter sets $\mathcal{S}_h = (\delta, \varepsilon, \beta, \alpha, \kappa)^T$ and $\mathcal{S}_d = (\gamma, m)^T$ define the shape of the yield surface in the hydrostatic ($\mathcal{S}_h$) and deviatoric plane ($\mathcal{S}_d$).

Following the concept of non-associated plasticity for frictional geomaterials, a suitable plastic potential, which describes the contractant and dilatant behaviour of the soil, is given by

$$G = \sqrt{\psi_1\mathrm{II}^D + \frac{1}{2}\alpha\,\mathrm{I}^2 + \delta^2\mathrm{I}^4} + \psi_2\beta\,\mathrm{I} + \epsilon\,\mathrm{I}^2. \tag{10}$$

Therein, ψ_1 and ψ_2 are material parameters, which serve to relate the dilatation angle to experimental data. The flow rule governing the plastic strain rate $(\varepsilon_{Sp})'_S$ reads

$$(\varepsilon_{Sp})'_S = \Lambda\frac{\partial G}{\partial \mathbf{T}^S_E}. \tag{11}$$

Therein, Λ is the so-called plastic multiplier, which in the framework of viscoplasticity using the overstress concept of *Perzyna* [22] is determined from

$$\Lambda = \frac{1}{\eta}\Big\langle\frac{F}{\sigma_0}\Big\rangle^r, \tag{12}$$

where $\langle\cdot\rangle$ are the *Macaulay* brackets, η is the relaxation time, σ_0 is the reference stress and r is the viscoplastic exponent. Note that the overstress concept also regularises ill-posed problems, for instance during the onset of shear bands (cf. [23] and the references therein), through a careful choice of the parameters η and r.

Any dilatant or compactive behaviour of soils is accompanied by macroscopic softening or hardening effects resulting in a shrinkage or an expansion of the yield surface in the principal stress space. Therefore, suitable evolution laws $(p_i)'_S$ for the parameter subset $p_i \in \{\beta, \delta, \epsilon, \gamma\}$ of the yield surface are used (cf. [20]):

$$(p_i)'_S = (p_i^V)'_S + (p_i^D)'_S = (\overset{*}{p}_i - p_i)[C_{pi}^V\,(\varepsilon_{Sp}^V)'_S + C_{pi}^D\,\|(\varepsilon_{Sp}^D)'_S\|],$$
$$\text{where} \quad p_i(t=0) = p_{i0}. \tag{13}$$

Note that the yield-surface-parameter evolution is split into volumetric $(p_i^V)'_S$ and deviatoric parts $(p_i^D)'_S$, which are driven by the corresponding volumetric and deviatoric plastic strain rates, $(\varepsilon_{Sp}^V)'_S$ and $(\varepsilon_{Sp}^D)'_S$. Moreover, p_{i0} and $\overset{*}{p}_i$ denote the yield-surface parameters at the initial and the saturated state, respectively, where the latter are associated with the failure surface.

Having cyclic loading conditions in mind, one has to take care of the mutual interlocking of the grains as a consequence of a preloading and their release during a subsequent reloading at a lower isotropic stress state. This influence has been observed during triaxial experiments and is considered in the model through a stress-dependent failure surface (cf. [18] for details)

$$\overset{*}{\epsilon}(\mathrm{I}) = \overset{*}{\epsilon}_0(1 + \overset{*}{C}_\epsilon\,\mathrm{I}) \qquad \text{with} \qquad \overset{*}{\epsilon} \geq \overset{*}{\epsilon}_{lim}. \tag{14}$$

Therein, $\overset{*}{C}_\epsilon$ is a constant evolution parameter of the failure surface, $\overset{*}{\epsilon}_0$ theoretically defines the failure surface for the unloaded virgin material and $\overset{*}{\epsilon}_{lim}$ defines the limit of the failure-surface parameter.

3 Numerical Treatment

3.1 Spatial Discretisation

The spatial discretisation of the semi-infinite domain is based on the finite-element method (FEM). In this connection, following a variational approach of *Bubnov-Galerkin*-type, the governing strong forms are multiplied by test function and are integrated over the spatial domain yielding the weak forms. However, in contrast to the standard FEM, the semi-infinite halfspace is spatially split into the near field (domain of interest) and far field (extension towards infinity) discretised by finite elements (FE) and infinite elements (IE), respectively.

At first, the attention is drawn to the spatial discretisation of the near field, which is carried out by the FEM. Therein, the governing strong forms (2)–(4) are multiplied with the test functions $\delta\mathbf{u}_S$, $\delta\mathbf{v}_F$ and δp and are integrated over the spatial domain Ω. The particular weak forms are taken from [17] and read

$$
\begin{aligned}
0 = &\int_{\Omega} \delta \mathbf{u}_S \cdot \{\rho^S (\mathbf{v}_S)'_S + \rho^F (\mathbf{v}_F)'_S - (\rho^S + \rho^F)\mathbf{b}\} \, \mathrm{d}v \\
&+ \int_{\Omega} \operatorname{grad} \delta \mathbf{u}_S \cdot (\mathbf{T}^S_E - p\,\mathbf{I}) \, \mathrm{d}v - \int_{\Gamma_\mathbf{t}} \delta \mathbf{u}_S \cdot \bar{\mathbf{t}} \, \mathrm{d}a,
\end{aligned} \tag{15}
$$

$$
\begin{aligned}
0 = &\int_{\Omega} \delta \mathbf{v}_F \cdot [\rho^F (\mathbf{v}_F)'_S - \mathbf{b}] \, \mathrm{d}v + \int_{\Omega} \delta \mathbf{v}_F \cdot [\frac{(n^F)^2 \gamma^{FR}}{k^F} \mathbf{w}_F - p \operatorname{grad} n^F] \, \mathrm{d}v \\
&+ \int_{\Omega} \operatorname{div} \delta \mathbf{v}_F (-n^F p) \mathrm{d}v + \int_{\Gamma_{\mathbf{t}^F}} \delta \mathbf{v}_F \cdot \bar{\mathbf{t}}^F \, \mathrm{d}a,
\end{aligned} \tag{16}
$$

$$
\begin{aligned}
0 = &-\int_{\Omega} \operatorname{grad} \delta p \cdot \frac{k^F}{\gamma^{FR}} \{\rho^{FR} [\mathbf{b} - (\mathbf{v}_F)'_S] - \operatorname{grad} p\} \mathrm{d}v + \int_{\Omega} \delta p \operatorname{div} \mathbf{v}_S \, \mathrm{d}v \\
&+ \int_{\Gamma_v} \delta p \, \bar{v} \, \mathrm{d}a.
\end{aligned} \tag{17}
$$

Therein, $\bar{\mathbf{t}} = (\mathbf{T}^S_E - p\,\mathbf{I})\mathbf{n}$ and $\bar{\mathbf{t}}^F = -n^F p\,\mathbf{n}$ denote the external loading vectors acting on the *Neumann* boundaries $\Gamma_\mathbf{t}$ and $\Gamma_{\mathbf{t}^F}$ of the overall aggregate and the pore fluid, respectively, and $\bar{v} = n^F \mathbf{w}_F\,\mathbf{n}$ is the volume efflux draining through the *Neumann* boundary Γ_v with $\mathbf{n}$ as the outward oriented unit surface normal.

In contrast to the near field, the far field is discretised via infinite elements (IE). Additionally, in order to achieve energy-absorbing properties, dashpots are introduced at the FE-IE interface. This procedure is often referred to as visco-damped boundaries (VDB) and originates from [24]. According to [19], the governing weak form, composed of a quasi-static and a viscous damped part, is given by

$$
\begin{aligned}
&\underbrace{\int_{\Omega} \operatorname{grad} \delta \mathbf{u}_S \cdot (\mathbf{T}^S_E - p\,\mathbf{I}) \mathrm{d}v - \int_{\Omega} \delta \mathbf{u}_S \cdot \rho\,\mathbf{b} \, \mathrm{d}v}_{\text{quasi-static part}} + \underbrace{\int_{\Gamma_\mathrm{I}} \delta \mathbf{u}_S \cdot \mathbf{r} \, \mathrm{d}a}_{\text{visous-damped part}} = 0 \\
&\text{where} \quad \mathbf{r} = \mathbf{P}^T \begin{bmatrix} a\rho c_p & 0 \\ 0 & b\rho c_s \\ 0 & b\rho c_s \end{bmatrix} \mathbf{P} (\mathbf{u_S})'_S.
\end{aligned} \tag{18}
$$

Therein, $\rho = \sum_\alpha \rho^\alpha$ denotes the density of overall aggregate, Ω denotes the volume of the infinite element, Γ_I the area of the FE-IE interface and $\mathbf{P}$ a projection matrix relating the global solid velocity components to the local coordinate system (normal and shear direction) on Γ_I. Furthermore, $\mathbf{r}$ represents an area-weighted 3-dimensional force vector containing the nodal contributions of the dashpots to the nodes associated with the area at the FE-IE interface. They depend on the compression- and shear-wave velocities, $c_p = \sqrt{(2\mu_S + \lambda_S)/\rho}$ and $c_s = \sqrt{\mu_S/\rho}$ (cf. [19]), and on the dimensionless compression- and shear-wave damping coefficients a and b.

In a second step, the unknown fields ($\mathbf{u}_S$, $\mathbf{v}_F$, p) and the corresponding test functions ($\delta\mathbf{u}_S$, $\delta\mathbf{v}_F$, δp) of the weak forms (15–18) are approximated by suitable test and ansatz functions, which, in the present scope, for the sake of stable solution procedure, need to fulfil the *inf-sup* condition (*Ladyshenskaya-Babuška-Brezzi* (LBB) condition) [25]. In particular, $\mathbf{u}_S$ and $\delta\mathbf{u}_S$ are approximated by quadratic shape functions, whereas linear shape functions are used for $\mathbf{v}_F$, p, $\delta\mathbf{v}_F$ and δp. Note that the test and ansatz functions of the finite and the infinite elements are not given here, instead, the interested reader is referred to [26, 27] for the FE and IE approximation, respectively.

Following this, the spatially discretised formulation combining the near and the far field can be summarised as

$$\boldsymbol{F}^h = \begin{bmatrix} \boldsymbol{G}^h(t,\ \boldsymbol{y},\ \boldsymbol{y}',\ \boldsymbol{y}'',\ \boldsymbol{q}) \\ \boldsymbol{L}^h(t,\ \boldsymbol{q},\ \boldsymbol{q}',\ \boldsymbol{y}) \end{bmatrix} = \begin{bmatrix} \boldsymbol{M}\boldsymbol{y}'' + \boldsymbol{C}\boldsymbol{y}' + \boldsymbol{k}(\boldsymbol{y},\boldsymbol{q}) - \boldsymbol{f} \\ \boldsymbol{A}\boldsymbol{q}' + \boldsymbol{r}(\boldsymbol{q},\boldsymbol{y}) \end{bmatrix} \stackrel{!}{=} \mathbf{0}. \quad (19)$$

Therein, $\boldsymbol{y} = [\hat{\mathbf{u}}_S,\ \hat{\mathbf{v}}_F,\ \hat{\mathbf{p}}]^T$ is a vector containing the nodel degrees of freedom of the finite-element mesh (global system $\boldsymbol{G}^h$) and a vector $\boldsymbol{q} = [\varepsilon_{Sp},\ \Lambda,\ \mathbf{p}]^T$, which gathers the internal variables plastic strains (ε_{Sp}), plastic multiplier (Λ) and yield-surface evolution parameters ($\mathbf{p}$) at the *Gauss* points of the finite-element mesh (local system $\boldsymbol{L}^h$). Note that for the sake of convenience the abbreviation $(\cdot)' = (\cdot)'_S$ is used. Moreover, $\boldsymbol{M}$ and $\boldsymbol{C}$ are the generalised mass and damping matrices, $\boldsymbol{k}(\boldsymbol{y},\boldsymbol{q})$ and $\boldsymbol{r}(\boldsymbol{y},\boldsymbol{q})$ denote the static residual vectors of the global and local system, respectively, and $\boldsymbol{f}$ is the generalised force vector acting on the *Neumann* boundaries.

3.2 Temporal Discretisation

In the next step, the temporal discretisation of Eq. (19) is carried out. In order to account for the specific requirements regarding numerical properties (e.g. stability, numerical damping) of the global and the local system, different time-integration schemes are deployed. In particular, the global system benefits from a numerical-damping-free procedure, whereas unconditional stability is desired for the local system.

In this regard, $\boldsymbol{G}^h$ is discretised through the implicit *Hilber-Hughes-Taylor* (HHT) method (cf. [28]) which is a generalisation of *Newmark*s method (cf. [29]) but allows for the explicit control of the numerical damping. Note that the for dynamic problems desired explicit schemes, which are more efficient, are not applicable within the current setting, as the incompressibility of the constituents leads to a singular and, thus, a non-invertible mass matrix (cf. [17] for details). The time-discrete form of Eq. $(19)_1$ is given by

$$
\begin{aligned}
& \boldsymbol{M}\boldsymbol{y}''_{n+1} + (1+\alpha)(\boldsymbol{C}\boldsymbol{y}'_{n+1} + \boldsymbol{k}(\boldsymbol{y}_{n+1}, \boldsymbol{q}_{n+1}) - \boldsymbol{f}_{n+1}) \\
& \qquad - \alpha(\boldsymbol{C}\boldsymbol{y}'_n + \boldsymbol{k}(\boldsymbol{y}_n, \boldsymbol{q}_n) - \boldsymbol{f}_n) \overset{!}{=} \boldsymbol{0} \\
\text{with} \quad & \boldsymbol{y}_{n+1} = \boldsymbol{y}_n + \Delta t\, \boldsymbol{y}'_n + \Delta t^2\big((\frac{1}{2} - \beta)\boldsymbol{y}''_n + \beta \boldsymbol{y}''_{n+1}\big), \\
& \boldsymbol{y}'_{n+1} = \boldsymbol{y}'_n + \Delta t\big((1-\gamma)\boldsymbol{y}''_n + \gamma \boldsymbol{y}''_{n+1}\big).
\end{aligned}
\tag{20}
$$

Therein, the parameter α controls the numerical damping, on the one hand, by adding the quasi-static residual contributions from the previous state (at t_n) to the current residual (at t_{n+1}) and, on the other hand, by the parameters β and γ, which are inherit from *Newmarks* method, and are given by

$$
\beta = \frac{1}{4}(1-\alpha)^2, \ \gamma = \frac{1}{2}(1-\alpha). \tag{21}
$$

A suitable choice of the parameter α ranges from $\alpha = -1/3$ (significant damping) to $\alpha = 0$ (no damping), whereby, in the latter, the trapezoidal rule ($\beta = 1/4, \gamma = 1/2$) is obtained. Note that a value of $\alpha = -0.05$ is in general considered as good choice as the inevitably time-stepping-induced high-frequency noise is quickly removed without a significant effect on the low-frequency response of the system.

The local system $\boldsymbol{L}^h$, in order to ensure unconditional stability, the implicit (backward) *Euler* scheme is exploited. In this regard, the time-discrete representation of $(19)_2$ is given by

$$
\boldsymbol{A}\boldsymbol{q}'_{n+1} + \boldsymbol{r}(\boldsymbol{q}_{n+1}, \boldsymbol{y}_{n+1}) \overset{!}{=} 0 \quad \text{with} \quad \boldsymbol{q}'_{n+1} = \frac{\boldsymbol{q}_{n+1} - \boldsymbol{q}_n}{\Delta t}. \tag{22}
$$

3.3 Solution Procedure

The solution of the coupled system (19) is carried out with respect to its block-structured nature through a generalisation of the Block *Gauß-Seidel-Newton* method, which is also know as multilevel or, in this particular case, as two-stage *Newton* method. This procedure results in two nested *Newton* iterations. In this connection, at each global iteration, which seeks the solution of the global variables $\boldsymbol{y}^{n+1}$, the nonlinear local system is iteratively solved for the internal variables $\boldsymbol{q}_{n+1}$ at each *Gauss* integration point with frozen global variables (e.g. cf. [30] and reference therein).

The discretised system is implemented into the FE package PANDAS[1] and linked through a general interface to the commercial FE package Abaqus [31]. This coupling

[1]**P**orous media **A**daptive **N**on-linear finite-element solver based on **D**ifferential **A**lgebraic **S**ystems, www.get-pandas.com.

allows for the definition of complex initial-boundary-value problems in terms of features, such as kinematic coupling and tie constraints, and in terms of large-scale analyses through parallelisation.

4 Parameter Identification

In order to identify the solid-skeleton material parameters for the sand of the research-unit FOR 1136 the course of actions is basically following the procedure described in [18]. Therein, a staggered identification scheme has been carried, in which, at first, the elastic shear modulus μ_S and the compression-extension-ratio parameter of the failure surface $\overset{*}{\gamma}$ are determined straightforward from triaxial loading-unloading loops and from compression and extension experiments at different confining pressures. The remaining model parameters are found through a minimisation of the squared error between simulation and experiment, which is commonly known as *Least-Squares* optimisation method. In particular, a gradient-based constrained optimisation is used, in which the *Hesse*an matrix is approximated through BGFS (Broyden, Fletcher, Goldfarb, Shannon) (cf. [32] and references) and and the parameter constraints are considered via the sequential-quadratic-programming (SQP) technique [33]. The identified solid-skeleton material parameters of the research-unit sand FOR 1136 are summarised in the Appendix.

5 Simulations

5.1 Parametric Studies

The following section addresses several parametric studies related to the numerical treatment of the semi-infinite unbound-domain. In particular, in the first investigation the macroscopic damping properties of a fully-saturated soil, which is mainly governed by the *Darcy* permeability, is investigated. The second set of studies servers as a parametric study of the present far-field treatment, in particular, of its energy-absorbing capabilities in relation to certain wave types. In this regard, as the following simulation are solely related to the elastic wave-propagation problem, the elasto-plastic solid is simplified to a purely elastic material governed by the material parameters given in Table 1. Note that the material parameters μ_S and ${k_s}^0$ are chosen arbitrarily and result in comparatively low compression- and shear-wave propagation velocities. The remaining material parameters, in particular, k^F and γ^{FR}, are defined in the related sections.

Table 1 Material parameters of the numerical parametric study of the far-field treatment

μ^S	k_S^0	k_S^1	ρ^{SR}	ρ^{FR}	n_{0S}^S
$0.26\,\text{m}^2/\text{MN}$	$0.40\,\text{m}^2/\text{MN}$	$0.0\,\text{m}^2/\text{MN}$	$2700\,\text{kg/m}^3$	$1000\,\text{kg/m}^3$	0.6

5.1.1 Parametric Study of the Damping Characteristics of a Fluid-Saturated Soil

When subjecting a soil to rapidly changing loading conditions waves are emitted at the source and propagate through the domain. The resulting, in general, complex particle motion can be considered as a superposition of two fundamental wave types, in particular, compression or primary waves (p-waves) and shear or secondary waves (s-waves). By the assumption of a purely elastic solid and by neglecting viscous shear forces within the pore fluid, its macroscopic observed damping properties are solely related to the solid-skeleton-pore-fluid interaction, in particular, to the solid-fluid momentum exchange (cf. Eq. 3). Therein, the momentum exchange is mainly driven by the solid volume fraction, $n^S = 1 - n^F$, and, thus, by the volumetric deformations (cf. Eq. 6). As deviatoric deformations do not provide significant volumetric strains within the small strain regime, the dissipative properties due to pure shear deformations can be neglected in the following parametric study. Hence, the damping-property study is solely applied to compression waves.

In this regard, the displacement amplitude of the solid skeleton is investigated at different depths with varying *Darcy*-permeability. Note that the specific pore-fluid weight is set to $\gamma^{FR} = 10^4\,\text{N/m}^3$. The underlying initial-boundary-value problem (IBVP) is depicted in Fig. 1 (left). Therein, a 3-dimensional soil column, which as simplified to a 1-d problem via suitable boundary conditions, is subjected to a displacement impulse applied on the top of the soil column given by

$$\overline{u}(t) = \overline{u}_0 \sin(2\pi t/T_0)[H(t) - H(t-\tau)] \tag{23}$$

with $T_0 = 1s$, $\tau = T_0/2$, $\overline{u}_0 = 5 \cdot 10^{-3}\,\text{m}$ and $H(t-\tau)$ as the *Heaviside* step function.

The evolution of the solid-skeleton displacement amplitude of the triggered p-wave for different *Darcy* permeabilities $k^F = \{10^{-1}, 10^{-3}, 10^{-3}\}\,\text{m/s}$ is depicted in Fig. 1 (right). It can be seen that with decreasing permeability, which increase the viscous friction between the soil grains and the pore fluid, the amplitude of the compression wave reduces rapidly. Thus, already for a relatively high *Darcy* permeability of $k^F = 10^{-3}\,\text{m/s}$, the influence of the p-wave can be neglected already one meter below the surface.

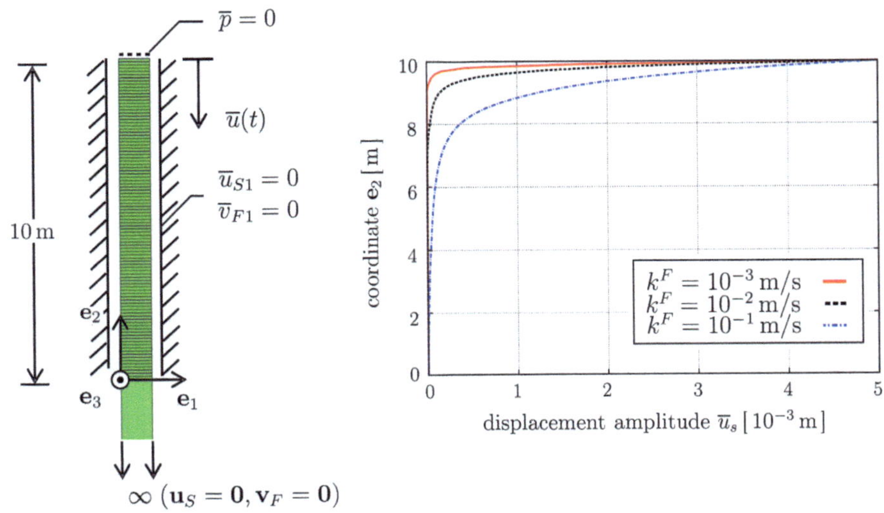

Fig. 1 IBVP to investigate the influence of the *Darcy* permeability on the p-wave penetration depth

Fig. 2 Initial-boundary-value problem of the far-field-treatment parametric study

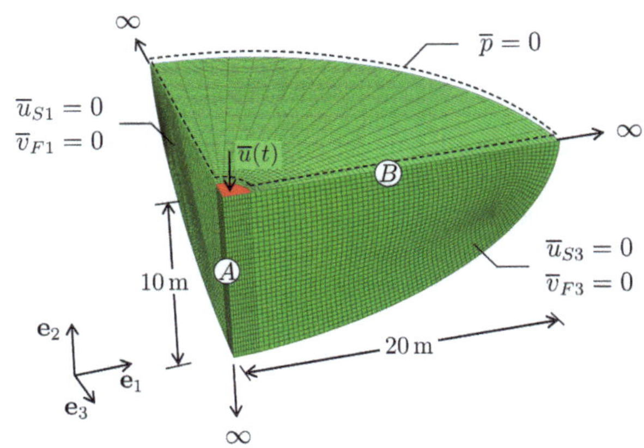

5.1.2 Parametric Study of Numerical Far-Field Treatment

The second example addresses a parametric study of the far-field treatment. In particular, the influence of the damping coefficients and of the quasi-static contribution on the energy-absorbing capabilities are investigated. The governing IBVP is depicted in Fig. 2.

Therein, a ellipsoidal domain (first and second minor axis: 20 m, third minor axis: 10 m) is extended towards infinity by use of infinite elements, whereas the FE-IE interface is described through viscous-damped boundaries. Note that the infinite elements are not depicted in Fig. 2. The resulting numerical model consists of approximately 40,000 elements, which results in approximately 700,000 degrees of

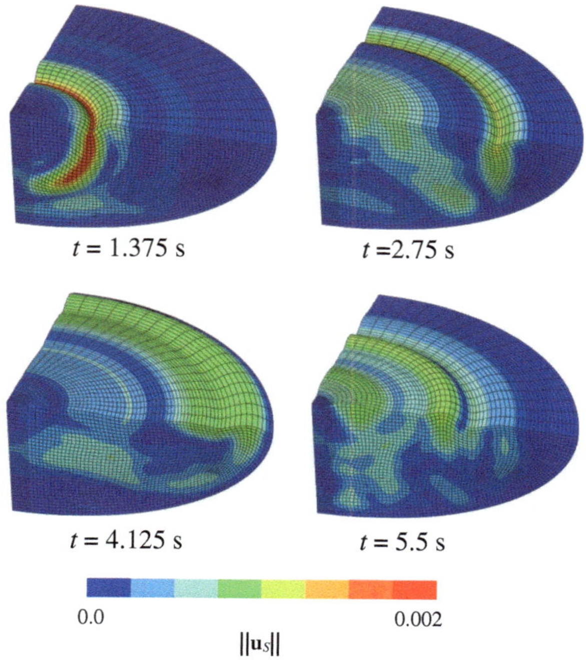

Fig. 3 Contour plot of the magnitude of the solid displacement $\|\mathbf{u}_S\|$ on the deformed geometry (scale factor: 500) of the truncated semi-infinite halfspace

freedom. Thus, due to the problem size, the simulations have been carried out in parallel on 40 cores. In order to trigger waves propagating through the domain, a displacement impulse, given by Eq. (23), is applied at the indicated area. Moreover, in order to judge the energy-absorbing capabilities, the vertical displacements of the solid skeleton at A (located at a depth of 5 m) and B (located 10 m from the vertical symmetry line) are evaluated.

In order to keep this parametric study within more general setting, the damping properties of the VDB with respect to compressional waves are investigated as well, although they are, as seen before, not relevant for most practical-oriented geotechnical scenarios. Thus, to allow for the induced wave to propagate nearly without a loss through the domain, the *Darcy* permeability and the specific weight are set to $k^F = 10^{-2}\,\text{m/s}$ and $\gamma^{FR} = 10^{-4}\,\text{N/m}^3$, respectively. Note that, hereby, the quasi-static contribution has been neglected and the normal and shear damping coefficients are set to $a = b = 1$.

The impact on the numerical solution of the specific far-field treatment is qualitatively illustrated in Figs. 3 and 4. The displacement impulse on top of the domain triggers a compression and a surface wave (*Rayleigh* wave) which, in case of the truncated domain, is reflected at introduced artificial domain boundaries (cf. Fig. 3). In contrast, using the special far-field treatment composed of infinite elements and an energy-absorbing layer at the near-field-far-field transition shows a significant improvement (cf. Fig. 4).

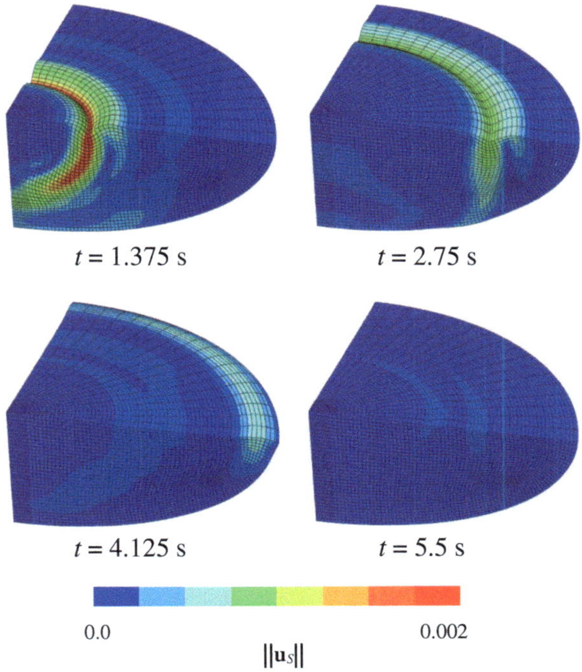

Fig. 4 Contour plot of the magnitude of the solid displacement $\|\mathbf{u}_S\|$ on the deformed geometry (scale factor: 500) of the semi-infinite halfspace incorporating VDB

5.1.3 Influence of the Damping Coefficients

The following simulations investigate the influence of the normal- and shear-damping coefficients, a and b, of the dashpots at the FE-IE interface on the energy-absorption behaviour. In particular, the proposed values of Lysmer and Kuhlemeyer (LK) [24] ($a = b = 1$), which give the best energy absorption if the wave-propagation direction is normal to the FE-IE interface, is compared to the approach of White et al. [34], which is based on the maximisation of the dissipated energy over different wave incidence angles. In the latter approach, the damping coefficients are computed via

$$a = \frac{8}{15\pi}(5 + 2c - 2c^2), \qquad b = \frac{8}{15\pi}(3 + 2c) \qquad \text{where} \qquad c = \sqrt{\mu^S/(\lambda^S + 2\mu^S)}. \tag{24}$$

By use of the material parameters of Table 1, the damping coefficients can be computed as $a = 0.93$ and $b = 0.69$. Note that the quasi-static contribution (cf. Eq. 18) has been neglected.

The results of the simulation are depicted in Fig. 5. Therein, the gray line is used as a reference (ref.) showing the result of the truncated domain. It can be seen that the approach of White et al. (W) gives slightly better results compared to Lysmer and Kuhlemeyer (LK). However, as the approach of White et al. is exploiting the linear elastic *Hooke* an law for the constitutive description of the solid, their proposal is tied

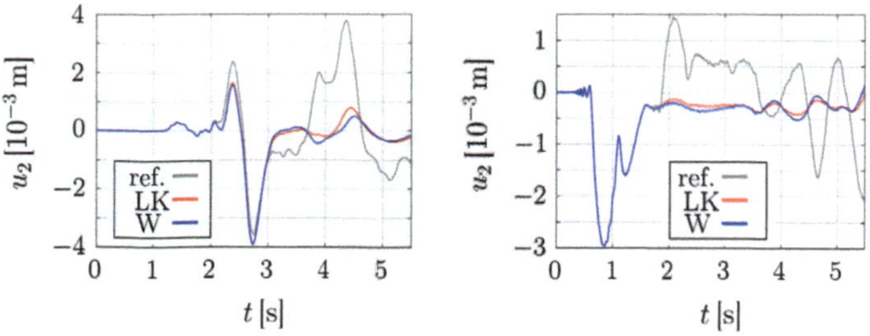

Fig. 5 Evolution of the vertical displacement of *A* (*right*) and *B* (*left*) for different damping parameters

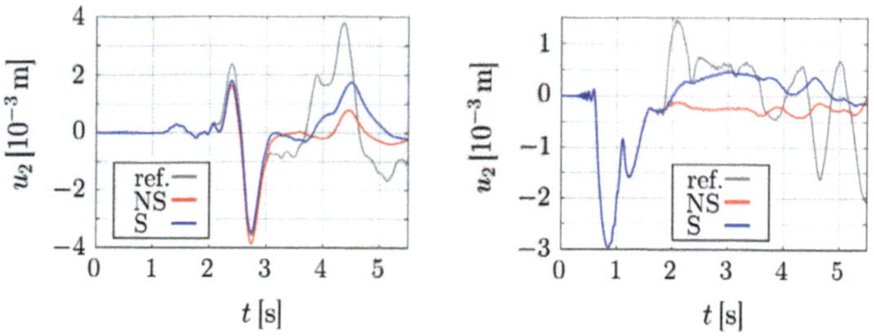

Fig. 6 Evolution of the vertical displacement of *A* (*right*) and *B* (*left*) for the quasi-static-contribution-influence study

to the first and second *Lamé* constants and, thus, may not be suitable for arbitrary solid material descriptions.

5.1.4 Influence of the Quasi-Static Contribution

The second example studies the influence of the quasi-static contribution. In this regard, a simulation involving the quasi-static contribution (S) is compared to a simulation without one (NS). Note that the damping parameters are set to $a = b = 1$.

The results of the simulation are depicted in Fig. 6. It can be seen that the best energy-absorbing capabilities are obtained if the quasi-static part is neglected (NS), which is in accordance with [35]. However, if the simulation contains quasi-static loading steps, for instance, if the transient load is preceeded by a consolidation step, the quasi-static contribution has to be considered, as it provides the necessary residual stiffness to the far-field.

5.2 *Liquefaction of Loose Sand*

The next example is related to the simulation of flow liquefaction in a loose, water-saturated sand under cyclic loading conditions, which is a common scenario, for instance, during earthquakes.

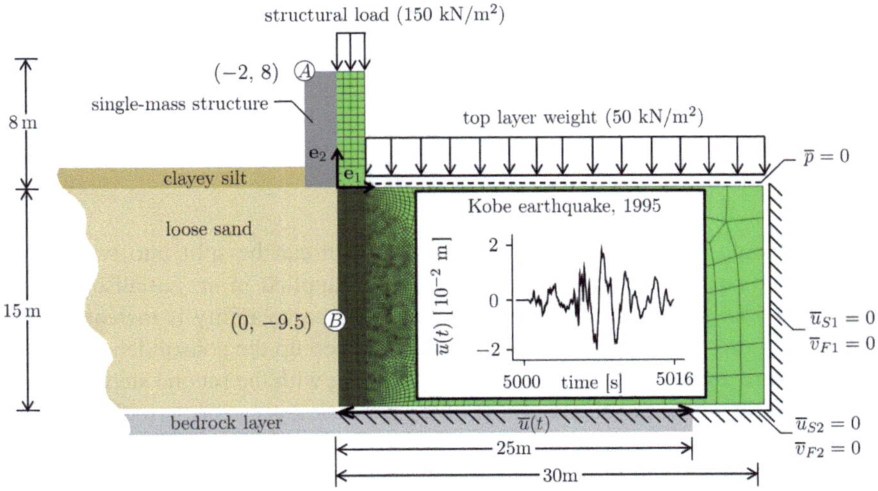

Fig. 7 Geometry (*left*) and deduced initial-boundary-value problem (*right*) of the liquefaction examples

The initial-boundary-value problem under consideration is inspired by the liquefaction-prone Wildlife Refuge area in Imperial Valley in southern California, where the layout of the domain of interest is as follows (Fig. 7, left). From top to bottom, the soil layers are a clayey silt, a liquefiable sand, a stiff clay and a bedrock layer. Based on that, a suitable numerical model is deduced (Fig. 7, right), where the weight of the single-mass structure and the top layer are replaced by uniformly distributed loads of 150 and 50 kN/m^2, respectively. Note that the replacement of the top layer by its corresponding load, avoids numerical difficulties as it ensures that the domain below is under compression during the simulation. Below that is a layer of a liquefiable sand, which is described as an elasto-(visco)plastic material with isotropic hardening and a stress-dependent failure-surface (cf. Sect. 2). The bedrock layer at the bottom of the modelled domain is subjected to lateral displacements according to the records of the Kobe earthquake in 1995 in Japan, which have been logged at the FUK station. Note that the modelled semi-infinite halfspace has been truncated without the use of a special far-field treatment. Due to the facts that, on one hand, the prescribed bedrock-layer displacements mainly initiate shear waves and that, on the other hand, the fluid-saturated soil exhibits significant damping properties with respect to compression waves, the size of the numerical problem size has been reduced, by discarding a special far-field treatment. Note that the finite-element mesh below structure has been refined in order to account for the expected strain localisation. Moreover, tie constraints have been imposed at the interface between the structure and the foundation soil to ensure kinematic compatibility. The resulting numerical model consists of approximately 24,000 elements with approximately 110,000 degrees of freedom.

Table 2 Comparison of the solid-skeleton material parameters for medium-dense FOR 1136 sand (set 1) to the estimated parameters for a loose sand (set 2)

Material	n_{0S}^S	ψ_1	ψ_2
Medium-dense FOR 1136 sand (1)	0.61	1.3	0.53
Loose sand (2)	0.4	0.4	0.1

The loading history of the liquefaction problem can be split into two stages. At first, the structural and top layer weights are applied in an initialisation step ($0\,\text{s} < t < 5000\,\text{s}$). Note that during that stage, the permeability is increased from $k^F = 10^{-5}$ m/s to $k^F = 10^{-3}$ m/s in order to speed up the consolidation process and to ensure a static equilibrium before proceeding with the second step ($5000\,\text{s} < t < 5016\,\text{s}$), in which the displacement of the bedrock layer is prescribed according to the records of the Kobe earthquake.

Note that in order to trigger liquefaction phenomena with the available material parameters, the prescribed displacements are scaled up by a factor of 15. Moreover, the material parameters of the medium-dense FOR 1136 sand are modified in order to describe a loose sand. In particular, the initial solid volume fraction n_{0S}^S and the material parameters governing the dilatation angle, ψ_1 and ψ_2, are varied such that liquefiable-prone loose sand can is mimicked (cf. Table 2). The initial solid volume fraction n_{0S}^S and the material parameters governing the dilatation angle, ψ_1 and ψ_2, are varied such that liquefiable-prone loose sand is mimicked (cf. Table 2).

A time sequence of contour plots of the norm of the accumulated plastic strain tensor $\|\varepsilon_{Sp}\|$ on the deformed finite-element mesh (unscaled) are depicted in Fig. 8. It clearly illustrates the failure of the loose soil foundation beneath the structure. This particular failure mode is known as punching shear failure [36].

As mentioned earlier, soil liquefaction is consequence of the pore-pressure build-up due to the contractant tendency of the soil, which reduces the intergranular normal forces, and thus, the intergranular frictional forces. To make this point clearer, the interplay between the pore pressure p and the effective volumetric solid stress $\mathbf{T}_E^S \cdot \mathbf{I}$, which is associated with the intergranular normal forces, is plotted at point B in Fig. 9 (left).

As can been seen due to the rapid cyclic motions that the slight pore-pressure build-up, approximately till $t \approx 5006$ s, continues to a dramatic pore-pressure increase resulting in a drop of the negative volumetric solid stress, which corresponds, according to the continuum-mechanical framework, to a decrease of the intergranular normal forces, and thus, reduces the intergranular frictional forces. Note that the effective volumetric solid stress may take slight positive values, which is non-physical in a cohesionless soil, but owed to the elasticity law (7) in case of positive volumetric solid strains. As a consequence of the reduced intergranular frictional forces, the foundation soil liquefies and does not recover into a static equilibrium any more. This is also illustrated in Fig. 9 (right), where the vertical displacement of point A located on top of the single-mass structure is depicted. Therein, the collapse of the

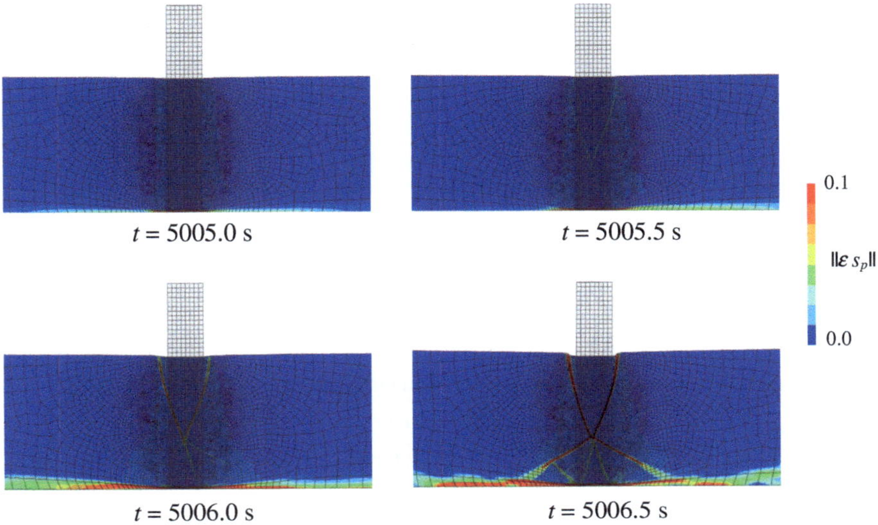

Fig. 8 Contour plots of the norm of the accumulated plastic strain tensor $\|\varepsilon_{Sp}\|$ on the deformed mesh (scale factor: 5) at different times illustrating flow liquefaction

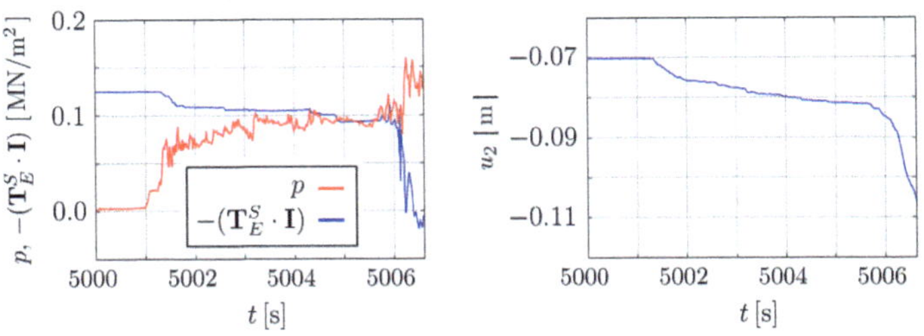

Fig. 9 Evolution of the pore pressure p and the effective volumetric solid stress $\mathbf{T}_E^S \cdot \mathbf{I}$ at point B (*left*) and the time history of the vertical displacement of point A (*right*) in the case of flow liquefaction

soil foundation is easily recognised by the rapidly increasing vertical displacement of the single-mass structure.

The computation terminates at approximately $t \approx 5006.5$ s due to extremely distorted finite elements located in the developing shear bands beneath the structure.

6 Conclusions

In this contribution, a modelling approach for the prediction of liquefaction phenomena in saturated soils has been presented. The underlying fluid-saturated soil model proceeds from a geometrically linear description based on the macroscopic TPM

Table 3 Elastic material parameters and initial solidity

μ^S	k_0^S	k_1^S	n_{max}^S	n_{0S}^S
190 MN/m^2	20 MN/m^2	47 MN/m^2	0.61	0.623

Table 4 Initial and saturation values of the parameters of the yield surface

δ_0	ϵ_0	β_0	γ_0	$\overset{*}{\delta}$	$\overset{*}{\epsilon}_0$	$\overset{*}{\beta}$	$\overset{*}{\gamma}$
0.0009 m^2/MN	0.1 m^2/MN	0.05	0	0.001 m^2/MN	0.01 m^2/MN	0.255	1.75

Table 5 Parameters of the yield-surface evolution and the failure surface

C_δ^V	C_ϵ^V	C_β^V	C_γ^V	C_δ^D	C_ϵ^D	C_β^D	C_γ^D	$\overset{*}{C}_\epsilon$	$\overset{*}{\epsilon}_{lim}$
−93 m^2/ MN	−150 m^2/ MN	−250	0	23 m^2/ MN	200 m^2/ MN	180	20	0.4 m^2/ MN	0.0001

Table 6 Additional yield surface, plastic potential and viscoplastic evolution parameters

κ	α	m	ψ_1	ψ_2	η	r	σ_0
0.0001 m^2/MN	0.01 m^2/MN	0.54	1.3	0.53	0.001 s	1.5	0.0001 MN/m^2

framework involving an elasto-(visco)plastic solid skeleton with isotropic hardening and a stress-dependent failure surface. The presented numerical results reveal the capability of the model to mimic the relevant physical behaviour necessary for the modelling of liquefaction phenomena under rapid cyclic loading conditions. In particular, the model accounts for the behaviour of granular assemblies undergoing volumetric strains under pure shear deformation and resulting a in pore-pressure build-up, which reduces the intergranular frictional forces, and thus, the strength of the whole soil. However, as the model has so far only been tested under rapid cyclic loading conditions, statements regarding the behaviour under quasi-static cyclic loading conditions can not made yet. This will be part of ongoing investigations.

Appendix: Material Parameters of the Elasto-Plastic Solid Describing Medium-Dense FOR 1136 Sand

Below the solid-skeleton material parameters of the research-unit sand FOR 1136[2] are summarised (Tables 3, 4, 5 and 6).

[2]Grain size: 0.1–1 mm; sieve retention: $d10 = 0.4$ mm, $d60 = 0.6$ mm.

References

1. Castro, G., Poulos, S.J.: Factors affecting liquefaction and cyclic mobility. ASCE. J. Geotech. Eng. Div. **103**, 501–506 (1977)
2. Casagrande, J.: The determination of the preconsolidation load and its practical significance. In: Proceedings 1st International Conference on Soil Mechanics and Foundation Engineering (1936)
3. Lin, C.-H., Borja, R.I.: Technical Report No. 137. The John A. Blume Eartquake Engineering Center, Stanford (2000)
4. Prevost, J.H.: Nonlinear transient phenomena in soil media. Mech. Eng. Mater. **30**, 3–18 (1982)
5. Prevost, J., Elgamal, A.M., Abdel-Ghaffar, A.M.: Earthquake-induced plastic deformation of earth dams. In: Proceedings: 2nd International Conference on Soil Dynamics and Earthquake Engineering, vol. 4, pp. 9–17 (1985)
6. Zienkiewicz, O.C., Bettes, P.: Soil Mechanics—Transient and Cyclic Loads. Wiley, Chichester (1982)
7. Zienkiewicz, O.C., Shiomi, T.: Dynamic behaviour of saturated porous media: the generalized Biot formulation its numerical solution. Int. J. Numer. Anal. Meth. Eng. **8**, 71–96 (1984)
8. Biot, M.A.: Theorie of propagation of elastic waves in a fluid-saturated porous solid. J. Acoust. Soc. Am. **28**, 168–178 (1956)
9. Manzari, M.T., Dafalias, Y.F.: A critical state two-surface plasticity model for sands. Geotechnique **47**, 255–272 (1997)
10. Roscoe, K.H., Burland, J.B.: On the generalized stress-strain behaviour of wet clay. In: Heyman, J., Leckie, F.A. (eds.) Engineering Plasticity. Cambridge University Press, Cambridge (1968)
11. Wichtmann, T., Niemunis, A., Triantafyllidis, T.: Strain accumulation in sand due to cyclic loading: drained cyclic tests with triaxial extension. Soil Dyn. Earthq. Eng. **27**, 42–48 (2006)
12. Zienkiewicz, O.C., Chan, A.H.C., Pastor, M., Schrefler, B.A.: Computational Mechanics—with Special Reference to Earthquake Engineering. Wiley, Chichester (1998)
13. Zienkiewicz, O.C., Chang, C.T., Hinton, E.: Non-linear seismics reponse and liquefaction. Int. J. Numer. Anal. Meth. Eng. **2**, 381–404 (1978)
14. Bowen, R.M.: Incompressible porous media models by use of the theory of mixtures. Int. J. Eng. Sci. **18**, 1129–1148 (1980)
15. De Boer, R.: Theory of Porous Media. Springer, Berlin (2000)
16. Ehlers, W.: Foundation of multiphasic and porous materials. In: Ehlers, W., Bluhm, J. (eds.) Porous Media: Theory, Experiments and Numerical Applications, pp. 3–86. Springer, Berlin (2002)
17. Markert, B., Heider, Y., Ehlers, W.: Comparision of monolithic and splitting solutions schemes for dynamic for dynamic porous media problems. Int. J. Numer. Meth. Eng. **82**, 1341–1383 (2010)
18. Ehlers, W., Avci, O.: Stress-dependent hardening and failure surface of dry sand. Int. J. Numer. Anal. Meth. Geomech. **37**, 787–809 (2012)
19. Heider, Y., Markert, B., Ehlers, W.: Dynamic wave propagation in infinite saturated porous media half spaces. Comput. Mech. **49**, 319–336 (2012)
20. Ehlers, W., Scholz, B.: An inverse algorithm for the identification and the sensitivity analysis of the parameters governing elasto-plastic micropolar granular material. Arch. Appl. Mech. **77**, 911–931 (2007)
21. Ehlers, W.: A single-surface yield function for geomaterials. Arch. Appl. Mech. **65**, 246–259 (1995)
22. Perzyna, P.: Fundamental problems in viscoplasticity. Adv. Appl. Mech. **9**, 243–377 (1966)
23. Ehlers, W., Graf, T., Ammann, M.: Engineering issues of unsaturated soil. In: Brinkgreve, R.B.J., Schad, H., Schweiger, H.F., Willand, E. (eds.) Geotechnical Innovations (Studies in Honour of Prof. Dr.-Ing. Pieter Vermeer on Occasion of his 60th Birthday). Verlag Glückauf, Essen, pp. 505–540 (2004)
24. Lysmer, J., Kuhlemeyer, R.L.: Finite dynamic model for infinite media. ASCE. J. Eng. Div. **95**, 859–877 (1969)

25. Brezzi, F., Fortin, M.: Mixed and Hybrid Finite Element Methods. Springer, New York (1991)
26. Marques, J.M.M.C., Owen, D.R.J.: Infinite elements in quasi-static materially nonlinear problems. Comput. Struct. **18**, 739–751 (1984)
27. Zienkiewicz, O.C., Taylor, R.L., Zhu, J.Z.: The Finite Element Method—Its Basis and Fundamentals. McGraw-Hill, New York (2005)
28. Hilber, H.M., Hughes, T.J.R., Taylor, R.L.: Improved numerical dissipation for time integration algorithms in structural dynamics. Earthq. Eng. Struct. Dyn. **3**, 283–292 (1977)
29. Newmark, N.M.: A method of computation of structural dynamics. ASCE. J. Eng. Div. **85**, 67–94 (1959)
30. Karajan, N.: An extended biphasic description of the inhomogeneous and anisotropic intervertebral disc, dissertation: Rep. No.: 09-II-19. Dissertation thesis, Report No. II-19, Institute of Applied Mechanics (Civil Engineering), University of Stuttgart (2009)
31. Schenke, M., Ehlers, W.: On the analysis of soils using an abaqus-PANDAS interface. Proc. Appl. Math. Mech. **11**, 431–432 (2011)
32. Spellucci, P.: Numerische Verfahren der nichtlinearen Optimierung. Birkhäuser, Basel (1993)
33. Spellucci, P.: A new technique for inconsistent QP problems in the SQP-method. Math. Meth. Oper. Res. **47**, 335–400 (1998)
34. White, W., Vallianppan, S., Lee, I.K.: Unified boundary for finite dynamic models. ASCE. J. Eng. Div. **103**, 949–964 (1977)
35. Häggblad, B., Nordgren, G.: Modelling nonlinear soil-structure interaction using interface elements, elastic-plastic soil elements and absorbing infinite elements. Comput. Struct. **26**, 307–324 (1987)
36. Day, R.W.: Geotechnical Earthquake Engineering Handbook. Mcgraw-Hill, New York (2002)

Experimental Strain Response-Envelopes of Granular Materials for Monotonous and Low-Cycle Loading Processes

S. Danne and A. Hettler

Abstract To look onto the stress-path-dependent strain behaviour of granular soils at low-cycle and monotonous loading processes as a basis for the development of new, improved or enhanced constitutive models, drained, stress-controlled triaxial-tests with a fine grained sand have been carried out. The focus was to investigate total as well as quasi-elastic strains for different stress-states by means of strain-response-envelopes. By subtracting the quasi-elastic strains from the total strains, a separate evaluation of plastic strains was also possible. For monotonous loading one separate soil-specimen has been used for each monotonous loading-direction. The shapes of the response-envelopes for initially isotropic stress-states were similar to ellipses, for initially anisotropic stress-states their shape was elongated and shifted away in the direction of the failure lines. For low-cycle loading, cycles of relatively small stress increments were applied in different directions in the stress-plane. It is found that quasi-elastic behaviour can already occur at a low number of cycles. The shapes of the obtained strain-response-envelopes were similar to symmetrical ellipses. It could be observed that the size of the ellipses decreases with increasing mean pressure p. The major axis of the ellipses rotates depending on the initial stress-state $\eta = q/p$, indicating a stress-induced anisotropy. Preloading seems to have little effect on the stiffness or the directions of the quasi-elastic strains. In the future it is intended to simulate the observed stress-strain behaviour by means of new, improved or enhanced constitutive models.

Keywords Triaxial tests · Strain response-envelopes · Incremental stress-strain behaviour · Granular soils

S. Danne (✉) · A. Hettler
Chair of Soil Mechanics and Foundation Engineering, Technical University of Dortmund, Dortmund, Germany
e-mail: stefanie.danne@tu-dortmund.de

Th. Triantafyllidis (ed.), *Holistic Simulation of Geotechnical Installation Processes*, Lecture Notes in Applied and Computational Mechanics 77,
DOI 10.1007/978-3-319-18170-7_12

1 Introduction

Due to quasi-static loading with cyclic progression there are plastic, i.e. irreversible, and elastic, i.e. reversible, deformations in the soil, without reaching fully elastic behaviour. In the quasi-elastic regime the material behaves asymptotically elastic. Goldscheider [10] describes this behaviour as "material shakedown".

Considering the number of cycles one can distinguish between high-cycle and low-cycle loading processes.

Effects of wind load on foundations of wind energy plants, vehicle crossing on foundation constructions, vibrating of foundation-elements e.g. retaining-wall-elements or grouted piles can be related to high-cycle loading. The number of cycles N during these processes is very high ($N \gg 50$). Due to accumulation of numerical errors and high computing time an implicit calculation of displacements, where the deformations during one cycle are calculated separately and accumulated, is not adequate. Instead, deformations due to high-cycle loading are calculated by using explicit models. Here the calculation of irreversible strains can be treated similar to creep deformations under constant loads, Niemunis et al. [20].

Low cycle loading processes can be defined for a lower number of cycles with ($N \leq 50$), see Danne and Hettler [6]. Deformations in this case are usually calculated implicitly, i.e. separately for each cycle and then accumulated.

Subject of this paper are low-cycle loading processes, of which it is assumed that inertial forces are negligible [12]. Un- and reloading for example, occurring during the construction phase of multiple braced excavation walls, produce stress paths quite similar to those of cyclically loaded systems at the first cycles before reaching shakedown. Therefore, these processes are also included within the scope of low-cycle loading.

An external cyclic load on a foundation for example does not lead to cyclic behaviour right from the beginning. This is the case only after a certain number of cycles. Some examples for low-cycle loading processes and related un- and reloading processes are:

- construction stages of multiple braced or anchored excavation walls
- braced excavation with force-controlled struts (to control deformations)
- temperature exposure of struts
- filling and emptying of locks or silos during first utilisation phase
- summer-/winter position of abutments of integral bridges due to temperature differences

The simplified consideration of a soil element behind a strutted retaining wall for example shows, that monotonous stress-paths as well as repeated low-cycle loading process with various directions can occur, Fig. 1.

In front of the embedded part of the wall stress-paths are similar, but extension may be important instead of compression.

Figure 2 shows typical stress-paths of a soil element beneath a watergate's bottom during construction and first utilisation phase of the lock. Stress-states in extension may appear here, too.

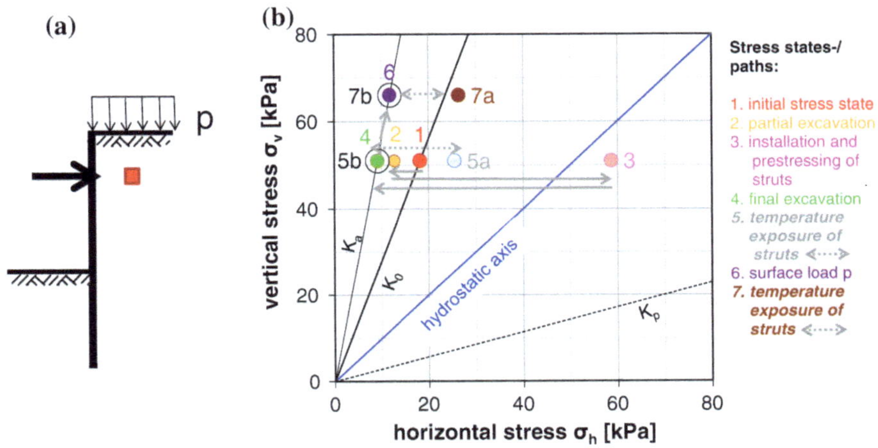

Fig. 1 Typical stress-paths (**b**) in a soil-element behind an excavation-wall (**a**)

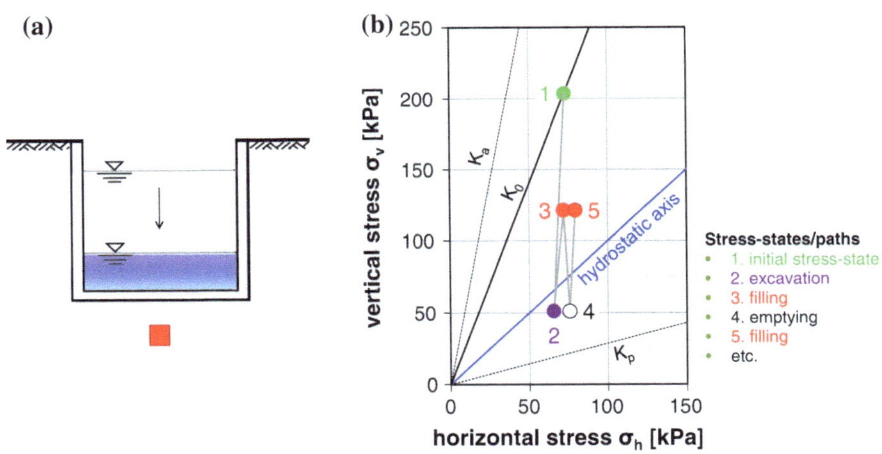

Fig. 2 Possible stress-paths during construction and first utilisation phase of a watergate (**b**) in a soil element beneath a watergate's bottom (**a**)

Element tests investigating the stress-strain behaviour of soils must therefore take into account any stress-path and monotonous as well as repeated un- and reloading processes when contributing successfully to the development of new or enhanced constitutive equations. It is also obvious, that stress-states in compression as well as in extension have to be considered.

2 Response-Envelopes

2.1 Stress- and Strain-Components in Triaxial Testing

Figure 3 shows the axisymmetric stress- and strain-components referred to in this paper.

In axial symmetric conditions $\sqcup_1$ denotes the axial component and $\sqcup_3$ the lateral component of stress or strain respectively.

The Roscoe invariants, mean pressure p and the deviatoric stress q, are defined as:

$$p = (\sigma_1 + 2\sigma_3)/3, \tag{1}$$

$$q = \sigma_1 - \sigma_3. \tag{2}$$

2 vectors, which are orthogonal to each other in the 3-dimensional principal stress space σ_1-σ_2-σ_3, are not perpendicular to each other anymore in the p-q-plane and their lengths get distorted. To avoid this, stresses and strains in this paper are presented in the rendulic plane, which is isomorphic. Its horizontal axis is $\sqrt{2}\sigma_3$ and $\sqrt{2}\varepsilon_3$ respectively and the vertical axis σ_1 and ε_1 (Fig. 4). The stress-ratio

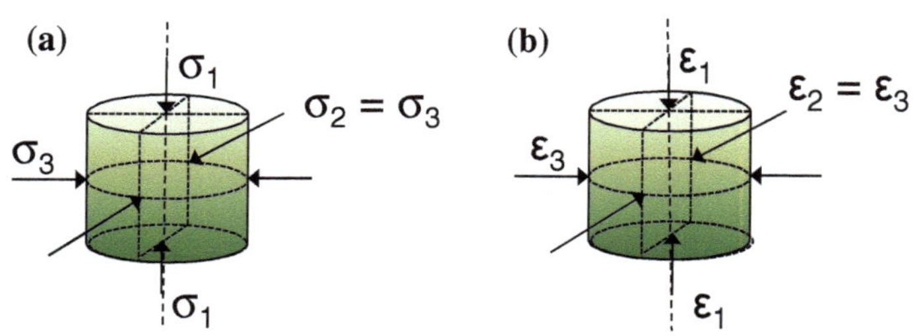

Fig. 3 Stress-(**a**) and strain (**b**)-components in triaxial testing

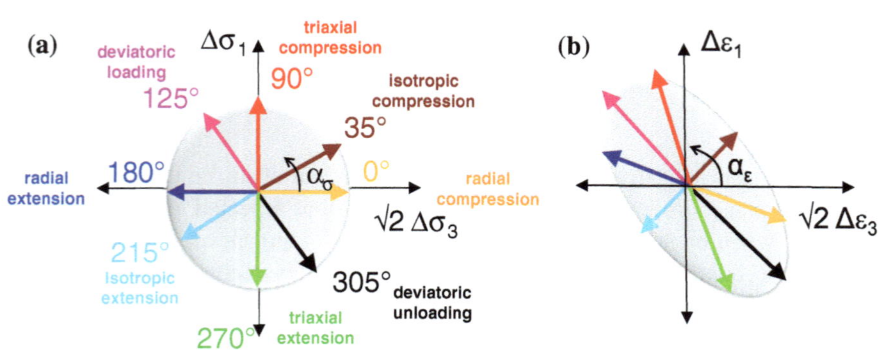

Fig. 4 Applied stress $\Delta\sigma$ increments (**a**) and corresponding strain responses (**b**) for different directions α_σ

$$\eta = q/p \tag{3}$$

describes the stress-state's position in the p-q-plane.

For all tests described in this paper were carried out drained, all stresses referred to are effective stresses ($\sigma = \sigma'$).

2.2 Concept

New or improved constitutive models need to be validated and calibrated. This is often done with the aid of experimental as well as numerical element tests, for example triaxial or oedometer tests. So called response-envelopes are a useful tool for calibrating, validating and comparing constitutive equations [1, 7, 15, 22, 23].

First basics of response-envelopes were presented in the 1970s by Lewin and Burland [17]. A few years later Gudehus [11] used this concept in context with the development of constitutive equations.

To obtain a response-envelope, a soil element is subjected to a certain stress- or strain-increment. The corresponding "response" of the soil in terms of either strain or stress is determined and presented graphically. The direction of the implied stress or strain increment with a constant absolute value is varied and leads to different stress or strain responses, endpoints of which are connected to a response-envelope. Considering the concept of strain-response-envelopes dealt with in this paper, a constant stress increment

$$\Delta\sigma = \sqrt{\Delta\sigma_1^2 + 2\Delta\sigma_3^2} \tag{4}$$

for at least 8 different stress probe directions α_σ is shown in Fig. 4a.

The strains too are plotted in the isomorphic rendulic diagram, where the resulting total strain-increment is

$$\Delta\varepsilon = \sqrt{\Delta\varepsilon_1^2 + 2\Delta\varepsilon_3^2}. \tag{5}$$

The 2 angles, α_σ and α_ε, shown in Fig. 4, are used herein to quantify the direction of incremental quantities where α_σ is the angle between stress probe vector and the positive $\sqrt{2}\sigma_3$-axis, (6), and α_ε is the angle between the strain increment vector and the positive $\sqrt{2}\Delta\varepsilon_3$-axis, (7). The angles are measured counter clockwise [1]

$$\alpha_\sigma = \arctan\frac{\Delta\sigma_1}{\sqrt{2}\Delta\sigma_3} \quad \text{with } \alpha_\sigma \in [0; 360) \tag{6}$$

$$\alpha_\varepsilon = \arctan\frac{\Delta\varepsilon_1}{\sqrt{2}\Delta\varepsilon_3} \quad \text{with } \alpha_\varepsilon \in [0; 360). \tag{7}$$

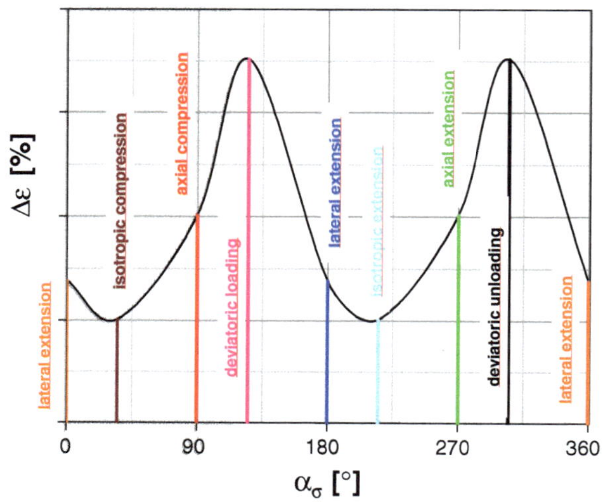

Fig. 5 Absolute value of strain increment $\Delta\varepsilon$ versus stress probe direction α_σ

As $\Delta\sigma$ = const. for all directions α_σ, one gets a circle in isomorphic diagrams, Fig. 4a. The red and the green lines in Fig. 4 for example represent triaxial compression/extension with $\Delta\sigma_3 = 0$. The brown and the turquoise lines stand for isotropic compression/extension with $\Delta q = 0$. A strain response in solely vertical direction ($\Delta\varepsilon_3 = 0$), i.e. a vertical arrow-line in Fig. 4b, would result from a K_0-stress path, which is—for the examined density with Jaky-$K_0 \approx 0.37$—an arrow line inclined under $\alpha_\sigma \approx 62°$ in Fig. 4a.

To show the correlation between the stress probe direction α_σ and the absolute value of the strain increment vector $\Delta\varepsilon$ defined in (5), an alternative evaluation method is to plot the strains $\Delta\varepsilon$ versus the angle of the stress-probe direction α_σ, Fig. 5.

From Fig. 5 it can be seen, which loading directions lead to the largest and smallest absolute strains respectively.

Performing stress-probe-experiments and evaluating the results with the concept of response-envelopes is a convenient tool to investigate the incremental stress-strain behaviour during first loading as well as during un- and reloading-processes.

In this paper the total as well as the quasi-elastic part of the strains due to a relatively low number of un- and reloading-processes ("low-cycle loading") is investigated and evaluated by means of strain-response-envelopes.

2.3 Literature

Some experimental tests to investigate the incremental stress-strain behaviour by means of response-envelopes can be found in literature. On the one hand, there are experiments which focus on either **the total** or **the plastic strains**. On the other

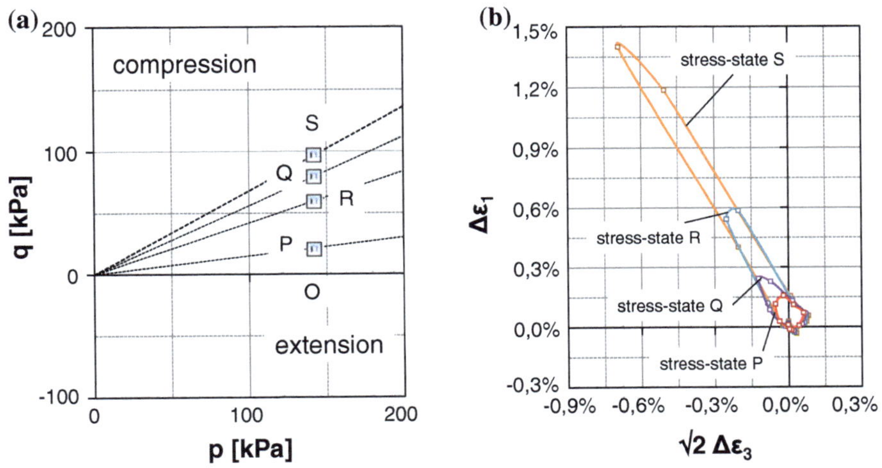

Fig. 6 Initial stress-states (**a**) and strain response-envelopes (**b**) of Lewin and Burland's experiments [17]

hand, there are experiments where very small stress- or strain-cycles are applied in different directions e.g. to investigate the cross-anisotropic **quasi-elastic properties** of the soil.

2.3.1 Stress- or Strain Probe Experiments Investigating Total or Plastic Strains

In the 1970s Lewin and Burland [17] performed stress-controlled triaxial tests on a remoulded, saturated, powdered slate dust. Their tests were performed from 5 initial stress-states shown in Fig. 6a. Lewin and Burland showed that using the normality condition of plasticity theory provides a reasonable basis for the prediction of shear strain behaviour. They found evidence that the direction of the plastic strain increment vector depends on the direction of the stress probe vector $\Delta\sigma$.

Anandarajah et al. [1] performed a series of stress-probe experiments on dense and medium dense Ottawa sand to investigate the dependence of magnitude and direction of incremental plastic strain on direction of incremental stress. 6 different initial stress-states were chosen and stress increments from $\Delta\sigma = 9$–52 kPa in up to 10 different directions were applied on triaxial sand specimens, Fig. 7. The focus was set on plastic strains, which were evaluated by subtracting the elastic strains from the total strains. The elastic strains again were either calculated "by using suitable elastic properties" [1] or determined by applying a stress cycle and measuring the elastic strains during reversal. The results conform to the theory of plasticity for stress-states close to the failure line. It differs from it for stress-states close to the isotropic axis.

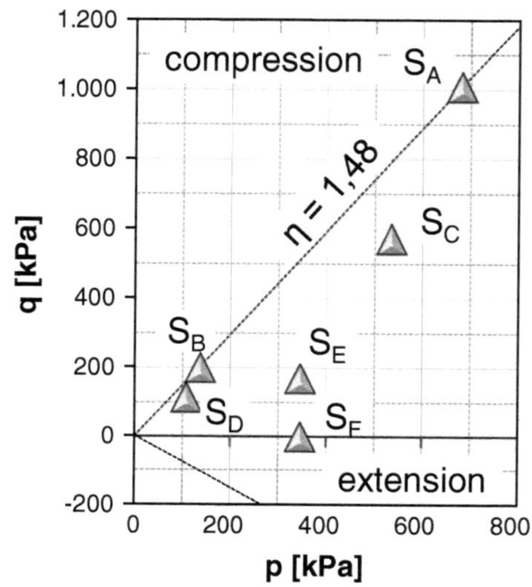

Fig. 7 Initial stress-states of Anandarajah's experiments [1]

Royis and Doanh [21] reported of strain-response-envelopes carried out at 3 initial stress-states for dense Hostun sand, Fig. 8a. The strains due to $\Delta\sigma = 10$ kPa were determined and presented graphically by strain-response-envelopes, Fig. 8b, c. For each stress probe direction, one soil-sample was used, so that the determined strain increments can be interpreted as total strains at first loading. Quasi-elastic strains were not investigated.

Calvetti et al. [4] used Doanh's results for a large numerical study with their Discrete-Element-Model (DEM) and compared the obtained with the experimental results.

Costanzo et al. [5] performed several triaxial tests on a silty clay to obtain strain-response-envelopes at 2 different initial stress-states, Fig. 9a. The strains were investigated and plotted for stress-increments between $\Delta\sigma = 20$ and 90 kPa, Fig. 9b, c (envelope for $\Delta\sigma = 90$ kPa not shown for sake of clarity). They found the response-envelopes to be consistent with each other, indicating an inelastic and irreversible material response, i.e. a strong dependence on the stress increment direction, also at relatively small strain levels. Quasi-elastic strains were not considered explicitly here either.

2.3.2 Stress- or Strain Probe Experiments Investigating Quasi-Elastic Strains

There is hardly any literature where "quasi-elastic" strain-response-envelopes due to low or high-cycle loading are presented. There are some papers though, in which quasi-elastic stress-strain-behaviour is investigated after applying very small axial and radial stress- or strain amplitudes. The object of these experiments is mainly to investigate the soil's inherent and stress-state-induced anisotropy at small

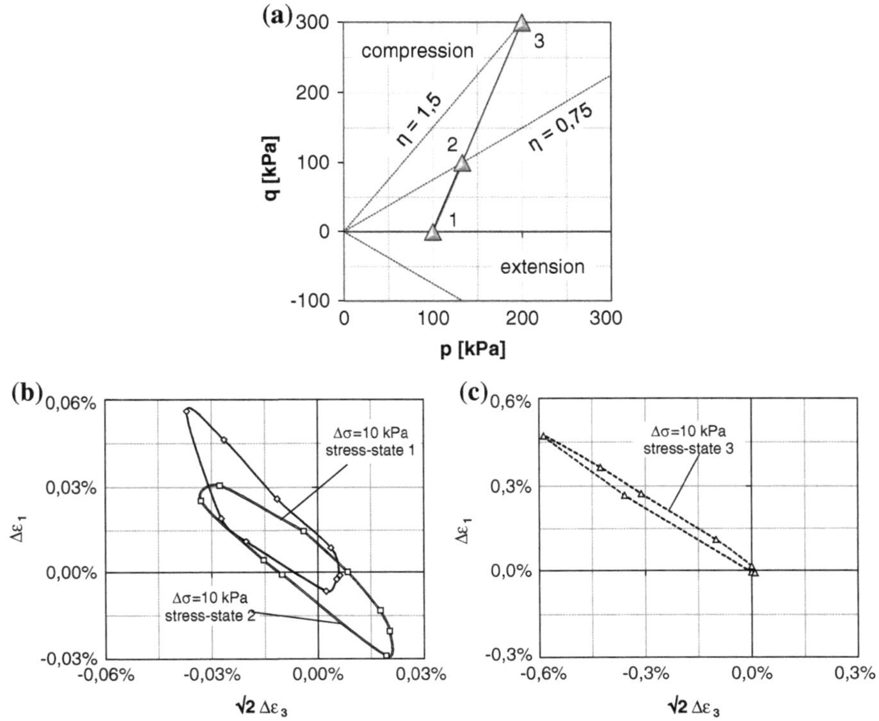

Fig. 8 Initial stress-states (**a**) and response-envelopes (**b**, **c**) of Royis and Doanh's experiments [21]

strain-regime. Hoque and Tatsuoka [14] for example investigated inherent and stress-state-induced anisotropy of different sands. They applied very small strain-amplitude cyclic normal stresses in vertical as well as horizontal direction at different initial isotropic and anisotropic stress-states, see Sect. 5.6 of this paper.

Ezaoui and Di Benedetto [9] carried out experiments to determine the quasi-elastic properties of dry Hostun sand. Very small axial cyclic static loadings and piezoelectric sensor-waves at different levels of the stress-strain curve were applied. The authors also investigated the influence of 3 different sample preparation-methods on anisotropic elastic behaviour, see Sect. 5.6 of this paper.

Kuwano et al. [16] imposed small shear waves to cylindrical sand samples at different stress-states to investigate the effect of stress ratio on anisotropic quasi-elastic properties.

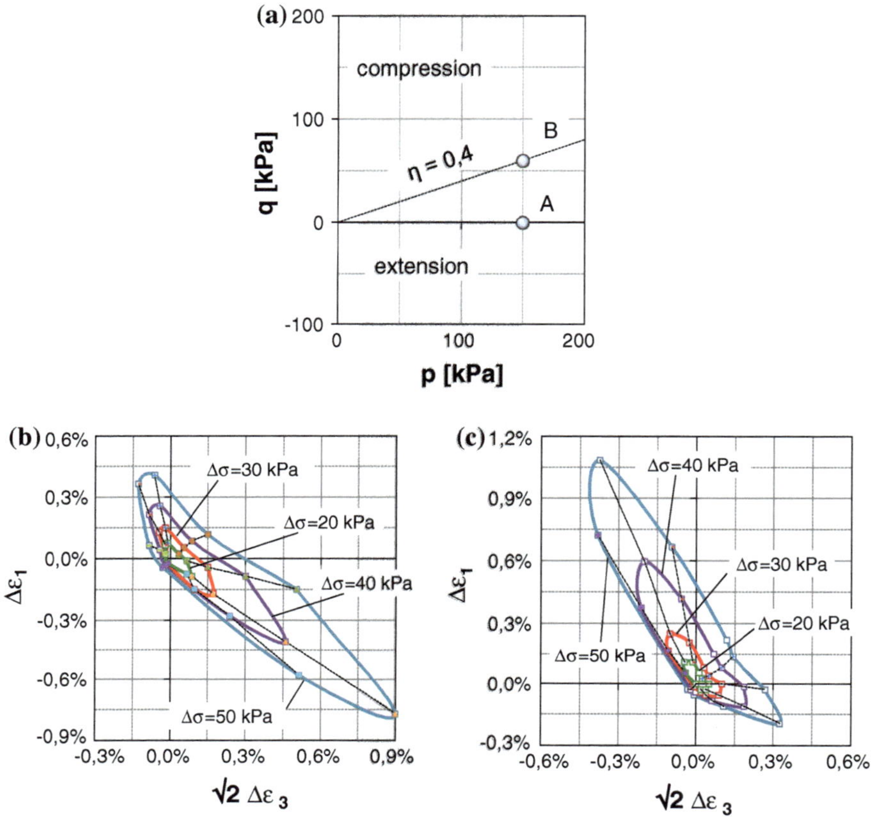

Fig. 9 Initial stress-states (**a**) and corresponding strain-response-envelopes at isotropic stress-state (**b**) and anisotropic stress-state (**c**) of Costanzo's experiments [5]

3 Experimental Fundamentals

3.1 Triaxial Device and Measuring Technique

The triaxial device used for the presented experiments is equipped with high-resolution measurement- and control-technology. It is equipped with an external force transducer to measure the axial strain and a volume-measuring device (burette) with a highly sensitive pressure sensor to measure the volumetric strain. The force transducer enters the cell through a ball bushing linear bearing and the force is measured internally. To avoid tilting, it is connected rigidly to the cell top, Fig. 10.

The confining pressure as well as the axial force can be controlled independently, so that any stress-path from any initial stress-state can be performed, either monotonously or cyclically. Height and diameter of the soil specimen are 10 cm.

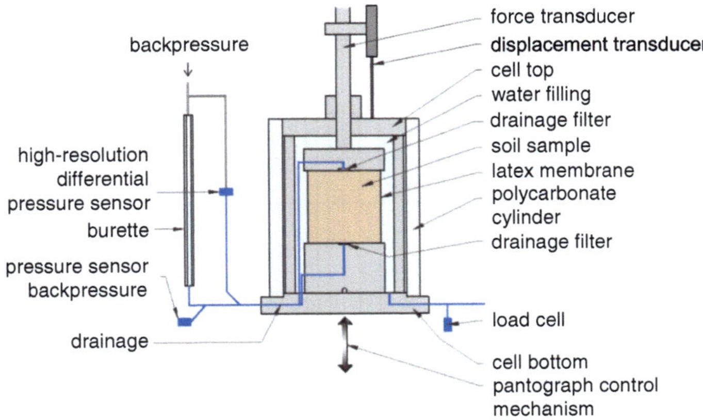

Fig. 10 Triaxial device

Radial strains are determined indirectly from measured axial and volumetric strains. For stress-increments $\Delta\sigma \leq 20\ldots50$ kPa, this procedure has led to plausible and reproducible experimental results of sufficient accuracy, Costanzo et al. [5], Anandarajah et al. [1].

To minimise end-friction, the specimen's end plates were first lubricated with a thin layer of silicon grease and then covered with a thin latex membrane ($t_M = 0.35$ mm). This method also contributes to a homogenous distribution of stress within the specimen.

3.2 Tested Sand

As the volumetric strains ε_v are measured via in- and out coming pore water and not only the axial stress σ_1, but also the cell pressure is varied, effects of membrane penetration also have to be taken into account, Nicholson et al. [18] and Baldi and Nova [2]. To minimise possible errors a fine-grained sand with a low uniformity-index C_U was used, Fig. 11.

3.3 Testing Procedure

The soil sample was fabricated by pluviating the dry sand, thereby maintaining a constant initial height. This specimen-preparation-method was kept constant for all tests. The achieved relative densities I_D were well reproducible with little deviation (±0.1).

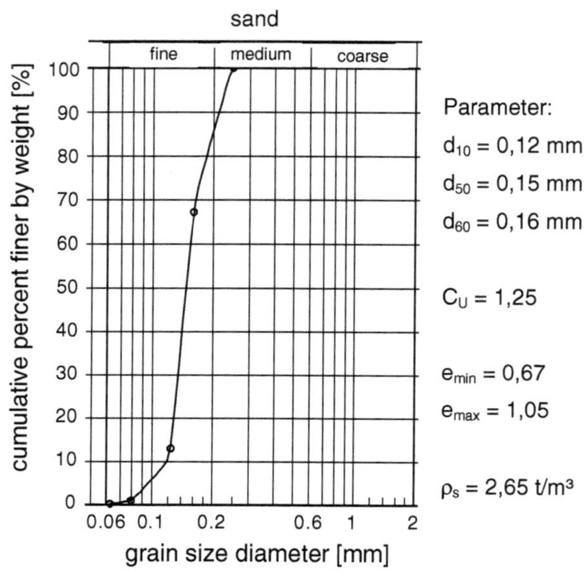

Fig. 11 Grain size distribution curve of the sand

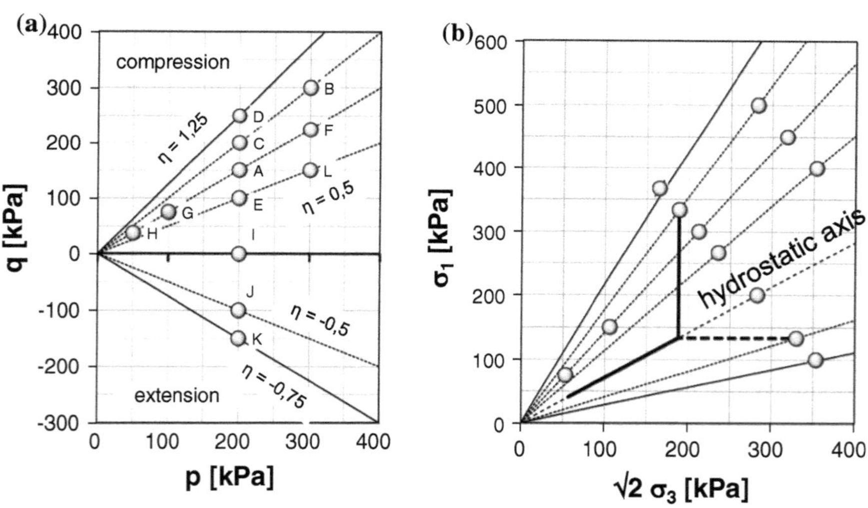

Fig. 12 Investigated stress-states (**a**) and reaching them (**b**)

The soil sample was then flushed with carbon dioxide (CO_2) for about 1 h and saturated with deaired water afterwards. After that a predefined initial stress-state (Fig. 12a) was reached, either by increasing the vertical stress (for stress-states in compression, solid line in Fig. 12b) or the horizontal stress (for stress-states in extension, dashed line in Fig. 12b). The backpressure was kept constant at 200 kPa.

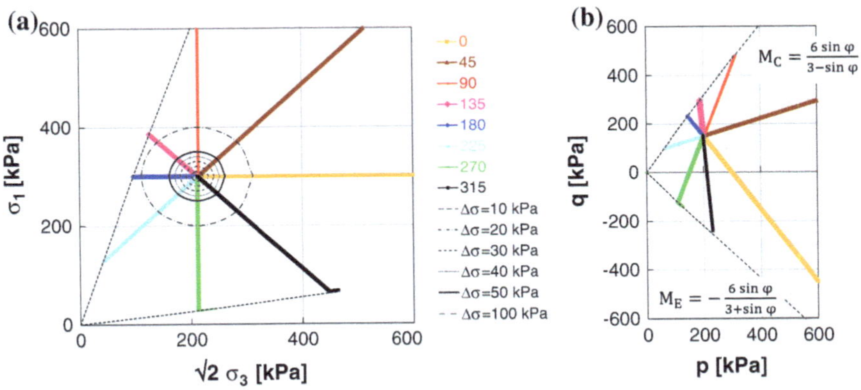

Fig. 13 Applying monotonous stress probes (here from stress-state A) in the rendulic-plane (**a**) and in the p-q-plane (**b**)

4 Experimental Results from Monotonous Loading

To obtain strain response-envelopes from monotonous loading paths, once a chosen initial stress-state is reached and consolidation is finished, a stress path in one direction is applied until failure, Fig. 13.

Total strains $\Delta\varepsilon$ are evaluated for different stress increments $\Delta\sigma = 20, 30, 40, 50$ and 100 kPa (circles in Fig. 13a).

Figure 14 shows different strain response-envelopes determined experimentally for 3 different stress-states located in compression (A), extension (J) and on the isotropic axis (I), which have the same mean pressure $p = 200$ kPa, but different stress-ratios η, Fig. 12.

From the stress-state located on the isotropic axis (I), the shape of the strain response-envelopes for $\Delta\sigma \leq 50$ kPa is almost similar to a symmetrical ellipse, Fig. 14b. For the 2 other stress-states (A and J), the strains due to stress probes indicating towards the failure-lines are significantly larger and the envelopes get elongated, Fig. 14a, c.

5 Experimental Results from Low Cycle Loading

5.1 Test Procedure to Determine Quasi-Elastic Strain-Response-Envelopes

To obtain quasi elastic strains for different stress probe directions, after consolidation at a chosen stress-state (Fig. 12), stress cycles of relatively small stress increments $\Delta\sigma \leq 50$ kPa are applied in a certain direction α_σ (Fig. 3). To avoid pore water pressure the frequency of the cycles was kept low.

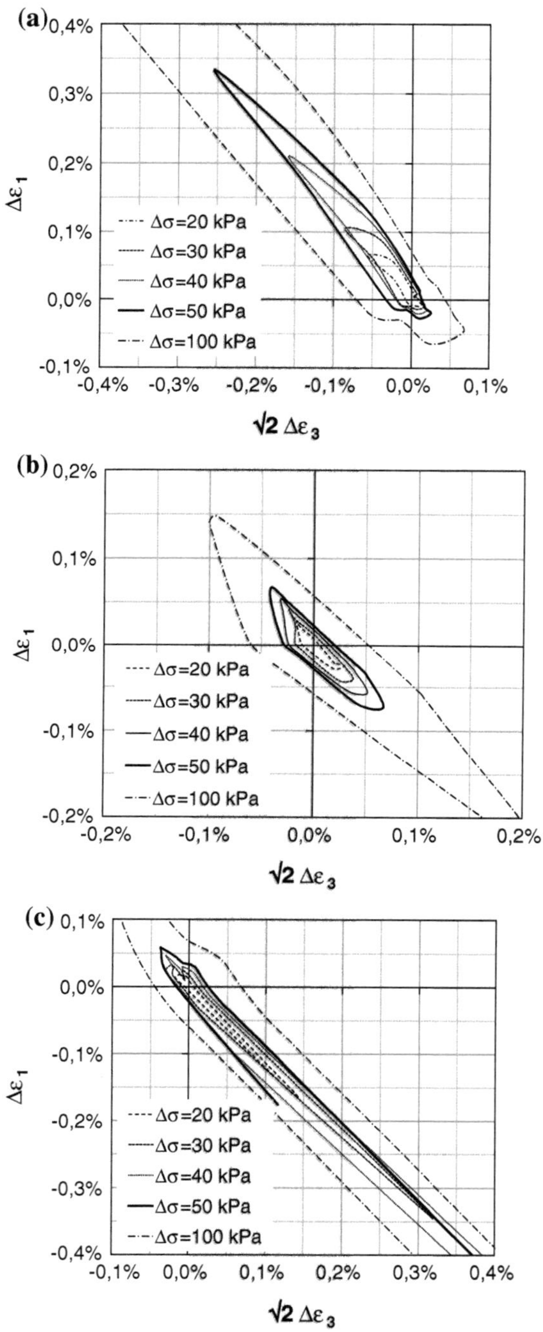

Fig. 14 Strain response-envelopes for stress-states A (**a**), I (**b**) and J (**c**) due to monotonous loading

The cyclic load in the first direction is repeated until the measured strains are practically reversible or rather quasi-elastic. The definition of "quasi-elasticity" implies that during one cycle the plastic strains are less than 1...3 % of the total strains, see Hettler and Danne [13]. It turns out that quasi-elastic behaviour can occur after a low number of cycles. The strain response of the last cycle of one direction is evaluated and plotted. After that, the test is continued by applying the same size of stress increment $\Delta\sigma$ in a different direction α_σ in the stress-space until quasi-elastic behaviour occurs again. The corresponding strains of the last cycle are plotted in a diagram, Fig. 15b.

The investigation of

1. the influence of the sequence of the directions,
2. the mean pressure p,
3. a monotonous isotropic prestress and
4. the stress-ratio η

on the shape, size and inclination of the response-envelope, i.e. the direction-dependent quasi-elastic stiffness, is described in the following.

5.2 *Shape of the Quasi-Elastic Strain-Response-Envelopes*

Figure 15 shows the quasi-elastic strain response-envelopes evaluated for different stress increments $\Delta\sigma$. In general, the envelopes seem to have the shape of symmetrical ellipses for different stress increments $\Delta\sigma$ and seem to be similar to each other, which may indicate a certain approximate linearity for investigated strain regime.

The highest absolute values of quasi-elastic strains always occur in the directions $\alpha_\sigma \approx 125° \ldots 135°$ and $\alpha_\sigma \approx 305° \ldots 315°$ (deviatoric un- and reloading with $\Delta p = 0$); the smallest absolute values result from directions $\alpha_\sigma \approx 35° \ldots 45°$ and $\alpha_\sigma \approx 215° \ldots 225°$(isotropic un- and reloading with $\Delta q = 0$).

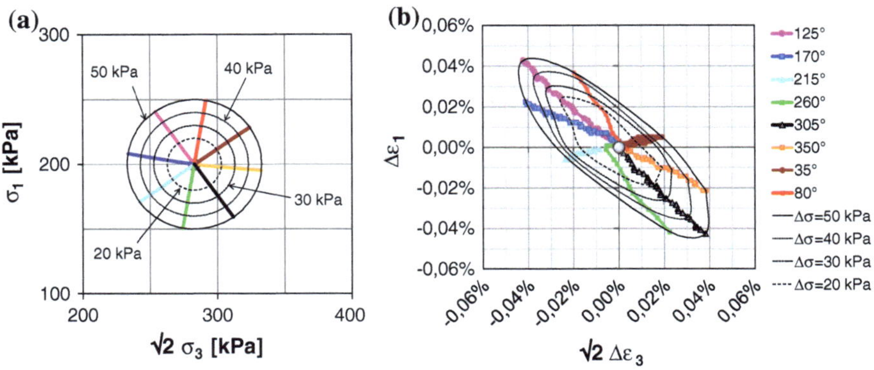

Fig. 15 Construction of the strain response envelope (here at stress-state I): **a** application of $\Delta\sigma$ = 50 kPa in 8 different directions and **b** corresponding strain-responses

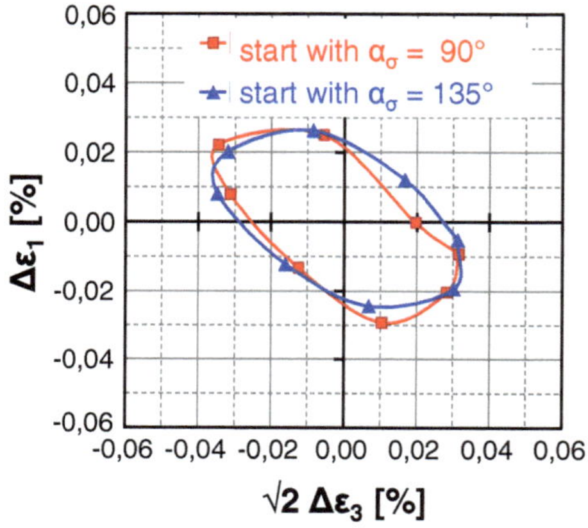

Fig. 16 Response-envelopes due to $\Delta\sigma = 50$ kPa for 2 different sequences of stress-paths from the same initial stress-state

5.3 Different Sequences of Stress Paths

To investigate the influence of the sequence of the applied stress-paths on the quasi-elastic strain-responses, the testing procedure described in Sect. 5.1 was applied for different sequences of directions α_σ. The rotational direction was also varied (clockwise and counter clockwise). It was found, that—for the investigated initial stress-states—neither the sequence nor the rotational direction of the applied stress-paths lead to a substantial influence on the strain response-envelopes, Fig. 16.

Further tests have been carried out which seem to confirm these results.

5.4 Stress-Dependent Stiffness

To investigate the stress-dependency of the quasi-elastic stiffness at low-cycle loading tests at 3 different initial stress-states with constant stress-ratios η and different mean pressures p are performed. The quasi-elastic strains due to stress increments $\Delta\sigma = 50$ kPa were determined and plotted by means of response-envelopes, Fig. 17.

As shown in Fig. 17b the size of the ellipses decreases with increasing mean pressure p and shows an increasing quasi-elastic stiffness for decreasing mean pressure p. This is especially evident at the stress-probe directions at deviatoric un- and reloading ($\alpha_\sigma = 125°$ and $\alpha_\sigma = 305°$). The influence of p on the elastic moduli at the directions $\alpha_\sigma = 35°$ and $\alpha_\sigma = 215°$ (isotropic un- and reloading) is much lower.

Figure 18 shows the absolute value of total strains $\Delta\varepsilon$ depending on the stress increment direction α_σ for 3 different stress-states at the same initial stress-ratio η = const. = 0.75.

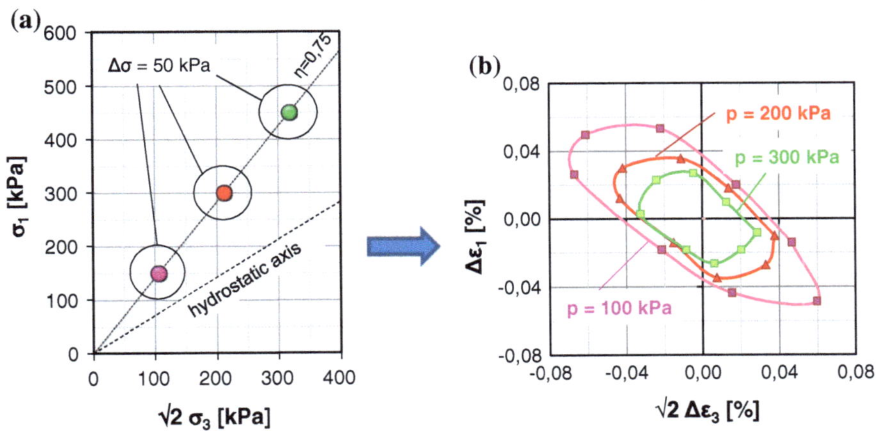

Fig. 17 Construction of the strain response envelope (here at stress-state I): **a** application of $\Delta\sigma = 50$ kPa in 8 different directions and **b** corresponding strain-responses

Fig. 18 Dependence of the absolute value of $\Delta\varepsilon$ on the stress increment direction α_σ (here for stress-states G, A and F with $\eta = \text{const.} = 0.75$)

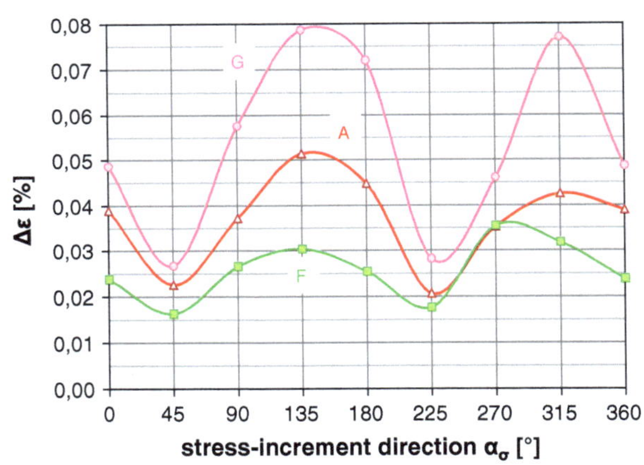

5.5 Isotropic Prestress

To examine the influence of a static isotropic preloading on the size and shape of the quasi-elastic response-envelopes, different tests were carried out starting at the same initial stress state with and without preloading, Fig. 19.

It seems that the influence of an isotropic preloading is negligible, Fig. 20.

Similar observations were made when applying an anisotropic preloading. It will be also necessary to investigate the influence of stress histories in general ("historiotropy") on quasi-elastic properties.

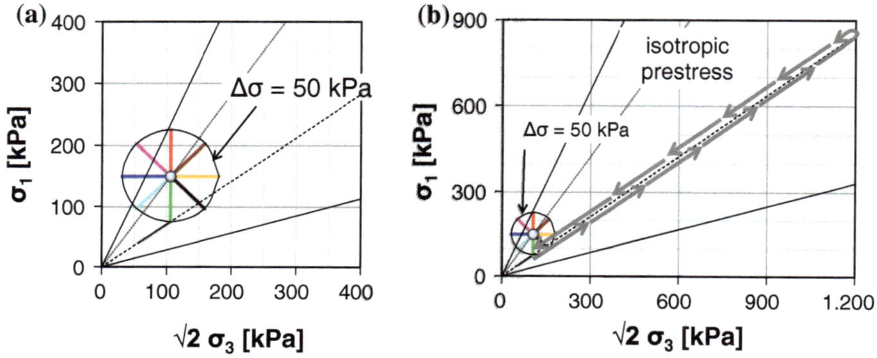

Fig. 19 Investigation of an isotropic prestress at stress-state G: experiment without (**a**) and with isotropic prestress (**b**)

Fig. 20 Response-envelope due to $\Delta\sigma = 50$ kPa with and without prestress at stress-state G

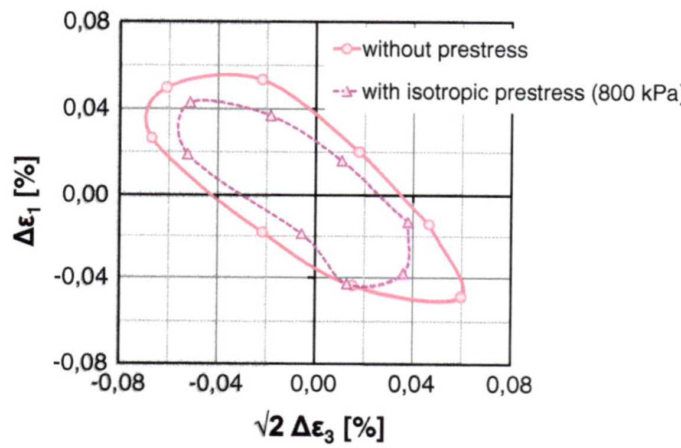

5.6 Anisotropy

To investigate anisotropic material properties, tests were carried out for different initial stress-ratios η. Figure 21 shows the corresponding envelopes plotted in the p-q-plane.

A considerable rotation of the main axes of the response-envelopes depending on the initial stress ratio η can be observed. This influence can be quantified. Figure 22 shows the ratio E_v/E_h of the vertical quasi-elastic stiffness $E_v = \Delta\sigma_v/\Delta\varepsilon_v$ and the horizontal quasi-elastic stiffness $E_h = \Delta\sigma_h/\Delta\varepsilon_h$ as a function of the stress-ratio η. For this purpose data were analysed for stress-probes $\alpha_\sigma = 90°$ and $270°$ (axial compression and extension) and $\alpha_\sigma = 0°$ and $180°$ (radial compression and extension).

The dependence of the ratio E_v/E_h on the initial stress-ratio η can be interpreted as a **stress-induced anisotropy**. Similar observations are also made when investigating much smaller stress- or strain-cycles, e.g. Ezaoui and Di Benedetto [9] or Hoque and Tatsuoka [14]. The coarser the sand, the more distinctive is the difference between E_v

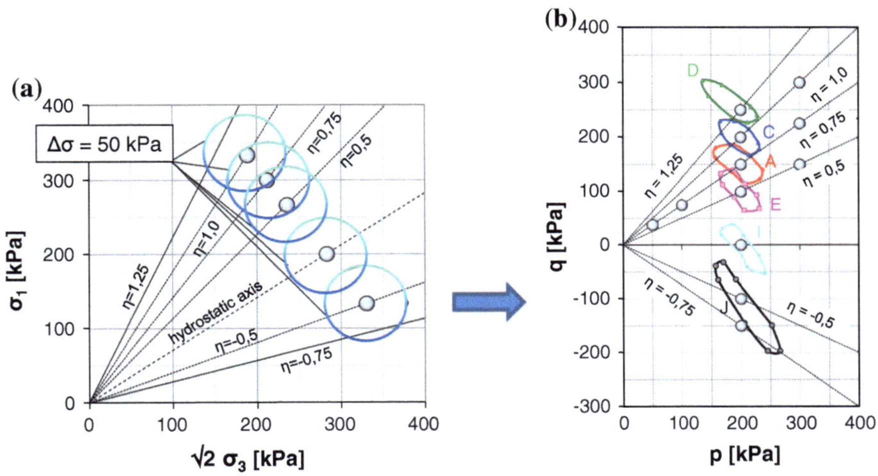

Fig. 21 Tests with different initial stress-ratios η and p = const. = 200 kPa: rotation of axes of the response-envelopes depending on the stress-ratio η

Fig. 22 Ratio of the quasi-elastic moduli E_v/E_h depending on the initial stress-ratio η

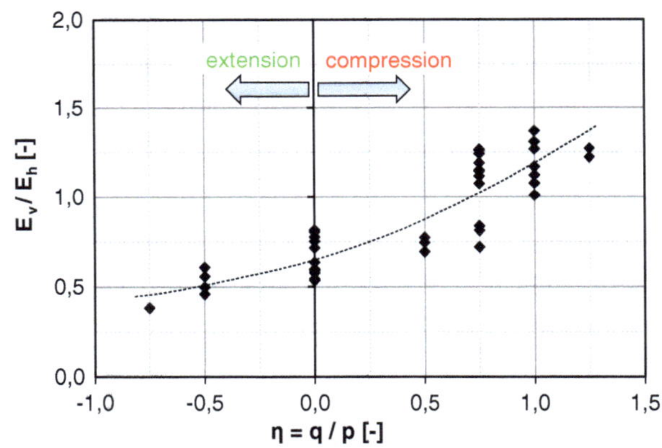

and E_h, i.e. the ratio E_v/E_h increases, Hoque and Tatsuoka [14]. A detailed analysis shows a stronger influence of the stress-ratio η on the vertical than on the horizontal stiffness, see Bellotti et al. [3].

Figure 22 does not only show a stress induced anisotropy. At the isotropic stress-state with $\eta = 0$ the ratio E_v/E_h is $\neq 1$. This means, that there are no isotropic properties at an initial isotropic stress-state, i.e. there also is an **inherent anisotropy**. Most authors come to similar conclusions. While Kuwano [16] and Hoque and Tatsuoka [14] find out $E_v/E_h \geq 1$ for all tested sands at isotropic stress-states, Ezaoui and Di Benedetto [9] also find ratios $E_v/E_h < 1$ for the preparation-methods pluviation and vibration and thus demonstrates a dependence of this ratio on the specimen preparation-method. These discrepancies seem to be due to several factors e.g. the grain-size distribution, the shape of the specimen and the preparation-method.

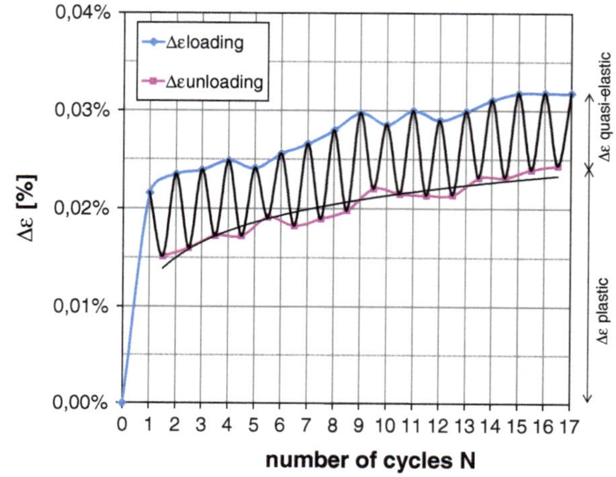

Fig. 23 Example of accumulation of strains $\Delta\varepsilon$ due to stress-probe-direction $\alpha_\sigma = 35°$ and $\Delta\sigma = 50$ kPa (initial stress-state I)

5.7 Plastic Strains

To investigate the accumulation of strains for small stress increments ($\Delta\sigma \leq 50$ kPa) one can look separately onto each stress probe direction, where stress cycles are performed. Figure 23 exemplarily shows the development of strains from loading and unloading with a stress-amplitude of $\Delta\sigma = \pm 25$ kPa in the direction $\alpha_\sigma = 35°$ (isotropic un- and reloading).

As expected, the strains increase with increasing number of cycles N. In Fig. 23 the accumulation of total strains becomes approximately 0 after about N = 15 cycles. Considering the pure quasi-elastic strains, i.e. the difference between $\Delta\varepsilon_{\text{loading}}$-$\Delta\varepsilon_{\text{unloading}}$, it seems that there is hardly any change of the absolute value of the quasi-elastic strains with the number of cycles N. For the construction of the strain response-envelopes shown in the previous sections, this means that the size of the quasi-elastic envelopes remains approximately constant and is independent of the number of stress-cycles.

6 Summary and Further Hints

Producing experimental or numerical response-envelopes is a convenient tool to investigate the soil's incremental stress-strain behaviour and to compare, calibrate or validate constitutive equations.

Most available experimental data are typically obtained by testing a soil in few loading conditions. In this paper an attempt is made to evaluate the incremental stress-strain behaviour for several loading directions, starting from many different initial stress-states.

From monotonous loading paths it could be shown, that from the stress-state located on the isotropic axis, the shapes of the strain response envelopes for

$\Delta\sigma \leq 50$ kPa are almost similar to symmetrical ellipses. For other stress-states the strains due to stress probes indicating towards the failure-lines, are significantly larger and the envelopes get elongated in this direction.

The investigation of the incremental stress-strain behaviour of sand at low-cycle loading in triaxial testing shows, that for investigated stress-increments $\Delta\sigma \leq 50$ kPa quasi-elastic behaviour can occur after a low number of cycles. While the influence of the sequence of the stress-paths as well as isotropic preloading on the quasi-elastic strains seems to be negligible, a significant influence of the mean pressure p and the initial stress-state η on the size and the inclination of the strain-response-envelopes can be observed.

Further triaxial tests are necessary in order to investigate e.g. the influence of the void ratio, K_0-preloading and the stress history in general (historiotropy). It may also be relevant to find out whether there is an influence of the direction of sedimentation during pluviating the sand sample on the anisotropic properties.

It is known, that some common constitutive models show deficits when predicting deformations due to high and low-cycle loading processes, e.g. ratcheting in hypoplasticity, elastic behaviour after the first un- and reloading in elastoplastic constitutive models, missing anisotropy. It is intended to use the presented results together with future tests as a basis for calibrating and validating more complex constitutive equations especially developed for low-cycle loading processes [8, 19].

Acknowledgments The work presented in this paper was supported by the German Research Foundation (DFG) as subproject 8 “Incremental stress-strain-behaviour of sand at low-cycle loading and application on excavation-models” of the interdisciplinary research group FOR 1136 “Simulation of geotechnical construction processes with holistic consideration of the stress strain soil behaviour (GeoTech) “Incremental stress-strain behaviour”. The authors appreciate the financial support from the DFG.

References

1. Anandarajah, A., Sobhan, K., Kuganenthira, N.: Incremental stress-strain behaviour of granular soil. J. Geotech. Eng-ASCE **1**, 57–68 (1995)
2. Baldi, G., Nova, R.: Membrane penetrations effect on triaxial testing. J. Geotech. Eng-ASCE **110**(3), 403–420 (1984)
3. Bellotti, R., Jamiolkowski, M., Lo Presti, D.C.F., O’Neill, D.A.: Anisotropy of small strain stiffness in Ticino sand. Géothechnique **46**(1), 115–131 (1996)
4. Calvetti, F., Viggiani, G., Tamagnini, C.: A numerical investigation of the incremental behaviour of granular soils. R.I.G. 11–19 (2003)
5. Costanzo, D., Viggiani, G., Tamagnini, C.: Directional response of a reconstituted fine-grained soil—Part I: experimental investigation. Int. J. Numer. Anal. Meth. **13**, 1283–1301 (2006)
6. Danne, S., Hettler, A.: Verhalten von nichtbindigen Böden bei niederzyklischer Belastung. Geotechnik **36**, 19–29 (2013)
7. Doanh, T.: Strain response envelope: a complementary tool for evaluating hostun sand in triaxial compression and extension: experimental observations. In: Constitutive Modelling of Granular Materials, pp. 375–396. Springer, Berlin (2000)
8. Ehlers, W., Avci, O.: Stress-dependent hardening and failure-surfaces of dry sand. Int. J. Numer. Anal. Meth. 1–23 (2011)

9. Ezaoui, A., Di Benedetto, H.: Experimental measurements of the global anisotropic elastic behaviour of dry hostun sand during triaxial tests, and effect of sample preparation. Géotechnique **59**(7), 621–635 (2009)
10. Goldscheider, M.: Shakedown and incremental collapse of structures in dry sand bodies. In: Proceedings of Dynamical Methods in Soil and Rock, Plastic and Long-Term Effects in Soils (1977)
11. Gudehus, G.: A comparison of some constitutive laws for soils under radially symmetric loading and unloading. In: Balkema, Proceedings of the 3rd International Conference on Numerical Methods in Geomechanics, pp. 1309–1323. W. Wittke (2001)
12. Hettler, A.: Verschiebungen starrer und elastischer Gründungskörper in Sand bei monotoner und zyklischer Belastung. Heft 90, Veröffentlichungen des Instituts für Bodenmechanik und Felsmechanik der Universität Fridericiana in Karlsruhe (1981)
13. Hettler, A., Danne, S.: Strain response-envelopes for low-cycle loading processes. In: Proceedings of the 18th International Conference on Soil Mechanics and Geotechnical Engineering, pp. 1491–1494. Paris (2013)
14. Hoque, E., Tatsuoka, F.: Anisotropy in elastic deformation of granular materials. Soils Found. **38**(1), 163–179 (1998)
15. Kolymbas, D.: Response-envelopes: a useful tool aus "Hypoplasticity then and now". In: Kolymbas, D., (ed.) Constitutive Modelling of Granular Materials, pp. 57–105. Springer, Berlin (2000)
16. Kuwano, R., Jardine, R.J.: On the application of cross-anisotropic elasticity to granular materials at very small strains. Géothechnique **52**(10), 727–749 (2002)
17. Lewin, P.I., Burland, J.B.: Stress-probe experiments on saturated normally consolidated clay. Géothechnique **20**(1), 38–56 (1970)
18. Nicholson, P.G., Seed, R.B., Anwar, H.A.: Elimination of membrane compliance in undrained triaxial testing. I. Measurement and evaluation. Can. Geotech. J. **30**, 727–738 (1993)
19. Niemunis, A., Prada-Sarmiento, L., Grandas-Tavera, C.: Paraelasticity. Acta Geotech. **6**(2), 147, 67–80 (2011)
20. Niemunis, A., Wichtmann, T., Triantafyllidis, T.: A high-cycle accumulation model for sand. Comput. Geotech. **32**(4), 245–263 (2005)
21. Royis, P., Doanh, T.: Theoretical analysis of strain response-envelopes using incrementally non-linear constitutive equations. Int. J. Numer. Anal. Meth. **22**, 97–132 (1998)
22. Sibille, L.: Directional stress probes to exhibit constitutive behaviour of discrete element models. Olek Zienkiewicz Course 2011—Discrete Mechanics of Geo-materials, Grenoble, June 27th-July 1st (2011)
23. Tamagnini, C., Masýn, D., Costanzo, D., Viggiani, G.: An evaluation of different constitutive models to predict the directional response of a reconstituted fine-grained soil. In: Modern Trends in Geomechanics, pp. 143–157. Springer, Berlin (2006)

MIX
Papier aus verantwortungsvollen Quellen
Paper from responsible sources
FSC® C105338

If you have any concerns about our products,
you can contact us on
ProductSafety@springernature.com

In case Publisher is established outside the EU,
the EU authorized representative is:
Springer Nature Customer Service Center GmbH
Europaplatz 3, 69115 Heidelberg, Germany

Printed by Libri Plureos GmbH
in Hamburg, Germany